Geoinformatics in Citizen Science

Geoinformatics in Citizen Science

Special Issue Editor

Gloria Bordogna

MDPI • Basel • Beijing • Wuhan • Barcelona • Belgrade

MDPI

Special Issue Editor
Gloria Bordogna
CNR IREA
Italy

Editorial Office
MDPI
St. Alban-Anlage 66
4052 Basel, Switzerland

This is a reprint of articles from the Special Issue published online in the open access journal *ISPRS International Journal of Geo-Information* (ISSN 2220-9964) from 2017 to 2018 (available at: https: //www.mdpi.com/journal/ijgi/special_issues/citizen-science)

For citation purposes, cite each article independently as indicated on the article page online and as indicated below:

LastName, A.A.; LastName, B.B.; LastName, C.C. Article Title. *Journal Name* **Year**, *Article Number, Page Range.*

ISBN 978-3-03921-072-5 (Pbk)
ISBN 978-3-03921-073-2 (PDF)

Cover image courtesy of Carlo Gerelli.
The Guest Editor thanks the Italian Ministry of Education through the project of national interest (PRIN) Urban-Geo Big Data (project code: 20159CNLW8 - PE10) for the financial support.

Contents

About the Special Issue Editor

Gloria Bordogna is a senior research scientist of CNR—IREA (Consiglio Nazionale delle Ricerche—Istituto per il Rilevamento Elettromagnetico dell'Ambiente). She has been with CNR since 1986, carrying out research on modeling uncertainty and imprecisions of textual and geospatial information by fuzzy logic and soft computing. From 2003 to 2010, she taught information retrieval systems and geographic information systems at Bergamo University. In 2017, she was awarded a fellowship in IFSA, the International Fuzzy System Association. She has organized several scientific events on the themes of information modeling and management, among which is the special track "Information Access and Retrieval" at the ACM Symposium on Applied Computing, which has been ongoing since 2008. Currently, she is researching multisource heterogeneous geospatial information collection, querying, retrieval, management, and fusion.

isprs International Journal of
Geo-Information

MDPI

Editorial

Geoinformatics in Citizen Science

Gloria Bordogna

CNR—IREA, via A. Corti 12, 20133 Milano, Italy, bordogna.g@irea.cnr.it; Tel.: +39-02-23699-299

Received: 29 November 2018; Accepted: 6 December 2018; Published: 11 December 2018

Abstract: This editorial introduces the special issue entitled "Geoinformatics in Citizen Science" of the *ISPRS International Journal of Geo-Information*. The issue includes papers dealing with three main topics. (1) Key tasks of citizen science (CS) in leveraging geoinformatics. This comprises descriptions of citizen science initiatives where geoinformation management and processing is the key means for discovering new knowledge, and it includes: (i) "hackAIR: Towards Raising Awareness about Air Quality in Europe by Developing a Collective Online Platform" by Kosmidis et al., (ii) "Coupling Traditional Monitoring and Citizen Science to Disentangle the Invasion of *Halyomorpha halys*" by Malek et al., and (iii) "Increasing the Accuracy of Crowdsourced Information on Land Cover via a Voting Procedure Weighted by Information Inferred from the Contributed Data" by Foody et al. (2) Evaluations of approaches to handle geoinformation in CS. This examines citizen science initiatives which critically analyze approaches to acquire and handle geoinformation, and it includes: (iv) "CS Projects Involving Geoinformatics: A Survey of Implementation Approaches" by Criscuolo et al., (v) "Obstacles and Opportunities of Using a Mobile App for Marine Mammal Research" by Hann et al., (vi) "OSM Data Import as an Outreach Tool to Trigger Community Growth? A Case Study in Miami" by Juhász and Hochmair, and (vii) "Experiences with Citizen-Sourced VGI in Challenging Circumstances" by Hameed et al. (3) Novel geoinformatics research issues: (viii) "A New Method for the Assessment of Spatial Accuracy and Completeness of OpenStreetMap Building Footprints" by Brovelli and Zamboni, (ix) "A Citizen Science Approach for Collecting Toponyms" by Perdana and Ostermann, and (x) "An Automatic User Grouping Model for a Group Recommender System in Location-Based Social Networks" by Khazaei and Alimohammadi.

Keywords: geoinformation in citizen science; VGI in citizen science; crowdsourced geoinformation collection and analysis

---

1. Introduction

The idea of editing this special issue was motivated by the observation of the increasing number of academic papers focused on the characteristics of volunteered geographic information (VGI) and crowdsourced geoinformation within citizen science (CS) projects, and on evaluations of the potential for VGI to help scientists, policy makers, and business companies in conceiving and launching new scientific projects [1–8]. VGI and crowdsourced geoinformation from social networks are being investigated as a novel opportunity to launch research projects with widespread ground data, including monitoring of natural, environmental, human-driven, and social changes and events. In these contexts, VGI appears as a relevant aspect of CS. Nevertheless, collecting VGI, filtering crowdsourced geoinformation from its sources, and analyzing it implies the adoption and application of geoinformatics techniques which were first developed for managing traditional geodata in GIS environments. Thus, the appropriateness, coverage, adaptability, and completeness of traditional geoinformation technologies to manage VGI and crowdsourced information in CS deserve an investigation.

The vast literature describing CS initiatives do not specifically focus on the geoinformatics algorithms and technologies applied in relation to the activities and tasks of the projects. This may be due to the fact that the community of researchers in CS is generally very heterogeneous, spanning from experts in various CS application domains, to social scientists studying crowd participation and volunteers' characteristics, and finally, to computer scientists who are often involved in CS activities as mere executors and implementers of solutions. The objectives of this special issue were to overview the latest geoinformation processing approaches used in CS initiatives to investigate CS activities and tasks that can benefit from the analysis of geoinformation, to envisage ongoing technological solutions and trends for geoinformatics in CS, and finally, to outline problems and unsolved issues.

This special issue received a total of 13 submitted papers with 10 papers accepted [9–18].

The authors' affiliations are distributed in the following countries: Austria, Germany, Greece, Italy, Netherlands, Norway, the United Kingdom, and the United States. The described CS initiatives span several geographic areas: Indonesia, Germany, Norway, Italy, the United States, and Iraq.

Topics covered include three main parts: (1) CS key tasks in leveraging geoinformatics, (2) evaluations of approaches to handle geoinformation in CS, and (3) novel geoinformatics research issues. The three topics and accepted papers are briefly described below.

2. CS Key Tasks in Leveraging Geoinformatics

Within this section, we examine descriptions of CS initiatives where geoinformation management and processing are the key means needed for pursuing the objectives of the CS projects (i.e., for discovering new knowledge on the specific application domain of the projects, or for performing some relevant activity of the project, such as reliable geodata filtering, management, analysis, synthesis, sharing, and visualization. This topic includes the following papers: (i) "hackAIR: Towards Raising Awareness about Air Quality in Europe by Developing a Collective Online Platform" by Kosmidis et al., (ii) "Coupling Traditional Monitoring and Citizen Science to Disentangle the Invasion of *Halyomorpha halys*" by Malek et al., and (iii) "Increasing the Accuracy of Crowdsourced Information on Land Cover via a Voting Procedure Weighted by Information Inferred from the Contributed Data" by Foody et al.

(i)   "hackAIR: Towards Raising Awareness about Air Quality in Europe by Developing a Collective Online Platform" by Kosmidis et al. well exemplifies some geoinformatics techniques which can be useful for crowdsourced multimedia data filtering and geolocating, multisource geoinformation merging in order to provide improved and more complete information in areas with partial and missing measurements, and personalized recommendations to citizens based on their profiles and areas. Motivated by the observation that air quality data are often scarce, the paper proposed a centralized air quality data hub with a loosely coupled service-oriented architecture. They applied up-to-date methods to collect multisource information from low-cost sensors and official measurement stations and consolidated technologies to merge these data with crowdsourced information filtered from social media (i.e., geotagged sky-depicting photos from Flicker and official webcam images). To automatically detect the presence of sky in an image, a visual concept detection model using deep convolutional neural networks was applied. Then, the location of the depicted sky was identified by applying a rule-based approach which was evaluated as yielding greatest performance with respect to using a fully convolutional network. Citizens can contribute to air quality monitoring by building and using low-cost sensing devices that optically determine air particles by means of a light scattering method. Finally, a data fusion algorithm based on geostatistics (i.e., universal kriging for interpolating the observations in space using model information as a spatial proxy) interpolated the point-based observations in space such that air quality estimates were available at any point within the domain. Since the final aim of the project was to provide personalized tips on how citizens can reduce their ecological footprint or personalized advice on how individuals may respond to existing atmospheric conditions, ontologies and semantic web technology were used for structuring and semantically integrating data.

(ii) "Coupling Traditional Monitoring and Citizen Science to Disentangle the Invasion of *Halyomorpha halys*" by Malek et al. In describing the "BugMap" science initiative to complement traditional ecological surveys and assist researchers in breaking down the behavior of invasive pests via a user-friendly and freely available mobile application, this paper well illustrates how social media, mobile platforms, and GIS can aid in recruiting and training volunteers to create observations and building species distribution models. The models were built by locality data geocorrelation with environmental variables extracted from authoritative geodata using GIS technologies. Specifically, the MaxEnt software package, a machine learning algorithm that applies the principle of maximum entropy, was used to predict the probability of the spatial distribution of species from presence-only data, represented by a Gaussian kernel function and environmental variables. Sensitivity analysis was performed by varying parameters and computing the receiver operating characteristic (ROC) curve to compare the area under the ROC curve (AUC) of all the models in order to identify the best bias treatment solution for the case study. The paper also reports an interesting geotemporal analysis of the characteristics of both the locations where volunteers created their observations and the species distribution.

(iii) "Increasing the Accuracy of Crowdsourced Information on Land Cover via a Voting Procedure Weighted by Information Inferred from the Contributed Data" by Foody et al. faces the critical issue of filtering reliable VGI to determine an ensemble classification of contributions which could be considered as the agreed classification of the crowd regarded as a unique contributing entity. In this work, the wisdom of the crowd was extrapolated by applying consensus dynamic models taking into account the geolocation of volunteers and their contributions; specifically, the paper explored how to increase the accuracy of crowdsourced data on land cover identified from satellite remote sensing images through the use of weighted voting strategies. Different consensus strategies were tested: the simple majority voting approach and several weighted voting strategies, in which both contributors' skills and models' parameters were considered. The results show that consensus approaches can aid in filtering reliable crowdsourced data and contributors with high agreement, so as to yield an ensemble classification that is more accurate than that achieved by any individual contributor.

3. Evaluation of Approaches to Handle Geoinformation in CS

This section includes CS initiatives, the focus of which is to analyze and critically evaluate approaches to create and manage geoinformatics that can be adopted for a given task in CS. It includes: (iv) "CS Projects Involving Geoinformatics: A Survey of Implementation Approaches" by Criscuolo et al., (v) "Obstacles and Opportunities of Using a Mobile App for Marine Mammal Research" by Hann et al., (vi) "OSM Data Import as an Outreach Tool to Trigger Community Growth? A Case Study in Miami" by Juhász and Hochmair, and (vii) "Experiences with Citizen-Sourced VGI in Challenging Circumstances" by Hameed et al.

(iv) "CS Projects Involving Geoinformatics: A Survey of Implementation Approaches" by Criscuolo et al. As stated in the title, this work tackled the objective of analyzing diversified ongoing CS initiatives from the perspective of geoinformation approaches they adopted for the various tasks of a CS project in action. To this end, they first proposed a common conceptualization of the CS activity workflow, from data generation and delivery, data visualization and access, data processing, to data qualification and validation. Then, a multidimensional classification of the selected CS initiatives was proposed in which each dimension, corresponding to a phase of the CS workflow, was categorized with respect to several main implementation approaches that can be applied. The final aim is to understand which are the most common and used approaches of geoinformatics actually employed in CS and how they evolved.

(v) "Obstacles and Opportunities of Using a Mobile App for Marine Mammal Research" by Hann et al. tackles the up-to-date issue of critically investigating how the use of a mobile application called Whale mAPP (www.whalemapp.org) for recording georeferenced opportunistic marine mammal sighting data in southeast Alaska impacts both the recruitment and commitment of contributors and the quality of VGI. Besides these objectives, the paper also included evaluating the potential educational and scientific benefits and limitations of mobile application use for the purpose of improving future CS projects. To achieve the educational objectives, citizen scientists completed a questionnaire before and after using the mobile app to assess participants' motivations, general experience, and educational outcomes of using the app. Technological glitches and participant retention added additional insight.

(vi) "OSM Data Import as an Outreach Tool to Trigger Community Growth? A Case Study in Miami" by Juhász and Hochmair presents the results of a study that explored if and how an OpenStreetMap (OSM) data import tool can contribute to OSM community growth. The software tool implements a hybrid approach for the building import task that consists of an automated bulk upload of buildings and a manual community review of the remaining buildings. A custom workflow using JOSM editor was developed and explained in a detailed tutorial to three targeted OSM user groups, namely, existing OSM members, local mappers, and students recruited to this purpose. The paper analyzed the spatiotemporal user contributions of the target groups of volunteers. Results revealed differences in editing patterns between newly recruited users and already-established mappers. More specifically, long-term engagement of newly registered OSM mappers did not succeed, whereas already-established contributors continued to import and improve data. In general, they found that an OSM data import tool can add valuable data to the map, but also that encouraging long-term engagement of new users, within or outside the academic environment, proves to be challenging.

(vii) "Experiences with Citizen-Sourced VGI in Challenging Circumstances" by Hameed et al. explores the process of VGI collection by assessing the relative usability and accuracy of a range of different means and methods for data collection among different demographic and educational groups and in different geographical contexts within a study area: smartphone with a GPS app installed for locating land parcel corners and attributing the resultant polygon; portable iPad Tablet PC with the official cadastral map uploaded and overwriting and annotating capability provided through the open source QGIS; and finally, paper-printed aerial or satellite images, with clipboard and pencil for demarcation and annotation. Assessments were made of positional accuracy, completeness, and the experiences of citizen data collectors with reference to the official cadastral data and the land administration system. Ownership data were validated by crowd agreement. The outcomes of this research show the varying effects of volunteers in relation with data collection method, geographical area, and application field.

4. Novel Geoinformatics Research Issues

This section groups three articles that exhibit novelty with respect to the geoinformatics approach they apply, analyze, or propose to perform regarding a specific task in a CS initiative. It includes: (viii) "A New Method for the Assessment of Spatial Accuracy and Completeness of OpenStreetMap Building Footprints" by Brovelli and Zamboni (ix) "A Citizen Science Approach for Collecting Toponyms" by Perdana and Ostermann, and (x) "An Automatic User Grouping Model for a Group Recommender System in Location-Based Social Networks" by Khazaei and Alimohammadi.

(viii) "A New Method for the Assessment of Spatial Accuracy and Completeness of OpenStreetMap Building Footprints" by Brovelli and Zamboni. Although tackling the very common topic of spatial accuracy evaluation of OSM data, it proposes an original artificial intelligence geoinformatics approach which mimics human behavior when making comparisons of maps. Specifically, the assessment of the spatial accuracy is based on the evaluation of the distance between points representing the same features in two different maps (or layers) depicting the same area. The implemented algorithm works on vector layers considering the vertices of the map featured as a set of coordinates. In detecting the homologous entity (the study case, the building footprint), it compares the position, shape, and semantics of the features on the two maps like a human being would. Finding such a correspondence is not trivial, since the two maps could both have slightly different scales and not exactly the same level of details. The comparison must then cope with vagueness and imprecision.

(ix) "A Citizen Science Approach for Collecting Toponyms" by Perdana and Ostermann. This research article starts from the assessment that crowdsourced geographic information and citizen science approaches can offer a new paradigm of toponym collection and addresses issues in advancing toponym practices. It starts by systematically examining the current state of the art of toponym collection and handling practices by multiple stakeholders and then identifies a recurring set of problems. Furthermore, it develops a citizen science approach, based on a crowdsourcing level of participation, to collect toponyms. The proposal identifies the minimum requirements that future mobile and web applications should have for collecting toponyms; specifically, nine essential functionalities are deemed important: navigation, marking GPS coordinates, tracking, displaying a map, taking geo-tagged photos, recording audio, ability to create geo-tagged notes or the generation of forms, offline functionality, and user friendliness and simple user interface. Finally, the implementation of the proposal in the context of an Indonesian case study is discussed.

(x) "An Automatic User Grouping Model for a Group Recommender System in Location-Based Social Networks" by Khazaei and Alimohammadi considers the problem of spatial group recommendations for suggesting places to a given set of users. In a group recommender system, members of a group should have similar preferences in order to increase the level of satisfaction. In this paper, an automatic user grouping model is introduced that obtains information about the preferences of the users, proximity of the places the users have visited in terms of spatial range, users' free days, and the social relationships among users automatically from location histories and users' profiles. These factors are combined to determine the similarities among users. The users are partitioned into groups based on these similarities. Notice that CS could leverage spatial group recommendation for several purposes, for example, for making suggestions of new areas to visit to contributors based on areas visited by others with similar preferences, so as to encourage user long-term commitment, which was identified as one major weak point of CS initiatives.

5. Conclusions

When I undertook the editing of this special issue, I expected to receive many contributions relative to VGI and sensor data interoperable web sharing, semantic representation and management of volunteers' contributions, and credibility/reliability/accuracy assessments of both volunteers and their contributions. Only the last topic is covered by the received papers, probably hinting at the fact that interoperability and semantic issues are solved problem. Many of the papers investigate or discuss the use of mobile applications as a suitable means for both collecting high-quality contributions and engaging long-term contributors. This testifies to the fact that mobile technologies are pervading our life habits, and thus, CS initiatives are investigating if and how mobile applications can constitute a potential to empower CS.

Some unexpected topics were also covered by the papers, such as the use of both machine learning algorithms and artificial intelligence, probably on the wave of popularity of these approaches.

I want to express my congratulation to the authors of the papers for their interesting works; my gratitude to the anonymous referees, whose excellent work made it possible to improve the contents of the papers; and finally, my thanks to the editorial staff of the IJGI for the assistance in producing this special issue.

Funding: This work was supported by URBAN-GEO BIG DATA, a Project of National Interest (PRIN) funded by the Italian Ministry of Education, University and Research (MIUR)—ID. 20159CNLW8.

Conflicts of Interest: The author declares no conflict of interest.

References

1. Goodchild, M.; Aubrecht, C.; Bhaduri, B. Special Issue Role of Volunteered Geographic Information in Advancing Science. *Trans. GIS* **2016**. Available online: http://onlinelibrary.wiley.com/doi/10.1111/tgis.12242/full (accessed on 4 July 2017).
2. Bordogna, G.; Carrara, P. (Eds.) *Mobile Information Systems Leveraging Volunteered Geographic Information for Earth Observation*; Earth Systems Data and Models Series; Springer: Heidelberg, Germany, 2018.
3. Mooney, P.; Zipf, A.; Jokar, J.; Hochmair, H.H. AGILE Workshop on VGI-Analytics. 2017. Available online: http://www.cs.nuim.ie/~{}pmooney/VGI-Analytics2017/ (accessed on 4 July 2017).
4. Mooney, P.; Zipf, A.; Jokar, J.; Hochmair, H.H. Special issue on Volunteered Geographic Information (VGI)-Analytics. Forthcoming in Geo-spatial Information Science, late 2017. 2017. Available online: http://explore.tandfonline.com/cfp/est/gsis/si3 (accessed on 4 July 2017).
5. Pfoser, D.; Voisard, A. *GEOCROWD Workshop Report: The Second Int'l Workshop on Crowdsourced and Volunteered Geographic Information 2013: (Orlando, FL—Nov. 5, 2013)*; SIGSPATIAL Special; ACM: New York, NY, USA, 2014; Volume 6, p. 11.
6. See, L.; Fritz, S.; de Leeuw, J. Special Issue Collaborative Mapping. *ISPRS Int. J. Geo-Inf.* **2013**, *2*, 955–958. [CrossRef]
7. Zipf, A.; Jonietz, D.; Antoniou, V.; See, L. Special Issue Volunteered Geographic Information. *ISPRS Int. J. Geo-Inf.* **2017**. Available online: http://www.mdpi.com/journal/ijgi/special_issues/VGI (accessed on 4 July 2017).
8. Zipf, A.; Resch, B. Special Issue on GeoWeb 2.0. *ISPRS Int. J. Geo-Inf.* **2015**. Available online: http://www.mdpi.com/journal/ijgi/special_issues/geoweb-2.0 (accessed on 4 July 2017).
9. Criscuolo, L.; Bordogna, G.; Carrara, P.; Pepe, M. CS Projects Involving Geoinformatics: A Survey of Implementation Approaches. *ISPRS Int. J. Geo-Inf.* **2018**, *7*, 312. [CrossRef]
10. Brovelli, M.A.; Zamboni, G. A New Method for the Assessment of Spatial Accuracy and Completeness of OpenStreetMap Building Footprints. *ISPRS Int. J. Geo-Inf.* **2018**, *7*, 289. [CrossRef]
11. Perdana, A.P.; Ostermann, F.O. A Citizen Science Approach for Collecting Toponyms. *ISPRS Int. J. Geo-Inf.* **2018**, *7*, 222. [CrossRef]
12. Kosmidis, E.; Syropoulou, P.; Tekes, S.; Schneider, P.; Spyromitros-Xioufis, E.; Riga, M.; Charitidis, P.; Moumtzidou, A.; Papadopoulos, S.; Vrochidis, S.; et al. hackAIR: Towards Raising Awareness about Air Quality in Europe by Developing a Collective Online Platform. *ISPRS Int. J. Geo-Inf.* **2018**, *7*, 187. [CrossRef]
13. Malek, R.; Tattoni, C.; Ciolli, M.; Corradini, S.; Andreis, D.; Ibrahim, A.; Mazzoni, V.; Eriksson, A.; Anfora, G. Coupling Traditional Monitoring and Citizen Science to Disentangle the Invasion of *Halyomorpha halys*. *ISPRS Int. J. Geo-Inf.* **2018**, *7*, 171. [CrossRef]
14. Hann, C.H.; Stelle, L.L.; Szabo, A.; Torres, L.G. Obstacles and Opportunities of Using a Mobile App for Marine Mammal Research. *ISPRS Int. J. Geo-Inf.* **2018**, *7*, 169. [CrossRef]
15. Juhász, L.; Hochmair, H.H. OSM Data Import as an Outreach Tool to Trigger Community Growth? A Case Study in Miami. *ISPRS Int. J. Geo-Inf.* **2018**, *7*, 113. [CrossRef]
16. Foody, G.; See, L.; Fritz, S.; Moorthy, I.; Perger, C.; Schill, C.; Boyd, D. Increasing the Accuracy of Crowdsourced Information on Land Cover via a Voting Procedure Weighted by Information Inferred from the Contributed Data. *ISPRS Int. J. Geo-Inf.* **2018**, *7*, 80. [CrossRef]
17. Khazaei, E.; Alimohammadi, A. An Automatic User Grouping Model for a Group Recommender System in Location-Based Social Networks. *ISPRS Int. J. Geo-Inf.* **2018**, *7*, 67. [CrossRef]
18. Hameed, M.; Fairbairn, D.; Speak, S. Experiences with Citizen-Sourced VGI in Challenging Circumstances. *ISPRS Int. J. Geo-Inf.* **2017**, *6*, 385. [CrossRef]

International Journal of
Geo-Information

isprs

MDPI

Article

CS Projects Involving Geoinformatics: A Survey of Implementation Approaches

Laura Criscuolo *, Gloria Bordogna, Paola Carrara and Monica Pepe

Institute for Electromagnetic Sensing of the Environment, National Research Council of Italy, Via Bassini 15, I-20133 Milano, Italy; bordogna.g@irea.cnr.it (G.B.); carrara.p@irea.cnr.it (P.C.); pepe.m@irea.cnr.it (M.P.)
* Correspondence: criscuolo.l@irea.cnr.it; Tel.: +39-02-23699599

Received: 30 March 2018; Accepted: 25 June 2018; Published: 2 August 2018

Abstract: In the last decade, citizen science (CS) has seen a renewed interest from both traditional science and the lay public as testified by a wide number of initiatives, projects, and dedicated technological applications. One of the main reasons for this renewed interest lies in the fact that the ways in which citizen science projects are designed and managed have been significantly improved by the recent advancements in information and communications technologies (ICT), especially in the field of geoinformatics. In this research work, we investigate currently active citizen science projects that involve geoinformation to understand how geoinformatics is actually employed. To achieve this, we define eight activities typically carried out during the implementation of a CS initiative as well as a series of approaches for each activity, in order to pinpoint distinct strategies within the different projects. To this end, a representative set of ongoing CS initiatives is selected and surveyed. The results show how CS projects address the various activities, and report which strategies and technologies from geoinformatics are massively or marginally used. The quantitative results are presented, supported by examples and descriptions. Finally, cues and critical issues coming from the research are discussed.

Keywords: citizen science; geoinformatics; projects survey

---

1. Introduction

Citizen science (CS) is currently arousing a great deal of interest from both the public and the scientific community. This is due to the unprecedented potential offered to CS by information and communications technologies (ICT), at a more rapid growth rate and at a larger scale than ever before. In fact, the Internet, smart mobile devices, global navigation satellite system (GNSS) sensors, broadband networks, cloud computing, and service-oriented and distributed-processing architectures are widespread technologies that are available for use. Additionally, geoinformatics is a mature discipline offering a geoenabling framework for the aforementioned technologies, benefiting those CS projects that are most sensitive to the geographic dimension of data. An investigation of the current state of the application of geoinformatics to CS is necessary to better understand the phenomenon as well as to envisage possible evolutions and challenges. This is the main objective of the present work, and is addressed by (i) defining a representation framework for the analysis of CS projects; (ii) collecting a significant set of CS initiatives; and finally (iii) examining the sample projects according to the proposed framework. In the following two subsections, we recall the notion of geoinformatics and cite literature dealing with CS characteristics.

1.1. Geoinformatics: A Tentative Definition

Geoinformatics is a term that was introduced in 2000, referring to the words "geo" (i.e., "geospatial"), and "informatics" (which stands for "information science"). It focuses on geoenabling modern information technologies (e.g., databases, decision support systems, the Internet), communication technologies (e.g., wireless networks, cell phones), and interconnection solutions (e.g., protocols, standards, compatibility, interoperability) [1]. In Figure 1, a diagram taken from the Geoinformatics Laboratory of Pittsburgh University shows the main components of the discipline. Geoinformatics is often misunderstood as geomatics. The term geomatics was first coined in 1981 by Michel Paradis, a Canadian photogrammetrist. It refers to "geo", which stands for "geodesy", and "matics", which stands for "mathematics" [1]. It is an engineering discipline using mathematics and engineering for geodesy and mapping. It embraces the more specific disciplines of surveying, geodesy, photogrammetry, remote sensing, cartography, and positioning.

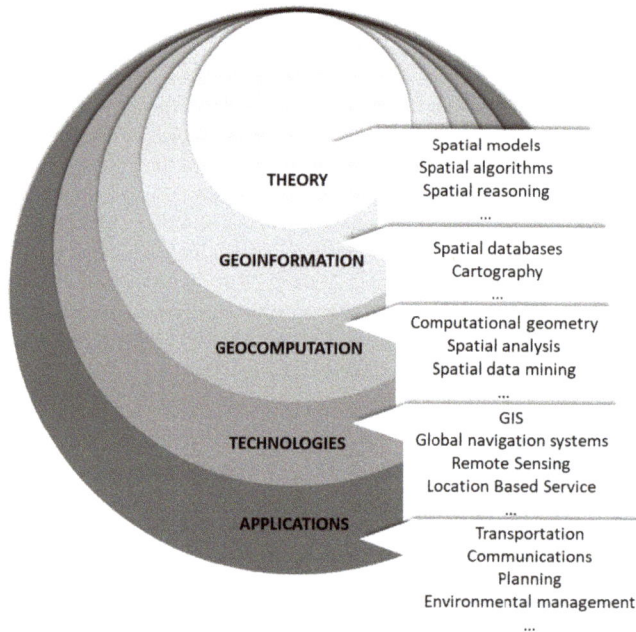

Figure 1. The traditional representation of the geoinformatics layers (from reference [2], modified). GIS: geographical information systems.

Geoinformatics saw a widespread diffusion after the introduction of the Digital Earth concept [3] and the evolution in the use of geographic content by the technological sector and by society at large. The diffusion of digital globe geo-browsers (e.g., NASA World Wind, Google Earth, and Microsoft's Bing Maps), together with an increase in the availability of satellite data, mobile devices, and navigation systems have been part of a digital revolution of geography. The interest in Digital Earth itself hence increased, with the introduction not only of enabling tools and technologies, but also of new concepts and perspectives, put forward by an international group of scientists [4] under the umbrella of Next-Generation Digital Earth. This vision fostered several important activities, also dedicated to education, such as the Vespucci Initiative for the Advancement of Geographic Information Science. In this framework, one of the key developments—in addition to geo-browsers, sensor networks, and spatial data infrastructures—is represented by volunteered geographic information (VGI) [5]. Geographical and Earth sciences are currently increasingly relying on digital spatial data acquired from smart phones, social media application programming interfaces, and remotely sensed images, analyzed by means of geographical information systems (GIS) or cloud-based applications, and distributed through complex infrastructures to target an ever-increasing variety of users. The technologies supporting these processes are at the core of current geoinformatics topics. On the one hand, all of the aforementioned changes and related technologies have familiarized citizens with geographic information, and on the other hand, they have changed their role from mere consumers to producers of geographic content.

All of these developments and progresses led to the so-called neogeography [6] to crowdsourced geographic knowledge, and, when created by a community of amateurs for scientific purposes, to geographic citizen science [7]. A good state-of-the-art as regards to data type, definitions, models, trends, and relationships with CS and VGI can be found in reference [8]. CS projects make use of VGI exploit geoinformatics, as we discuss in this paper, and thus need functionalities and strategies for geo-data acquisition, validation, storage, management, analysis, and portrayal, among others.

1.2. Reference Framework

CS is currently a hot topic, as proved by the literature, with several journals' Special Issues dedicated to the technologies adopted in different application domains (e.g., sustainability [9], public health [10], disaster management [11], geoweb [12]), and a growing number of papers [13] and projects. There are dedicated university courses and classes, seminars, conferences, and educational activities in general (e.g., Conference on Human Factors in Computing Systems, and the Conference on Computer Supported Collaboration and Social Computing, both sponsored by the Association for Computing Machinery). It is also worth mentioning the establishment of national and international initiatives (e.g., the white paper of CS for Europe, the USA government website on CS [14]), as well as the creation of associations (e.g., the American, European, and Australian CS associations, SCA, ESCA, and ASCA [15]), and of international groups of scientists. Finally, the European Commission has founded the Citizen Science COST Action CS-EU-CA-15212 [16] to promote creativity, scientific literacy, and innovation throughout Europe. It explores the potential transformative power of CS for smart, inclusive, and sustainable ends, and provides frameworks for the exploitation of the potential of European citizens for science and innovation. Within this initiative, motivated by the great heterogeneity of CS projects, a working group (Working Group 5—Improve data standardization and interoperability) is dedicated to improving CS data standardization and interoperability. The aim is to define an ontology for CS projects and for the data they created in order to enable CS data sharing and reuse.

The literature regarding the plethora of CS projects and their description is massive. Here, we focus only on those papers dealing with the categorization and analysis of CS, not on the papers describing single initiatives. A few papers have aimed at organizing and providing a reference framework for CS. To this end, they have introduced schemes and models for defining it and describing its characteristics [7,17,18]. Many best practices and guidelines for the implementation of projects have been proposed [19–21]. This is particularly important because CS is a multi-disciplinary domain including several different sectors ranging from ecology to social sciences, as outlined in reference [22], where the authors identified common terminology used in CS initiatives, particularly as related to different types of CS contexts. Although CS is a broad domain, very often, CS activities involve tools and processes which properly belong to geomatics, geography, and geoinformation in general. This paper focuses only on the branch related to geoinformatics. Geoinformatics and CS feature a mutual relationship and positive feedback. In principle, geoinformatics can, on one hand, provide CS with powerful tools and structured methodologies for geodata handling at all levels, potentially introducing a revolutionary spread of CS initiatives. On the other hand, CS can offer interesting use cases, asking for solutions in terms of user experience/interaction, publications and sharing, interoperability, semantic awareness, and the management of big, heterogeneous, incomplete, uncertain, and non-authoritative spatial datasets.

A reference work for the present study was reference [23]. Its perspective is broad, since it examines the past, present, and future of CS in terms of its research processes, program and participant cultures, and scientific communities. This paper, published in 2012, emphasizes the potential of emerging technologies in empowering CS projects. It foresees networks, open science, and the use of online computer/video gaming as important tools to engage non-traditional audiences. It stresses the role of mobile applications (apps), wireless sensor networks, and gaming as promising for advancing citizen science. Gaming genres include alternate- and augmented reality games, context-aware games, and games that involve social networking. It states that the use of these technologies will enhance the ability of scientists and practitioners to centrally consolidate scientific information across projects, promote collaborative writing, and create virtual forums and communities.

2. Materials and Methods

We set up our analysis framework and experimental evaluation methodology in order to consistently analyze the wide variety of CS projects from a geoinformatics perspective. We first identified the main activities that take place during the implementation of a CS initiative. For each activity, we identified the principal approaches that can be adopted for its accomplishment. The main activities and related approaches are detailed in Section 2.1. Once the analysis framework was set up, we selected a list of representative CS projects in terms of typologies, scientific domain, and features (this activity is described in Section 2.2). After pruning the initiatives that were not compliant with some criteria from the list (detailed in Section 2.2), we analyzed the remaining ones following the evaluation methodology, hence, by activities and approaches.

2.1. Schema of the Categorization

In this section, we describe the eight activities identified and detailed by means of a set of technical features which implement the distinct approaches. Such features are not limited to technologies for handling geoinformation, but, as far as possible represent, all the aspects covered in an operational CS project.

2.1.1. Recruitment

The recruitment of volunteers is often an awkward task. Some initiatives directly target small groups of experienced—or already trained—volunteers. Many other projects rely on a generic crowd of voluntary participants (crowd-sourcing). Different approaches and technologies are thus needed, depending on the number of participants, the foreseen participants' preparation, their technological skills, age, geographical provenance, etc. In this research, we considered seven possible recruitment approaches that can be applied individually or in combination. While some approaches have been adopted from the very beginning of citizen science (e.g., recruitment within gentleman associations or scientific networks), some others are quite novel, such as those exploiting Web 2.0 technologies. The main identified approaches were the following:

1. Project website-based: This happens when the broadcast of the initiative and the opportunity to join it are mainly entrusted to a project-dedicated website;
2. Smart app-based: The opportunity to join to the CS project is offered by means of a mobile application connected to the Internet;
3. Web platform-based: The initiative is promoted and supported by thematic web platforms or multi-project websites. Users and visitors of the platform are informed of the existence of many ongoing initiatives. Sometimes they are assisted in selecting suitable initiatives (e.g., based on the user preferences or location) and encouraged to take part. Sometimes the web platform also takes charge of data collection, management, and access, as well as user interactions;
4. Social media-based: The use of social networks is the channel to encourage recruitment of volunteers;
5. Local facility-based: Participation is not promoted on the web, but in real locations. This typically occurs when visitors of museums, natural oases, or public offices are informed of a CS initiative and encouraged to join it;

6. Association/network-based: The recruitment is proposed within associations or relies on the diffusion among collaborators—both professionals and amateurs;

7. Education/academia-based: Proposals of participation specifically designed for school or academic classes and their teachers.

2.1.2. Data Generation

The data generation activity regards data capture by human observation, by the collection of field material, by the acquisition of measurements from instruments or sensors, or by means of automatic or even unaware mechanisms. However, it could also include the compilation of meta-information regarding data capture (e.g., position, timespan, author profile, IP address, etc.). While more traditional CS projects were mainly based on a common process for data generation and collection that basically consisted of human observations, instrumental measurements, and written reporting, the current wave of CS initiatives is experiencing a great variety of different generation procedures, significantly relying on modern participative web and geoweb mechanisms. The generation of data for current CS projects takes place in several different ways; it can consist of the acquisition of samples or measurements from sensors and networked devices, or of the compilation of forms for collecting human observations. It includes the production of transcriptions, classifications, tags, geographical features, attributes, or boundaries. In some special cases, there is no direct input action as in participative grid computing projects, serious games, or health monitoring projects based on wearable devices. In order to analyze the data generation activity, we considered the following approaches, which are not mutually exclusive. They offer a significant glimpse into the technological skills required of volunteers and the different ways for acquiring or coding information.

1. Field activity: Participants are asked to perform activities in particular places or environments in order to generate data and information (e.g., sample collections, field observations, etc.);

2. Guided human observation: Volunteers are asked to report what they see, hear, feel, and experience in a given situation, aided by means of schemas, forms, protocols, conditioned data entry, etc.;

3. Transcription: Aimed at producing digital copies of documents (such as museum specimens' labels or old log-books), or translating documents in a different language, such that data generation consists of a transcription task;

4. Sensor observation: Typical for data acquired by sensors and transmitted directly or by volunteer intervention to a cyber-infrastructure;

5. Sampling: Used when real objects or specimens should be collected in the field by participants (soil, plants, animals, etc.) and analyzed subsequently by experts;

6. Multimedia data capture: Proposed to participants either as a way to create data (or metadata), or even as a token for proving the veracity of the volunteered observation. Multimedia may consist of photographs, videos, or audio files;

7. Human analysis or decision: Human skills, logic, or critical thinking are required. In these cases, data derive directly from human deduction or interpretation (e.g., pattern finding, sound/images recognition, object classification, etc.);

8. Serious game: Volunteers are enabled to produce data and information just by playing a game. Data are automatically extracted from users' interactions and decisions made during the game. Serious gamers are often aware of the contribution they are giving to research, but are involved by means of typical gaming mechanisms and environment: competitions, interactive interfaces, stimulating messages, amusing activities, etc. Occasionally, games are used to train volunteers (e.g., to recognize species) instead of producing data, but even in this case, this approach is useful for improving the initiative;

9. Georeferencing and geocoding: Used for relating elements to a geographic reference system. Georeferencing can be applied to physical entities (e.g., a lake, a measurement station),

to representations of physical entities (e.g., aerial photos), or even to abstract concepts and events as long as they they are related to a geographic location (e.g., nesting area of a bird, point of sighting of a cetacean). Geocoding transforms physical addresses (e.g., buildings centroids, postal code centroids, administrative boundary centroids) into geographic locations represented in numerical coordinates. They are often associated with gazetteers, and can be used effectively in CS to make data "human-readable", even those with complex spatial relations, to display them, and to make them spatially searchable.

2.1.3. Data Delivery

The delivery of data has undergone a deep transformation in recent decades, following the evolution of communication technologies from the traditional delivery by postal service to the use of emails, which requires network connection and allows almost immediate delivery of data in a digital format, until the more recent delivery strategies that lean on web communication protocols. We envisaged the following most-used data delivery strategies:

1. Postal delivery: Used for sending physical contributions in non-digital format (e.g., samples, paper documents, etc.);
2. Email delivery: Requires data to be in a digital format (images, numerical, categorical, textual, etc.), within given constraints on attachment dimension in bytes. Both an email address and an Internet connection are required. Unlike postal delivery, email delivery is almost immediate;
3. Web delivery: The exchange of digital data by means of standard communication protocols on the Internet, such as FTP and HTTP. In the case of HTTP, the volunteer is usually required to fill web forms or check boxes, or to select features or geographic areas. The volunteer performs this activity on websites or web platforms, or by means of specific web applications. Data delivery via social media messaging and sharing is included in this approach;
4. Smart app delivery: A special case of web delivery performed only via mobile applications, without using a web browser. It requires a mobile device (phone/tablet or watch) that is connected to the Internet;
5. Unaware delivery: Happens when the (web) contributor is not completely aware of creating data for the project, or is unaware of the specific kind of data he/she is creating. This strategy can be adopted in serious gaming, but also in other contexts, sometimes raising ethical questions on privacy and consensus.

2.1.4. Data Search

Various types of data search can be implemented, depending on a project's needs and kind of data. The search can be performed on archives of texts in natural language, which might have different structure and length; on multimedia archives of images, audio, and video files; and on archives of geospatial data, as commented by metadata. The search is generally performed by expressing queries either by free terms or controlled keywords, possibly within specific metadata fields (e.g., refined search on a timespan). Alternatively, users can autonomously perform their searches, scrolling through the pages of a website by navigating the links and interactive content, or they can be assisted by dedicated interfaces. Since it is impossible to generate an exhaustive list of all the possible search mechanisms, here, we list some general approaches:

1. Discovery service: Enables web users to search for spatial data sets and services by retrieval mechanisms on metadata. Search criteria can be expressed by keywords, geographic references, timespan, and author names, among others;
2. Full text search: Indicates the possibility of retrieving documents, web pages, or any piece of data containing text on the basis of the presence of the search terms within it;
3. Multimedia search: Allows the retrieval of multimedia information in different formats, such as images, audio files, video, etc. The query can be a text (in this case, the metadata are matched to

retrieve the multimedia object), or multimedia content, such as an image (which is used as an example to retrieve similar multimedia contents). It can be implemented by content-based visual or audio information retrieval systems;

4. Browsing & navigation: It is very common for users to explore data on the web by navigating dedicated websites guided by menus and interactive lists or maps, links, and action buttons. The browsing and navigation possibilities are encouraged by the availability of mutually linked interactive data and metadata;

5. Spatial search: Requires an interactive map, on which users can specify points or areas of interest (e.g., by clicking on a position of the map, or by drawing a bounding box, circle, or polygon) in which to perform the data search. Alternatively, a spatial search can be run by entering spatial queries by means of specific tools and interfaces, which often translate addresses into coordinates (e.g., geocoding tools exploiting gazetteers). The analysis differentiates the cases of punctual queries and range (areal) queries.

2.1.5. Data Visualization and Access

This activity covers the ways in which data and metadata are displayed and made accessible by web services. Indeed, the web enables a number of possibilities in selecting, presenting, organizing, unfolding, and accessing data, having direct consequences on their usability. The following approaches, which are not mutually exclusive, were identified:

1. Open access: Refers to the possibility for a generic web user to examine the collected data by any of the following approaches. The policy of open access lets any web user visualize the whole data collection. A restricted policy on data access instead imposes constraints on data consultation. There may be constraints on accessing certain data (e.g., sensitive data related to protected species), or on parts of data and metadata (e.g., occurrence can be provided, but not locations and timespan), or again, data can be made accessible only to logged-in users or those having special access permissions;

2. Web portal and tools access: Includes a broad range of web environments used by CS project designers to provide general or dedicated access to data. The dedicated environment can lean on predefined forms, web applications and platforms, specific hardware for 3D and virtual reality fruition, etc.;

3. Map visualization: Displays some data content in the form of a map or a virtual globe, depending on the geographic reference associated. Maps can represent each data item separately, for instance, as punctual or polygonal features, clustered as groups, or aggregated information (e.g., a density map). The map can be published as a simple image or as an interactive web map;

4. Download: When this feature is enabled, web users can access data by downloading it in one or more data format;

5. Standard web services access: It can transfer machine-readable file formats and support interoperable machine-to-machine interaction over the web. This means that multiple standard clients can access the same service, avoiding the duplication of data repositories and fostering their reuse. Specific standard web services for spatial data are the ones provided by the Open Geospatial Consortium (OGC) [24] to access maps, features, coverages, sensor observations, and metadata catalogues.

2.1.6. Operations on Data

This activity focuses on the mechanisms suitable for transforming, modeling, aggregating, or analyzing data and metadata to extract new information and knowledge. Among the many existing methods, we chose the following approaches that we consider to be particularly suitable for handling and displaying geoinformation in CS.

1.   (Geo)statistics and summaries: Comprehends all processes to organize geodata in order to offer an interpretation through summaries, graphs, indexes and trend indicators, interpolations, and more;

2.   Spatial analysis and spatial properties calculation: Includes a wide variety of techniques aimed at computing metric and topological properties of geodata sets;

3.   Spatial clustering: Refers to algorithms that identify groups of spatial data which share some spatial proximity and possibly, similar attributes;

4.   Map customization: Refers to enabling the customized or personalized representation of geodata on a map by the end user. The personalization can concern the legend styles, the selection/deselection of elements and layers, or other options that modify the data display without modifying the content;

5.   Map editing: Allows users to make changes directly to the dataset by interacting with the map client (e.g., adding features or modifying the geometry of spatial objects).

2.1.7. Qualification/Validation

There has been a long and still ongoing debate within the scientific community on the questionable quality—and thus usability—of CS data. This has stimulated the development of several techniques for checking and assessing the quality of contributions. Since the pioneering examples of the last century, some supervision has been employed by experts. Recently, however, CS has been able to rely on a great variety of strategies and supporting technologies for improving, assessing, and managing the quality of data and metadata. Quality control methods can be divided into two major groups: ex-ante and ex-post methods. The former acts on the preparation of the volunteer and on the assistance and correction in the data generation phase as a strategy for quality assurance. The latter operates selections, treatments, and fixing after the data delivery phase. Even after the publication of data on the web, mechanisms for enabling users for a collaborative revision can be employed to flag, comment on, and improve contributions. A more detailed description of the ex-ante and ex-post strategies can be found in reference [25,26], while an important analysis of their application in CS projects dates back to 2011 [27]. An interesting aspect of these technologies is the role of the one entrusted for controlling quality. More and more often, automatic mechanisms help technicians to check data and purge weak contributions. At the same time, the community of volunteers is frequently called upon to collaborate in the identification of vices, in the enrichment of the contributions, and in their revision and validation. The integration of automatic and collaborative data control mechanisms is particularly useful in CS projects where the large volume of voluntary contributions, or their geographical scattering, does not allow the expert staff to perform controls on their own. For instance, automatic algorithms can identify data that are missing some important information (e.g., timespan), and geostatistical techniques can identify outlier contributions or unrealistic observations for a given geographical area (e.g., the presence of alien species). The user community, despite having an amateur preparation, has a widespread presence in the territory and can greatly contribute to the validation of local reports. The following approaches can be used as unique strategies, or hybridized to better achieve consistent results:

1.   Learning material: Consists of providing volunteers with tutorials, interactive guides, or other types of instructions. It is an easy but effective ex-ante strategy to improve the quality of the contributions and prevent misreporting. Nevertheless, it is typically optional, so it does not guarantee a common preparation baseline for all volunteers;

2.   Compiling assistance: Gathers all techniques that help—and seldom constrain—the volunteer while compiling his/her contribution. They include the use of controlled vocabularies, geographic gazetteers, auto-completion, templates with automatic error-checking capabilities, checklist configuration tools, etc.;

3.   Assessment: Quality assessment can be performed after the delivery of the contributions by a panel of experts, by automatic techniques, or by the community of volunteers. We took into account who among experts, automatic agents, and community performed the assessment. A special kind of assessment is the auto-assessment. It asks contributors to report their level of confidence in the data generation. This kind of assessment is in fact an ex-ante strategy that enriches the metadata;

4.   Cross-comparison with authoritative data: An ex-post strategy that is commonly used to identify the accuracy of datasets and to validate geographic data and information. Cross-referencing is performed by comparing VGI with authoritative information from administrative or commercial datasets. Nevertheless, in some cases, the accuracy of crowd-sourced geographic information can exceed that of authoritative information, as reported in reference [28].

2.1.8. User Interaction and Participation

Recent web and mobile technologies have provided users with several tools to interact with. Some of these technologies are strategically employed to attract volunteers, to keep them engaged, and to reward them for their contributions. Web visitors can often explore the project resources, create custom maps, perform analyses and comparisons, share content on various social channels, add comments and ratings, or take part in discussions on message boards. The creation of customized virtual profiles is widely used in CS projects. It enables registered users to keep track of their activities and to apply to reach goals and awards. Virtual profiles can also be useful to the scientific team to identify and make decisions about the users (from rewarding, to credibility, to commitment assignment). Participation can be stimulated even by calling for volunteers for collaboration during the project design phase, and by recognizing their merits in the scientific publications derived from the project. Local meetings also provide positive feedback on participation; they act both by strengthening the community and attracting new volunteers. Serious games are emerging and they look very promising, since they can involve and entertain participants while providing useful information to the research [23]. Here, we summarize the most-used approaches found in CS projects. Even if they are not directly related to geoinformatics, they are worth surveying because some of them are particularly suitable for future integration with geoinformatic functionalities.

1.   User profiles: Often the participants are encouraged (or asked) to create virtual profiles in order to access data or to deliver contributions. Registered users can keep logs of their activities and reach goals and awards and can share their work both inside and outside the community. User profiles are also useful to the scientific team for contacting the users, requiring clarifications, involving them in local initiatives, sending periodic messages and newsletters, assigning them privileges or tasks, or for producing usage statistics. Encoding user position in a profile can aid the implementation of location-based functionalities;

2.   Scores and ranking: These techniques are aimed at motivating and honoring the most active and good contributors. They can be specialized in different tasks and different areas and can also be exploited for quality assessment. They require users to register in the project web infrastructure. This approach may also employ geoinformatic techniques, such as assigning the scores depending on the user position;

3.   Competitions and prizes: As with the previous ones, these produce some healthy competitiveness among the participants, motivating them and rewarding them for the quantity or quality of their contributions. Rewards can be symbolic, or can be award money and prizes;

4.   Forum: An online discussion site where the community of volunteers can share experiences, ask for help, and search for information. Often some mediators are selected within the administrative and scientific staff or among the most experienced volunteers;

5.   Social media: Social media act as a megaphone for many CS initiatives, managing to reach many contacts and visualizations immediately. They can even be used to collect contributions,

suggestions, or to keep users informed on the project status. Social media capability to share images and multimedia files is often a useful support for projects that would have low dissemination power otherwise. Additionally, in this case, geoinformatics technologies can be exploited to target specific areas;

6. Newsletter: An easy way to keep subscribers updated and engaged in the project. It reports project progress, calls for performing particular tasks, highlights important dates and events, and even publicly acknowledges the best contributors. When its automatic delivery is not available, it can be replaced by manual sending to a mailing list;

7. Community review: this group includes the different technologies that can be implemented to allow the community to comment, integrate, report, or review the resources shared by the project;

8. Meetings and events: periodic social events can be useful in CS projects, not only for performing training and data collection (e.g., during bio blitz), but also for reinforcing bonds among volunteers and to arouse the interest of the local communities;

9. Co-authoring: In some cases, especially when professionals, associations, or expert amateurs are involved as volunteers, contributions can be encouraged, recognizing co-authorship in scientific papers, magazine publications, etc.;

10. Project definition: This engages citizens right from the project design stage, or gives them the opportunity to independently develop sub-projects. Among the advantages, this strategy allows researchers to better understand the needs of communities, and to more easily obtain the favor of the public or private parties involved;

11. Games: The use of games in CS projects is part of the phenomena called gamification and serious gaming, namely, the use of typical game design elements and principles in non-game contexts, not only for entertainment. This use of games, in fact, strengthens participants' engagement and can be exploited by science both to train volunteers or to encourage them to perform certain tasks while having fun.

2.2. Selection of CS Projects

To perform the analysis following the methodological framework introduced above, a significant dataset of active CS projects had to be found and then filtered in order to obtain those in which geoinformatics plays a role. As no official or complete list of active CS projects exists, in order to find and select a set of projects useful for our purposes, we initially evaluated a couple of possibilities:

- Identifying CS projects by submitting queries such as "CS projects" to search engines and analyzing the first-ranked retrieved web pages;
- Adopting an existing-unofficial-list to start with and then refining it.

The first solution, performed by submitting several different queries to multiple search engines, gave poor results that were affected by weaknesses such as the scarce variety of retrieved projects, bulky presence of off-topic items ("best of" lists, multi-project platforms, CS associations, etc.), the loss of small or local projects, the absence of non-web-based projects, together with lexicon and temporal biases. We then decided to adopt the second approach and started with the projects listed in the dedicated English Wikipedia web page [29] (as released on 10 September 2017). The list of active projects includes 194 projects, disparate in discipline, aim, lifetime, country, and spread (both in terms of technologies and people involved). The voluntary nature of the Wikipedia system does not guarantee full representativeness of the enlisted projects but makes it possible to discover some small or local projects (some not even translated into English) that would not have been considered otherwise and which are not commonly known. This also enabled us to analyze some of the many smaller local realities together with famous and technologically advanced CS projects. Moreover, this choice comprised CS initiatives lacking official web pages and thus not indexed on the web. We initially randomized the items of the list and sorted the first 121 projects enlisted. Then, we filtered out and discarded the projects (i) not involving geoinformation and geoinformatics at all; (ii) apparently non-active and not analyzable; and finally (iii) the duplicated items in the list. The purged selection resulted in 87 projects. They varied—as was our intention—in discipline, geographic area, technology, involved volunteers, dimension, and lifetime (the first started in 1969, and the most recent in 2017). The analysis was carried out on these 87 projects by following the criteria described in Section 2.1 and by going through the projects with the tools and documentation provided by them—considering the projects' websites, web and mobile applications, user guides and tutorials, social media, press releases, descriptions on CS platforms, and any other non-technical documentation made available on the web. Direct experiencing of the projects was preferred with respect to analyzing scientific literature reporting descriptions. The lists of both accepted and discarded projects is annexed (Appendix A). Figure 2 illustrates the proportion among selected (71.9%) and discarded (28.1%) CS projects. When considering the discarded projects, 21.5% were rejected for being irrelevant to geoinformatics, 5.0% for being no longer active or currently unreachable, and 1.7% for being duplicated items.

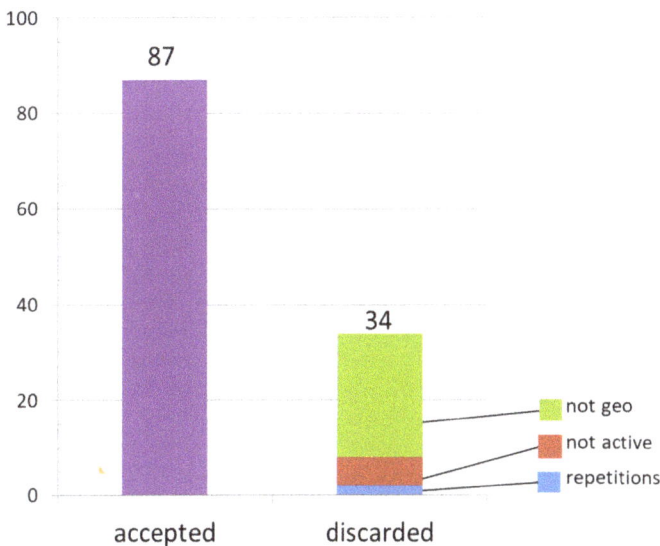

Figure 2. Chart representation of the number of selected and discarded projects within the initial subset.

The high percentage of projects accepted (71.9%) and discarded as not-geo, but still representative of current CS projects, (71.9 + 21.5 = 93.4%) indicates the good consistency and update status of the original Wikipedia list.

We are aware that the collection may be affected by some bias, namely

- Linguistic bias, due to the fact that we adopted the English list (which is the longest one but not necessary the most representative);
- Lexicon bias, due to the many initiatives that are not self-defined as "citizen science", even if they involve not-expert or not-professional volunteers in contributing to research. These initiatives tended to be excluded from the collection simply for lexical reasons;
- Methodological bias, because we chose to base the selection of the dataset on a single list which was not assumed to be statistically representative of the whole CS realm. We are also aware that non-web-based projects are not included in the selection, since the list requires the availability of some web description of the initiative. This is also necessary for our direct analysis.
- Temporal bias, because the dataset includes only active CS projects and is not representative of completed initiatives, or of past technologies to collect, process, and share geographical

information, even if they have been highly fruitful, popular, or significant. Moreover, the analysis was conducted during the period from September 2017 to March 2018 and reports information on the status of the projects at that time. It is possible that further initiatives or those advertised for a very limited timespan were missed.

Considering the above limitations, we do not pretend that our selection is fully representative of the complex setting of geoinformation in CS, nor of the most advanced technologies. Nevertheless, we still believe that the selected dataset reasonably represents a wide part of the CS scene in relation to geotechnology and geoinformation usage in its framework.

3. Results and Discussion

The results of the survey are presented in the form of bar charts and are summarized as frequencies of the different approaches for each activity belonging to the Section 2.1. As any project can adopt more than one approach at the same time (e.g., recruitment can be performed by both social media and smart apps), the sum of the percentages for each activity can exceed 100%. Some examples are reported to better describe the most significant features found.

3.1. Recruitment

The survey of the different recruitment approaches (Figure 3) highlighted how a large part of CS projects rely on a website to present their activities and to call for volunteer contributors (96.6% of the subset). Smart apps, web platforms, and social media were well-represented, but with far lower frequencies (30.3%, 25.8%, and 15.7% respectively). All of these approaches rely on the Internet to gather participants and to crowd-source the project tasks. The projects that addressed the recruitment within academic circuits, pre-existing associations or networks, or by local facilities mainly called for specific groups of users connected by particular interests or circumstances. They accounted for 12.3%, 6.7%, and 5.6%, respectively. Of the many projects considered, only a few used geoinformatics solutions to improve recruitment, despite the existence of several already viable and well-known possibilities. An example of such a solution is offered by SciStarter (https://scistarter.com/), a geoenabled website acting as a project hub where users can filter the CS projects that are active in their areas of interest. Furthermore, with the user's consent, the site can detect the position of the active IP and use it for suggesting and ranking a list of projects. The Atlas of Living Australia (Figure 4), which hosts geodata from some of the surveyed projects, uses an even more refined mechanism to filter suitable CS projects for its web users. It offers some filters based on administrative levels and the possibility to draw shapes and polygons on a basemap as well as to center the search on a marker or on the detected IP location (https://www.ala.org.au/, https://biocollect.ala.org.au). This kind of solution—a sort of proximity market applied to volunteerism—benefits both the website visitors and the hosted projects. The former are offered a personalized selection, while the latter have the opportunity to make themselves known and to recruit new participants. A similar functionality could even be offered by CS mobile apps, which could detect the GPS position of the device and offer suitable CS activities or tasks.

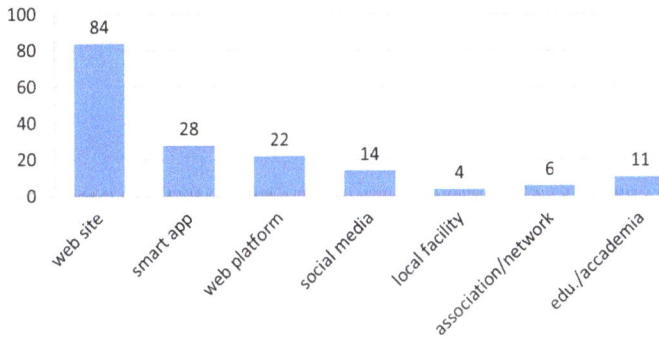

Figure 3. Summary of the results relative to the recruitment approaches.

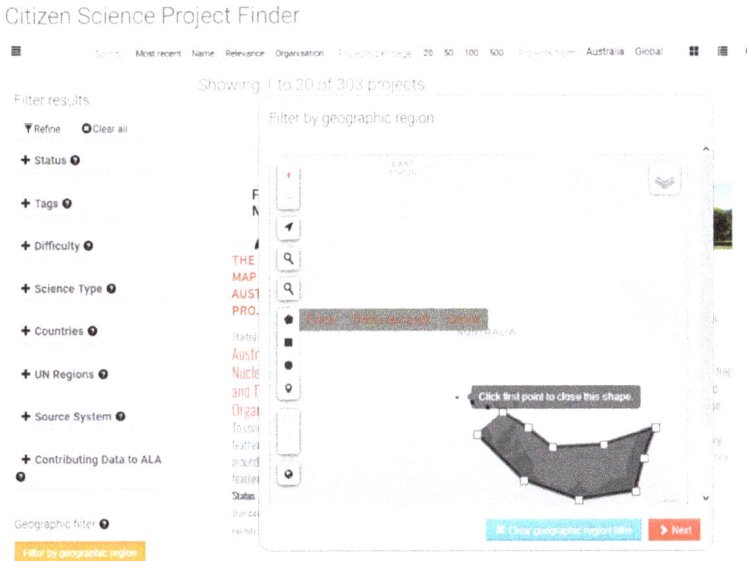

Figure 4. A view of the Citizen Science Project Finder interface implemented in The Atlas of Living Australia website.

3.2. Data Generation

The results (Figure 5) show how commonly CS projects ask volunteers to collect field data. In fact, data came from field activities in 78.6% of projects, and in 77.5% of cases, data were from human observations. In 61.8% of cases, data or metadata generation included the acquisition of multimedia files (photos, video, or audio). Georeferencing and geocoding practices are very well established in projects for the generation of geographic information (89.7%). Nevertheless, it is interesting to see that in a small amount of projects (13.5%), data was derived directly from human inference or interpretation—participants perform analyses and make decisions on their own. In order of frequency, the following infrequent approaches were also found to be used for data generation: transcription (5.6%), sampling (5.6%), serious gaming (5.6%), and sensor observation (2.2%).

Interesting cues regarding geoinformatics in data generation could be taken from the investigated projects. For instance, the TreeSnap project (https://treesnap.org/) randomly alters the GPS coordinates in order to face the issue of geodata privacy. As an additional option, the project allows volunteers to hide the geographic components of their sensitive contributions from the shared map. A gaming approach is used in some of the projects proposed in the Geo-Wiki platform (https://www.geo-wiki.org/). FotoQuest-Go (Figure 6, http://fotoquest-go.org), for instance, challenges the participants in a treasure hunt, in which the goal is to reach specific places and photograph them. The mobile application is a location-based service that visualizes the closer target points as markers on an interactive map and in augmented reality environments, leading the users to its destination. An interesting example of the sensor observation approach can be found in the Air Quality Egg (https://airqualityegg.wickeddevice.com/), a commercial initiative enabling CS. The promoting company sells egg-shaped wireless sensor devices to be used to create a community-led air quality sensing network. Data are generated by pollution sensors and shared on a web dashboard.

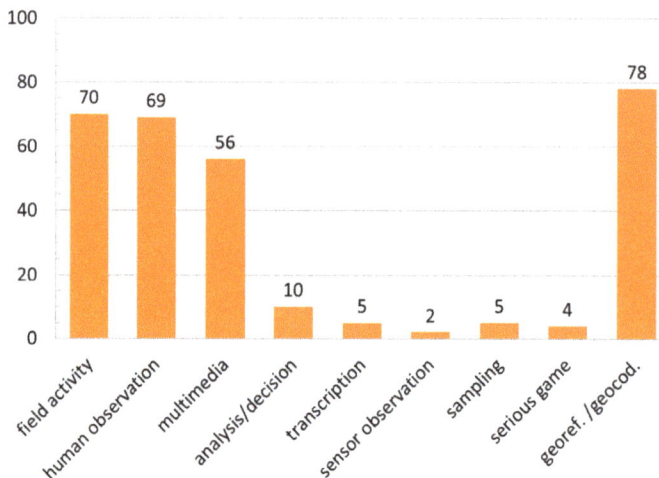

Figure 5. Summary of the results relative to the data generation approaches.

3.3. Data Delivery

The web was by far the preferred data delivery medium (86.5%) (see Figure 7), followed by smart applications (31.5%) and email (20.2%). Postal delivery reached fourth place in the ranking (16.9%). Finally, unaware delivery was present only in one project of the dataset. Looking deeper into the results, it can be noted that postal delivery is the only possible approach in projects based on specimen collection (e.g., the Backyard Bark Beetles and the Monarch Health projects, available at http://www.backyardbarkbeetles.org/ and http://www.monarchparasites.org/, respectively), and it was almost exclusively present in projects with a local or national scale, In fact, only one of the projects following this approach had a global dimension, while the other fourteen that used postal delivery were local or national projects. The only detected case of unaware delivery was that of Project Discovery (https://www.eveonline.com/discovery/). This looks like a minigame included in the MMORPG (massively multiplayer online role-playing game) EVE online. Players are called to solve problems for an independent police force, and find themselves indicating possible planetary transits on light curves (diagrams representing long-term measurements of luminosity of distant stars). They do not need to be aware that curves report data from the CoRoT telescope (COnvection ROtation and planetary Transits, operating in space since 2006), nor do they need to understand how and where their performances will be delivered, processed, and used. As stated in the description of the "unaware delivery" approach, contributors can be not completely aware that they are creating data for a scientific project, or not aware of which data they are creating. In a couple of CS projects, four delivery media—postal, email, web, and smart apps—are enabled together. For instance, the initiative launched by the B.C. Cetacean Sightings Network (http://wildwhales.org/sightings/) asks citizens to report cetacean and sea turtle sightings by sending back a hardcopy logbook, by describing the sighting in an email, or even by submitting a detailed web form, or the list of sightings recorded in the mobile apps even when out of mobile reception range. As for geoinformatics usage, in this project, only smart apps implement the automatic geolocation of the device, as well as the geocoding support and the "click on map" option. It must be noted that within the considered list, every project enabling the smart app data delivery includes georeferencing and geocoding functionalities.

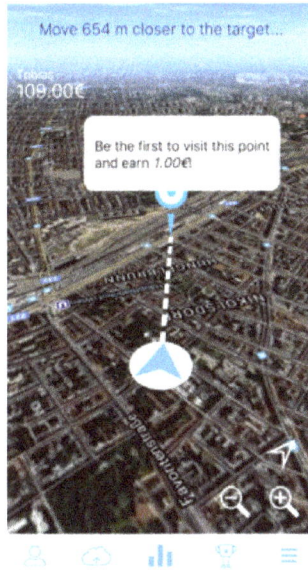

Figure 6. An image from the FotoQuest-Go mobile app.

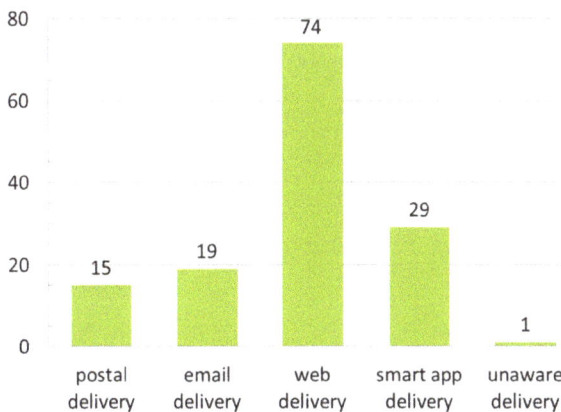

Figure 7. Chart representation of the results relative to the data delivery approaches.

3.4. Data Search, Visualization, and Access

The most frequent approaches for enabling the data search on the web were the navigation of the project website and the related applications (62.1% of projects), and the spatial search (56.2% of projects) (Figure 8). Spatial search was enabled mostly through punctual queries (58.3%). Often, such queries were flanked by mechanisms that allowed spatial areas to be queried (29.2%), for example, by selecting bounding boxes or search radius. Projects allowing areal queries only were less frequent (12.5%). Discovery services, which allow searching by exploiting metadata collected in catalogues, were not as frequent (30.7%), and full-text querying was rarely offered (14.6%). Regarding data visualization and access (Figure 9), open access was available in 37.1% of the cases and downloading was possible in about 29.2% of initiatives. Standard web services were poorly represented (in 8% of projects). The access (even partial or restricted) to data was provided by means of dedicated portals and tools in 69.0% of projects. A map visualization of the geographic data components was provided in 79.3% of projects, while 32.2% let users customize or style the map representation. Only 8.0% of the projects provided map editing tools.

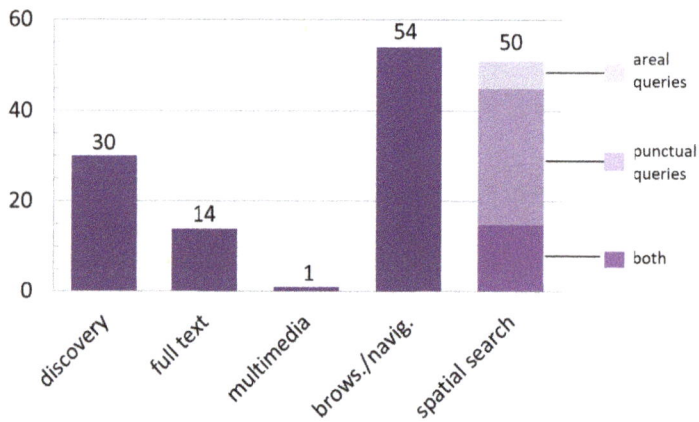

Figure 8. Chart representation of the results relative to the data search approaches.

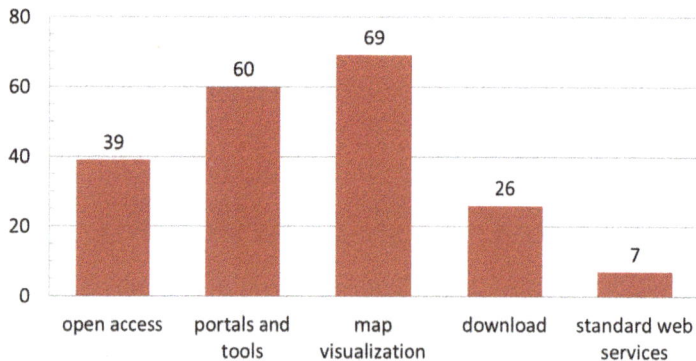

Figure 9. Chart representation of the results relative to the data visualization and access approaches.

In the surveyed list, some interesting examples were extracted that underline how geoinformation can contribute to the data search, visualization, and access activities. The Globe at Night project (http://globeatnight.fieldscope.org/) relies on the Fieldscope platform (http://www.fieldscope.org) to explore the contributed datasets. This application, developed with the support of the National Science Foundation and the National Geographic Society (USA), enables a number of geodata search, visualization, and graph options, depending on the customization chosen by the staff. Moreover, the main website of the project Globe at Night (http://www.globeatnight.org/) uses geo-based navigation, and provides users with a number of geoinformatic functionalities for searching, visualizing, and accessing geodata. For instance, it can provide web users with a customized star map, indicating to the volunteer which constellation to observe and which position to look for, depending on their IP location. IP localization and geocoding mechanisms allow web users to filter the contributions visualized on interactive maps (e.g., by setting the central point and the search radius). Herpmapper (http://www.herpmapper.org/) and PARS (https://paherpsurvey.org/) are herpetological CS projects standing out from the survey for their rich geosearch and visualization functionalities. The first one offers many search possibilities (at three different administrative regional levels, in addition to the more common search by species and taxa, by user, by date, etc.) and visualizes single or grouped results with hotspot maps, metrics, summaries, temporal graphs, etc. The second project provides, in its homepage, a simple and immediate geographic search system. In fact, it offers an interactive heatmap already divided into counties (as an option, blocks can be shown instead of counties), which can be selected by a click, and can be customized in the mapped data type (atlas data, recent activities, diversity, etc.). The legend underlying the heat map provides qualitative information at first glance. Artportalen (https://www.artportalen.se) is the Swedish Species Observation System, and Artsobservasjoner (http://www.artsobservasjoner.no/) is the analogue Norwegian system. They provide professional functionalities to perform web searches, data visualization, and access to their large observations dataset. Here, we recall only some of the spatial features provided: the drawing or uploading of polygons to perform "search by map"; eight different spatial search parameters (including map accuracy); and five different types of interactive map presentations as the search output. From the homepage, a simplified geographic search for the most recent sightings is made available to users which gives the possibility to filter observations by region, by species group, or by day. The portals collect thousands of sightings a day and publish them openly. Data are also forwarded to the Global Biodiversity Information Facility (GBIF, https://www.gbif.org/). Wakame Watch (http://wakamewatch.org.uk) and the Mitten Crab Watch (http://mittencrabs.org.uk/) are two CS projects fostered by the British Marine Biological Association. They deal with the monitoring of invasive species which currently have limited diffusion in the UK. The projects rely on the vast NBN Atlas (National Biodiversity Network, https://nbnatlas.org/), the country's largest collection of biodiversity information, so that even such small initiatives can benefit from its complex functionalities. In fact, the NBN Atlas allows project users to search by species record, environment, climate, and soil information, personal biological records, and habitat, either in single or combined database, and makes it possible to download or export maps, reports, and summaries (Figure 10). The NBN Atlas uses OGC web services for the deployment of spatial layers and lets users add further layers from OGC web map services (WMS) to the map client. The Australian Reef Life Survey project (https://reeflifesurvey.com/) coordinates surveys of rocky and coral reefs with the aim of improving the sustainable management of marine resources. The project enables open access to the collected data through its geoportal (http://reeflifesurvey.imas.utas.edu.au/portal/search). It is essentially a metadata catalogue enriched with a map viewer, deploying a wide list of spatial web services: OGC Catalog Service for the Web (CSW), OGC Web Coverage Service (WCS), OGC Web Feature Service (WFS), OGC Web Map Service (WMS), OPeNDAP, THREDDS, etc. These services let users consult metadata, have a preview of the data content, filter, and select the needed data collections and download them freely (Figure 11).

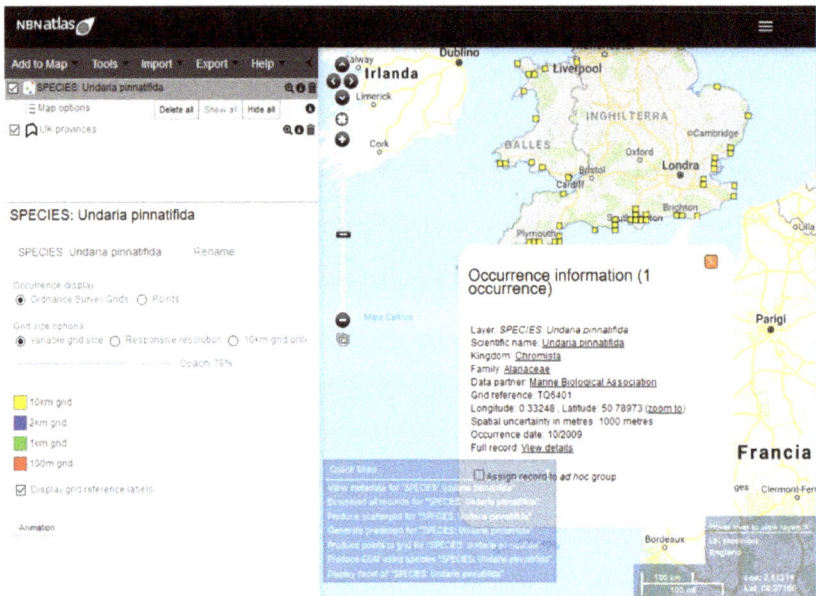

Figure 10. A view of the National Biodiversity Network (NBN) Atlas analysis portal, showing data collected within the Wakame Watch project.

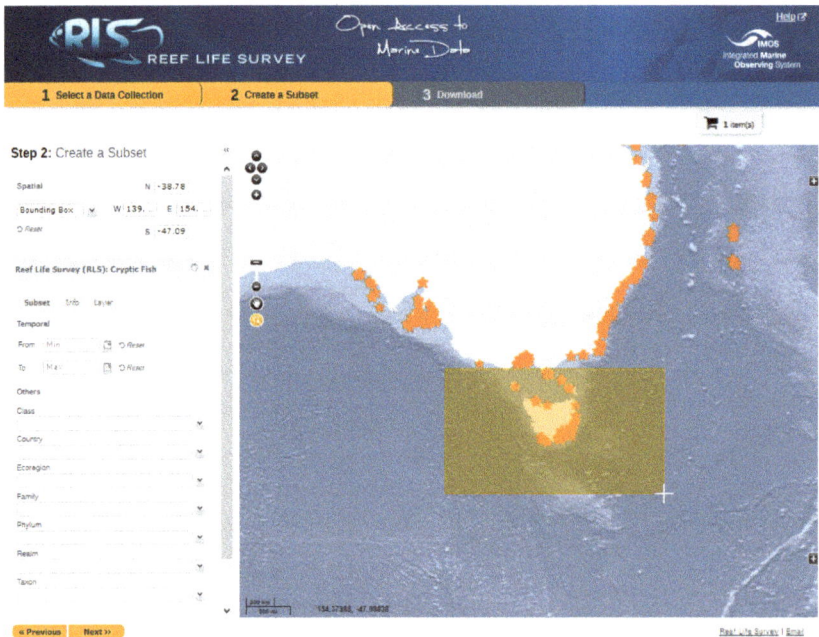

Figure 11. The selection of data subset on the Reef Life Survey data portal.

3.5. Operations on Data

Geostatistics and summaries are available for 58.6% of projects. Spatial analysis and spatial properties calculation occurred in 21.8% projects, while spatial clustering occurred in 23.0% (Figure 12). Many projects related to biology or life science disciplines (botany, entomology, ornithology, biodiversity, etc.) reported the results obtained from spatial analysis of the collected data on their websites and portals (e.g., zonal occurrence maps, heatmaps, hotspot or gap analysis, spatial and temporal series plots, distribution graphs, etc.). It is less frequent that web users were guided to generate customized analyses with the tools provided by the project. This happened for instance in projects relying on the already-cited Fieldscope platform, such as, for instance, FrogWatch USA (Figure 13, http://frogwatch.fieldscope.org). This system guides users in defining variables and parameters in order to generate a number of graphical analyses on data (scatter plots, histograms, time series plots, range comparison plots, calendars, etc.). Some projects provide toolkits to perform spatial analysis or interactive environments for specific tasks. For instance, Old Weather (https://www.oldweather.org/) has a rich set of navigation tools available online, aimed at assisting users in determining ships' positions, distance, course, and speed, using bearings and other geographic references. Data from the African MammalMap project (http://mammalmap.adu.org.za) can be accessed and reworked effectively by the Virtual Museum (http://vmus.adu.org.za), a CS platform developed by the Animal Demography Unit from the University of Cape Town. In the Virtual Museum, web visitors can generate customized summaries and distribution maps, coverage maps, four different types of hotspot analysis, and a gap analysis coverage map.

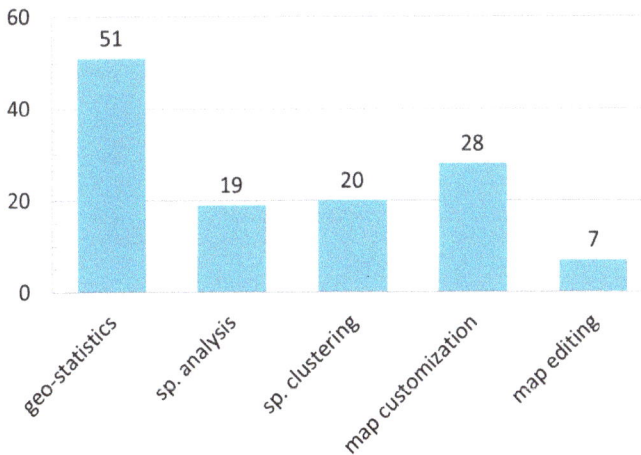

Figure 12. Summary of the results relative to the approaches for operations on data.

3.6. Qualification/Validation

Among the four approaches analyzed within this activity (Figure 14), the first two can be observed as ex-ante strategies for the qualification and validation of data, while the last two are ex-post strategies. As for the ex-ante (preventive) quality control approaches, the most frequent one relies on learning material to prepare and train volunteer collaborators (82.0%), while assistance in the data compilation—aimed to constrain data creation—was made available by fewer initiatives (27.0%). The Ontario Reptile and Amphibian Atlas is an example of an integrated and effective use of ex-ante strategies for data qualification. Its qualification system is made up of different components; its website and mobile app provide users with comprehensive guides, interactive maps useful to make them aware about the ranges of species, and it has a web form for submitting observations with many constrained or assisted compilation fields (automatic data and time detection, species codelist, assisted geocoding and automatic coordinates detection options, overlaying maps consultation, etc.). Concerning the ex-post strategies, in spite of the vast amount of related literature, the cross-comparison between voluntary and authoritative data is performed consistently only in 4.5% of the projects. Moreover, in the surveyed projects, this comparison does not involve the spatial component of the data, probably due to the lack of ground truth data. Nevertheless, in this context, it is worth mentioning a couple of comparison tools found in the survey. LACO-Wiki (https://laco-wiki.net), within the Geo-Wiki project (https://www.geo-wiki.org/), does not provide expert validation on amateur geoinformation. It allows the user to directly compare their own vector and raster maps with a variety of reference layers to generate validation samples and to obtain a customized accuracy assessment report. In this way, all citizens taking part in the Geo-Wiki project as well as any other web users are enabled to qualify their maps. Additionally, the Herbaria@home project (http://herbariaunited.org/atHome/) provides a parser, enabling users to upload their own taxa dataset and grid-references, and returning the comparison among the uploaded records and the known distribution. Regarding ex-post approaches, assessment evaluation of the contributions is used by about half (48.3%) of the projects, mainly relying on experts' revisions. Automatic data assessment is still a rare choice in CS (Figure 14), although it is a suitable strategy to guarantee uniform and objective evaluation. The iNaturalist platform (Figure 15, https://www.inaturalist.org/) is outstanding for the complex validation functionalities provided. iNaturalist charges the community with the identification/validation of taxa for each naturalistic observation contributed. At the same time, an automatic algorithm calculates the most probable taxon on the basis of those proposed by the community and assigns it to the observation. Moreover, a quality grade is associated to the observation, depending on the number of taxon validations and the fulfillment of a series of quality requirements. In order to assist the identification and the validation of observations, distribution maps, statistics, seasonality data, misidentification tips, and other information are shown for each species.

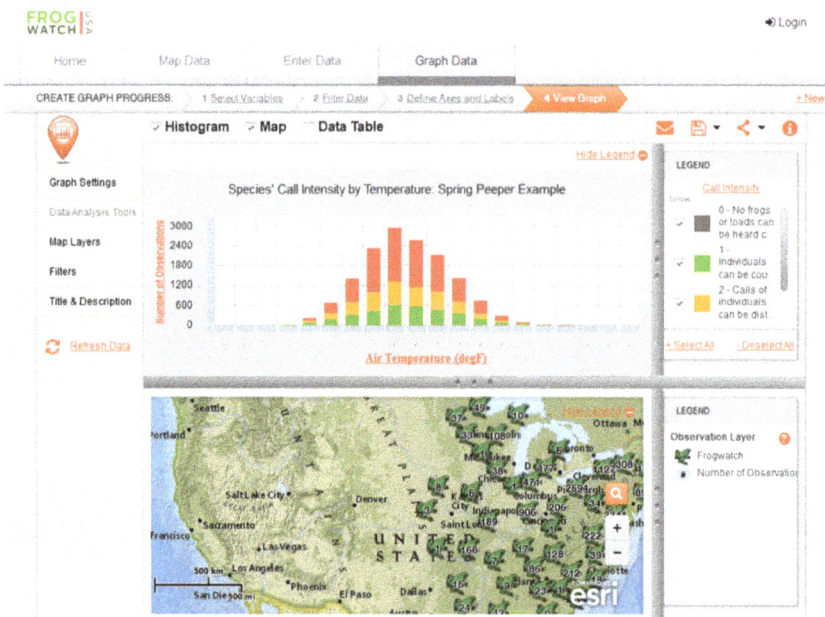

Figure 13. A view of the Fieldscope graphical analysis interface for the FrogWatch USA project.

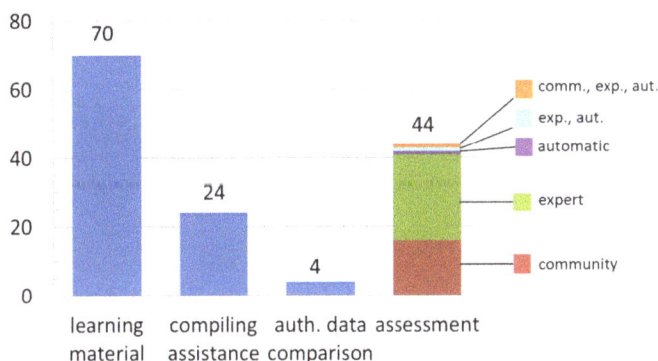

Figure 14. Summary of the results relative to data qualification and validation categories.

Figure 15. Some of the identification/validation functionalities provided by the iNaturalist platform.

3.7. User Interaction and Participation

Social media is the most frequent approach to interaction and participation activities (70.1%), followed by the introduction of user profiles (67.8%) (Figure 16). Community participation via reviewing and commenting on the project's contents (35.6%) and the attribution of scores and ranking (31.0%) appear to be rather widespread practices. Newsletters and local events are used to strengthen interactions and bonds in 29.9% and 14.9% of the projects, respectively. Quite rarely, non-traditional strategies such as competitions, games, user-based project definition, and co-authoring play a role in the user participation possibilities.

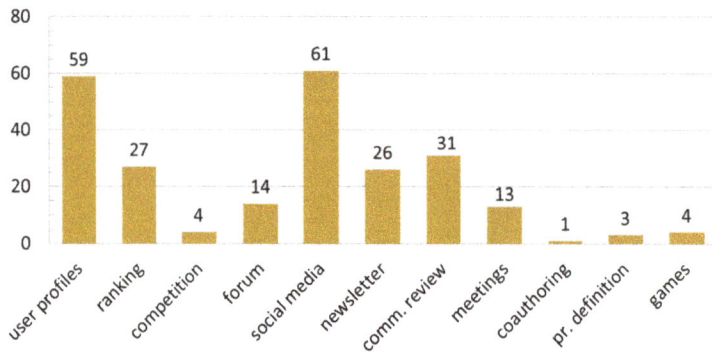

Figure 16. Summary of the results relative to the user interaction and participation approaches.

Looking deeper into the social media accounts of the projects (mainly Facebook and Twitter accounts), it can be seen that they are mostly used to share noteworthy photographs, news, events, thematic publications, achievements, or to make vital contacts with similar communities and stakeholders. The CS project from the North American Field Herping Association is technically a Facebook group (https://www.facebook.com/groups/NAFHA/). In addition to fostering discussions and photo sharing among the community members, it publishes a series of recommended places to perform recreational field herping. We could not report explicit usage of geoinformatic facilities on social media that are aimed at encouraging interaction and participation. Anyway, it is possible that users' positions are implicitly exploited in the projects. For instance, social account administrators can derive important information on the dissemination of the initiative by using web page usage statistics. We had no possibility in the present work to investigate this kind of usage. User profiles are often enriched with tools for monitoring and managing personal contributions. Among these tools, it is not rare to find personalized maps, metrics, and achievements. This is the case for Digivol (https://volunteer.ala.org.au/), a crowd-sourcing platform developed by the Australian Museum in collaboration with the Atlas of Living Australia (https://www.ala.org.au). Each user is provided with a rich personal webpage, where they can consult their own contribution lists, the current validation stage of contributions, the related summaries and statistics, their distribution on a interactive map, and more. The previously mentioned iNaturalist platform (as well as the customized versions, NatureWatch NZ and Natusfera) offers registered users similar advanced functionalities on personal pages. In addition, the users can subscribe to particular places (and taxa) of interest and receive personalized updates. This is also a nice example of the implementation of the community review approach. In iNaturalist, contributors are extremely active in commenting photos, suggesting identifications and data qualification tips, flagging contents, adding new "places" to the platform of where to focus personal studies (each place page displays all the known species from that place, including information about their abundance, conservation status, and first observers), etc. Moreover, each user can create personal projects, define their geographical extension, taxa list, membership, and observation rules, and customize the fields of the observation form, as well as the graphical items of the web page. Another interesting example in the survey, related to community review and user profile, comes from the Habitat Network project (Figure 17, http://www.habitat.network/, http://content.yardmap.org/), based on the Yardmap web application. The project literally calls on individuals and neighbour communities to draw digital maps of their backyards, parks, farms, schools, and gardens, and enables collaborative mapping of the local landscape. It is aimed at achieving better decisions about how to sustainably manage the local environment. Each registered user can map sites and not only manage their environmental descriptors (site characteristics, habitats, and objects) but also set the possible goals and the actions needed to reach them. A forum and messages let participants communicate and suggest improvements.

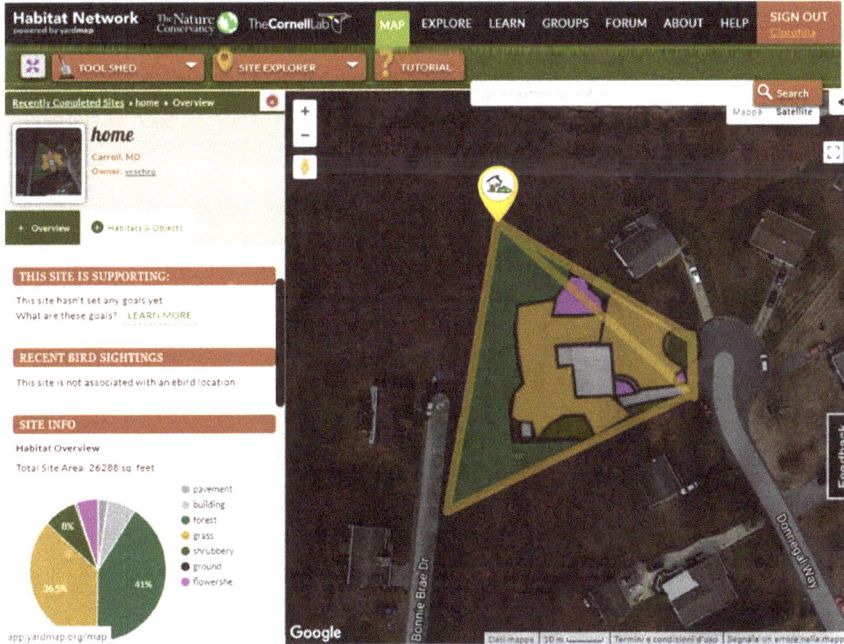

Figure 17. A view of the habitat network collaborative platform.

3.8. Concluding Remarks

From our analysis, websites emerged as the leading unavoidable framework for recruitment, data delivery, and user interaction activities in particular. The limited employment of smart applications and smartphone facilities is surprising. It could be interpreted as the initial phase of a new trend that is still developing. In contrast, it could be seen as the effect of the affirmation of new ubiquitous computing technologies alternative to the use of smart applications. The popularity of websites and portals as hubs for CS initiatives brings many advantages. Besides, we must consider that the digital divide, caused either by poor communication or poor training/knowledge [30], can limit the participation of many people worldwide in CS initiatives. To obtain statistics, it is interesting to consult the most recent reports provided by the ICT Data and Statistics Division of the Telecommunication Development Bureau, International Telecommunication Union [31] and to consider that the global Internet penetration rate is only 30%, with North America and Europe having 10–70 times more data than the developing world or global South [7]. Moreover, we cannot ignore that in conducting our analysis, mainly based on the availability of web pages, we may have potentially missed many local CS initiatives that do not rely on web technologies or on smart applications.

It is also apparent that contributors' training is massively performed via traditional methods (i.e., learning material distribution).

Regarding the access, visualization, and processing of geodata, the facilities offered are the basic WebGIS functionalities (access and visualization), and only rarely can implementations of typical features of mature GIS such as styling, analysis, and decision-making be found.

Standards are rarely used. This demonstrates a lack of attention to data exchange, data reuse, and interoperability issues. A lack of both standardization and common guidelines means uneven or duplicated vocabularies and redundant software libraries, applications, and tools, which could cause inefficiency and ineffectiveness, such as

1. The disorientation of volunteer contributors and their dispersion within too many proposals;
2. The lack of a shared knowledge base;
3. The low robustness of project implementation choices.

On the contrary, the most successful collaborative mapping initiatives, such as the extremely popular OpenStreetMap [32] and Google Earth [33], offer harmonized frameworks, where many applications can grow with similar underlying technologies and user interaction characteristics. The use of open source software and libraries also guarantees robustness, since large developer communities are usually involved. With respect to openness, open data proved to still be far from taking the scene with a low possibility of verification by the public and ultimately lower quality.

Users' interaction approaches can be strengthened and conveniently associated with quality improvement activities. For instance, the user profiling and scores assignment approaches that are typically introduced to increase interactions with users and to reward their participation can even help research staff to detect suitable contributors for performing quality improvement or validation tasks. Once the characteristics and the competencies of users are identified, the staff can assign them specific quality checks or data enhancement operations.

The overall view that emerged from the analysis suggests a landscape populated by initiatives with similar geoinformatics features, although in different application contexts. There has been no dedicated effort to design and propose novel tools or approaches with technological features specific for CS, or differing from the traditional ICT ones.

Many scientific publications [34], as well as reports on social media hitting the headlines [35], have shown that new technologies present still-critical issues, such as privacy and geoprivacy, licensing, intellectual property rights, lack of accuracy, standardization, and interoperability [36]. Another main critical and still disregarded topic is how to interpret "no data values" (i.e., lack of data in some geographic areas) and bias due to both volunteer attitudes and field logistics (e.g., shots of slow-moving animals are more likely to be contributed, because they are easy to catch, while fast-running animal shots are harder; more observations are provided in easily-accessible locations than in remote ones). Nevertheless, strategies and technological means have been proposed to manage such biases, incompleteness, and uncertainty [37–40].

In summary, the performed analysis shows that what was foreseen five years ago in reference [23] is still true—CS is still in need of improvement, which could be achieved by applying geoinformatics technological advances. Quality and reuse are still priorities. For example, syntactic interoperability can be obtained by adopting standard geoservices which are now uncommon in CS. Furthermore, the adoption of shared vocabularies (and/or domain ontologies) and common metadata schema could help normalize data creation and retrieval and enable semantic interoperability between projects. This would allow a semantic-aware fruition and reuse of CS data [16,37,41]. Extending data quality and interoperability of practices means shifting CS initiatives from the "ghetto" of amateurs to the level of authoritative science.

We advocate that future projects meet two requirements: (1) they have a choice of agreed and shared technological standards; and (2) there is wide international coordination of initiatives. In fact, a lesson could be learned from successful CS experiences—the use of standards, well-recognized platforms, and the adoption of general goals—internationally defined though connected to networks of local contributors—allows for both activities and data to be coordinated, potential contributors to be encouraged and engaged over time, and good practices to be rooted in local communities.

Author Contributions: L.C., G.B., P.C., and M.P. conceived and designed the survey, wrote the paper and revised it; L.C. performed the survey.

Funding: L.C.'s research fellowship was granted by the RITMARE Italian Flagship project of the Italian Ministry of University and Research and by the SATURNO project of Lombardy Region, funded by the European Agricultural Fund for Rural Development.

Acknowledgments: The authors would like to thank the anonymous reviewers for their valuable comments and suggestions to improve the manuscript. Special thanks goes to Cristiano Fugazza who performed the English revision.

Conflicts of Interest: The authors declare no conflict of interest. The founding sponsors had no role in the design of the study; in the collection, analyses, or interpretation of data; in the writing of the manuscript, and in the decision to publish the results.

Appendix A. Selected Projects

Tables A1 and A2 list the citizen science (CS) projects considered for the analysis, while Table A3 shows those that were discarded.

Table A1. First half of accepted and analyzed projects.

Project Name	Area	Began
Air Quality Eggs	Ithaca, NY, USA	
GeoTag-X	Worldwide	2012
Loss of the Night	Germany	2013
Globe at Night	Worldwide	
CidadĊo Cientista	Brazil	2004
Geo-Wiki	Worldwide	2008
BugGuide	USA, Canada	2003
Manta Matcher	Worldwide	2012
Whistler Biodiversity Project	Canada	2004
Citclops	Europe	2012
Monarch Larva Monitoring Project	Canada, USA	
Track a Tree	UK	2014
Backyard Bark Beetles	USA	2014
Track My Fish	Canada	2012
Wakame Watch	UK	2014
Ontario BioBlitz	Canada (Ontario)	2012
Natusfera	Europe	2016
DigiVol	Australia	2011
Herbonauten	worldwide, Germany	2016
Herbarium@home	UK	2006
Agent Exoplanet	Goleta, CA, USA	
Great World Wide Star Count	Worldwide	2007
Big Butterfly Count	UK	2010
Crowdcrafting	Worldwide	2011
Observation.org	Worldwide	
Cicada Watch	Northeastern America	2013
NatureWatch NZ	New Zealand	2006
CyanoTracker	Worldwide	2014
Pennsylvania Amphibian and Reptile Survey	Pennsylvania, US	
BeeSpotter	USA (IL, IN, MO, OH)	2007
North American Field Herping Association	North America	2007
Hare Survey	UK	2015
Big Moss Map	UK	2015
Hazelnut Project, The	USA	2000
Mitten Crab Recording Project	UK	
NatureWatch	Canada	
Landscape Watch Hampshire	Landscape Change Consortium	2015
Anecdata	Worldwide	2013
B.C. Cetacean Sightings Network	Canada (British Columbia)	1999
Project Discovery II	Worldwide	2017
Teatime4Science	Worldwide	2016
SciStarter	Worldwide	2011

Table A2. Second half of accepted and analysed projects.

Project Name	Area	Began
Striped AmBASSadors	NS, Canada	2010
Project Splatter	UK	2013
Pieris Project	Worldwide	
Garden Wildlife Health	UK	2013
Report-a-weed	Canada	
Cape Citizen Science	South Africa	2015
Portland Urban Coyote Project	Portland, Oregon, USA	2011
FrogWatch USATM	USA	1998
Cities at Night	Global	2014
Reef Life Survey	Australia	
eButterfly	USA, Canada	2010
Floodcrowd	UK	2015
TreeSnap	USA	2017
Big Bug Hunt	USA, UK	2016
Project Roadkill	Worldwide	2014
Species Observations System	Norway	2008
Artportalen	Sweden	1999
MammalMAP	Africa	2012
Old Weather	Worldwide	
Aquila Project	West Kimberley North Western Australia	2010
Monarch Health	Canada, USA	
Habitat Network	North America	2012
The Shore Thing Project	UK	2006
iNaturalist	Global	2008
AppEAR	Argentina, South America	2015
Bumble Bee Watch	Canada, USA	2014
Mosquito Alert	Spain	2013
Local Environmental Observer Network	Worldwide	2012
Ontario Reptile and Amphibian Atlas	Canada (Ontario)	
Massachusetts Herpetological Atlas	USA	1992
Vermont Reptile and Amphibian Atlas	USA (VT)	1995
Michigan Herp Atlas Project	USA (MI)	2004
Herpetological Education and Research Project	North America	2007
Turtle Survey and Analysis Tools	Australia	2014
Amphibian Migrations and Road Crossings	New York, USA	
HerpMapper	Global	2013
CrowdWater	Worldwide	2017
iSeahorse	Global	2013
Go Viral Study	USA	2013
Ontario Butterfly Atlas Online	Canada (Ontario)	1969
Reef Environmental Education Foundation	Key Largo, USA (FL), Worldwide	1990
Marine Metres Squared	New Zealand	
eOceans	Global	2014
Monarch Watch	Canada, USA	1992
eShark	Global	2005

Table A3. List of discarded projects.

Project name	Area	Began
Operation Wallacea	UK	1996
Identify animals	New Zealand	2015
BioNote	Worldwide	2016
Flying ant survey	UK	2012
Science Gossip Biodiversity Heritage Library	Worldwide	2015
Galaxy Zoo	Worldwide	2007
Galaxy Explorer	Australia	2015
Project Soothe	Worldwide	2014
ARTigo	Worldwide	2007
VerbCorner	Worldwide	2013
AgeGuess	Worldwide	2012
Diver Safety Guardian	Europe and Africa	1994
Project Dive Exploration	North America	
Radio Galaxy Zoo	Worldwide	2013
Orca Game	Worldwide	2013
Reading Nature's Library	UK	
Mark2Cure	USA (CA)	2012
Disk Detective	Worldwide	2014
Stardust@Home	Worldwide	2006
VT Fish Diaries	USA (VT)	2015
Artsobservasjoner	Norway	2008
Digital Access to a Sky Century @ Harvard	USA	2001
Weather Detective	Australia	2014
Clumpy	Worldwide	2012
theSkyNet	Worldwide	2011
Smithsonian Transcription Center	USA	2014
Doing It Together Science DITOs	Europe	2016
Notes from Nature	Worldwide	
Fraxinus	Worldwide	2013
Cochrane Crowd	Worldwide	2016
Plankton Project	Worldwide	2013
Foldit	Worldwide	2008
SETI@home	Worldwide	1999
Socientize	Worldwide	2012

References

1. Kemp, K. *Encyclopedia of Geographic Information Science*; SAGE: Thousand Oaks, CA, USA, 2008.
2. Geoinformatics Laboratory of Pittsburgh University, 2017. Available online http://gis.sis.pitt.edu/images/GeoinformaticsDiagram.jpg (accessed on 29 March 2018).
3. Gore, A. The Digital Earth. *Aust. Surv.* **1998**, 43, 89–91, doi:10.1080/00050348.1998.10558728. [CrossRef]
4. Gould, M.; Craglia, M.; Goodchild, M.F.; Annoni, A.; Camara, G.; Kuhn, W.; Mark, D.; Masser, I.; Maguire, D.; Liang, S.; et al. Next-generation digital earth: A position paper from the vespucci initiative for the advancement of geographic information science. *Int. J. Spat. Data Infrastruct. Res.* **2008**, 43, 146–167.
5. Goodchild, M.F. Citizens as sensors: the world of volunteery geography. *GeoJournal* **2007**, 69, 211–221. [CrossRef]
6. Turner, A. *Introduction to Neogeography*; O'Reilly Media, Inc.: Sebastopol, CA, USA, 2006.
7. Sui, D.; Elwood, S.; Goodchild, M. *Crowdsourcing Geographic Knowledge: Volunteered Geographic Information (VGI) in Theory and Practice*; Springer Science & Business Media: Berlin, Germany, 2012.
8. See, L.; Mooney, P.; Foody, G.; Bastin, L.; Comber, A.; Estima, J.; Fritz, S.; Kerle, N.; Jiang, B.; Laakso, M.; et al. Crowdsourcing, Citizen Science or Volunteered Geographic Information? The Current State of Crowdsourced Geographic Information. *ISPRS Int. J. Geo-Inf.* **2016**, 5, 55, doi:10.3390/ijgi5050055. [CrossRef]
9. Jokar Arsanjani, J.; Vaz, E. Special Issue Editorial: Earth Observation and Geoinformation Technologies for Sustainable Development. *Sustainability* **2017**, 9, 760, doi:10.3390/su9050760. [CrossRef]

10. Arsanjani, J.J. Remote Sensing, Crowd Sensing, and Geospatial Technologies for Public Health: An Editorial. *Int. J. Environ. Res. Public Health* **2017**, *14*, 405, doi:10.3390/ijerph14040405. [CrossRef] [PubMed]

11. Pirasteh, S.; Li, J. *Global Changes and Natural Disaster Management: Geo-Information Technologies*; Springer: Berlin, Germany, 2017.

12. ISPRS. IJGI, Special Issue Geoweb 2.0, 2015. Available online http://www.mdpi.com/journal/ijgi/special_issues/geoweb-2.0 (accessed on 29 March 2018).

13. Follett, R.; Strezov, V. An analysis of citizen science based research: Usage and publication patterns. *PLoS ONE* **2015**, *10*, e0143687. [CrossRef] [PubMed]

14. The USA Citizen Science Initiative. Available online https://www.citizenscience.gov/ (accessed on 29 March 2018).

15. Australian Citizen Science Association (ACSA). Available online http://csna.gaiaresources.com.au/ (accessed on 29 March 2018).

16. COST_Actions CA15212, Citizen Science to Promote Creativity, Scientific Literacy, and Innovation Throughout Europe, 2016. Available online http://www.cost.eu/COST_Actions/ca/CA15212 (accessed on 29 March 2018).

17. Wiggins, A.; Crowston, K. From conservation to crowdsourcing: A typology of citizen science. In Proceedings of the 2011 44th Hawaii International Conference on System Sciences (HICSS), Kauai, HI, USA, 4–7 January 2011; pp. 1–10.

18. Serrano Sanz, F.; Holocher-Ertl, T.; Kieslinger, B.; Sanz García, F.; Silva, C. *White Paper on Citizen Science for Europe*; Socientize Consortium, European Commission: Brussels, Belgium, 2014.

19. Roy, H.E.; Pocock, M.J.; Preston, C.D.; Roy, D.B.; Savage, J.; Tweddle, J.; Robinson, L. *Understanding cItizen Science and Environmental Monitoring: Final Report on Behalf of UK Environmental Observation Framework*; NERC/Centre for Ecology & Hydrology: Wallingford, UK, 2012.

20. Wiggins, A.; Bonney, R.; Graham, E.; Henderson, S.; Kelling, S.; LeBuhn, G.; Litauer, R.; Lots, K.; Michener, W.; Newman, G. *Data Management Guide for Public Participation in Scientific Research*; DataONE: Albuquerque, NM, USA, 2013; pp. 1–41.

21. Law, E.; Williams, A.C.; Wiggins, A.; Brier, J.; Preece, J.; Shirk, J.; Newman, G. The Science of Citizen Science: Theories, Methodologies and Platforms. In Proceedings of the Companion of the 2017 ACM Conference on Computer Supported Cooperative Work and Social Computing, Portland, OR, USA, 25 February–1 March 2017; ACM: New York, NY, USA, 2017; pp. 395–400.

22. Eitzel, M.; Cappadonna, J.L.; Santos-Lang, C.; Duerr, R.E.; Virapongse, A.; West, S.E.; Kyba, C.C.M.; Bowser, A.; Cooper, C.B.; Sforzi, A.; et al. Citizen science terminology matters: Exploring key terms. *Citiz. Sci. Theory Pract.* **2017**, *2*. [CrossRef]

23. Newman, G.; Wiggins, A.; Crall, A.; Graham, E.; Newman, S.; Crowston, K. The future of citizen science: Emerging technologies and shifting paradigms. *Front. Ecol. Environ.* **2012**, *10*, 298–304. [CrossRef]

24. Open Geospatial Consortium. Available online http://www.opengeospatial.org/ (accessed on 29 March 2018).

25. Bordogna, G.; Carrara, P.; Criscuolo, L.; Pepe, M.; Rampini, A. On predicting and improving the quality of Volunteer Geographic Information projects. *Int. J. Digit. Earth* **2016**, *9*, 134–155, doi:10.1080/17538947.2014.976774. [CrossRef]

26. Criscuolo, L.; Carrara, P.; Bordogna, G.; Pepe, M.; Zucca, F.; Seppi, R.; Oggioni, A.; Rampini, A. Handling quality in crowdsourced geographic information. In *European Handbook of Crowdsourced Geographic Information*; Capineri, C., Ed.; Ubiquity Press: London, UK, 2016; p. 57.

27. Wiggins, A.; Newman, G.; Stevenson, R.D.; Crowston, K. Mechanisms for data quality and validation in citizen science. In Proceedings of the 2011 IEEE Seventh International Conference on e-Science Workshops (eScienceW), Stockholm, Sweden, 5–8 December 2011; pp. 14–19.

28. Goodchild, M.F.; Li, L. Assuring the quality of volunteered geographic information. *Spat. Stat.* **2012**, *1*, 110–120. [CrossRef]

29. Wikipedia Page: List of Citizen Science Projects. Available online https://en.wikipedia.org/wiki/List_of_citizen_science_projects (accessed on 29 March 2018).

30. Ess, C.; Sudweeks, F. On the edge: Cultural barriers and catalysts to IT diffusion among remote and marginalized communities. *New Media Soc.* **2001**, *3*, 259–269. [CrossRef]

31. International Telecommunication Union, ICT Data and Statistics Division of the Telecommunication Development Bureau. Available online https://www.itu.int/en/ITU-D/Statistics/Documents/facts/ICTFactsFigures2017.pdf (accessed on 29 March 2018).
32. OpenStreetMap. Available online https://www.openstreetmap.org/ (accessed on 29 March 2018).
33. Google Earth. Available online https://earth.google.com/web (accessed on 29 March 2018).
34. Campelo, C.; Elízio, C.; Bertolotto, M.; Corcoran, P. *Volunteered Geographic Information and the Future of Geospatial Data*; IGI Global: Hershey, PA, USA, 2017.
35. Mirowski, P. Against Citizen Science, 2017. Available online https://aeon.co/essays/is-grassroots-citizen-science-a-front-for-big-business (accessed on 29 March 2018).
36. Sturm, U.; Schade, S.; Ceccaroni, L.; Gold, M.; Kyba, C.; Claramunt, B.; Haklay, M.; Kasperowski, D.; Albert, A.; Piera, J.; et al. Defining principles for mobile apps and platforms development in citizen science. *Res. Ideas Outcomes* **2017**, *3*, e21283. [CrossRef]
37. Bordogna, G.; Frigerio, L.; Kliment, T.; Brivio, P.A.; Hossard, L.; Manfron, G.; Sterlacchini, S. Contextualized VGI Creation and Management to Cope with Uncertainty and Imprecision. *ISPRS Int. J. Geo-Inf.* **2016**, *5*, 234. [CrossRef]
38. Billiet, C.; de Weghe, N.V.; Deploige, J.; Tré, G.D. Visualizing and Reasoning With Imperfect Time Intervals in 2-D. *IEEE Trans. Fuzzy Syst.* **2017**, *25*, 1698–1713, doi:10.1109/TFUZZ.2016.2633363. [CrossRef]
39. Rocchini, D.; Foody, G.M.; Nagendra, H.; Ricotta, C.; Anand, M.; He, K.S.; Amici, V.; Kleinschmit, B.; Förster, M.; Schmidtlein, S.; et al. Uncertainty in ecosystem mapping by remote sensing. *Comput. Geosci.* **2013**, *50*, 128–135. [CrossRef]
40. Karam, R.; Favetta, F.; Laurini, R.; Chamoun, R.K. Uncertain Geoinformation Representation and Reasoning: A Use Case in LBS Integration. In Proceedings of the 2010 Workshops on Database and Expert Systems Applications, Bilbao, Spain, 30 August–3 September 2010; pp. 313–317, doi:10.1109/DEXA.2010.68. [CrossRef]
41. Bastin, L.; Schade, S.; Mooney, P. Volunteered Metadata, and Metadata on VGI: Challenges and Current Practices. In *Mobile information Systems leveraging Volunteered Geographic Information for Earth Observation*; Bordogna, G., Carrara, P.P., Eds.; Springer: Berlin, Germany, 2018.

International Journal of
Geo-Information

MDPI

Article

A New Method for the Assessment of Spatial Accuracy and Completeness of OpenStreetMap Building Footprints

Maria Antonia Brovelli * and Giorgio Zamboni

Department of Civil and Environmental Engineering (DICA), Politecnico di Milano, P.zza Leonardo Da Vinci 32, 20133 Milano, Italy; giorgio.zamboni@polimi.it
* Correspondence: maria.brovelli@polimi.it

Received: 29 April 2018; Accepted: 20 July 2018; Published: 24 July 2018

Abstract: OpenStreetMap (OSM) is currently the largest openly licensed collection of geospatial data, widely used in many projects as an alternative to or integrated with authoritative data. One of the main criticisms against this dataset is that, being a collaborative product created mainly by citizens without formal qualifications, its quality has not been assessed and therefore its usage can be questioned for some applications. This paper provides a map matching method to check the spatial accuracy of the building footprint layer, based on a comparison with a reference dataset. Moreover, from the map matching and a similarity check, buildings can be detected and therefore an index of completeness can also be computed. This process has been applied in Lombardy, a region in Northern Italy, covering an area of 23,900 km^2 and comprising respectively about 1 million buildings in OSM and 2.8 million buildings in the authoritative dataset. The results of the comparison show that the positional accuracy of the OSM buildings is at least compatible with the quality of the reference dataset at the scale of 1:5000 since the average deviation, with respect to the authoritative map, is below the expected tolerance of 3 m. The analysis of completeness, given in terms of the number of buildings appearing in the authoritative dataset and not present in OSM, shows an average percentage in the whole region equal to 57%. However, worth noting that the opposite, namely the number of buildings in OSM and not in the reference dataset, is not zero, but corresponds to 9%. The OSM building map can therefore be considered to be a valid base map for direct use (territorial frameworks, map navigation, urban analysis, etc.) and for derived use (background for the production of thematic maps) in all those cases where an accuracy corresponding to 1:5000 is required. Moreover it could be used for integrating the authoritative map at this scale (or smaller) where it is not complete and a rigorous quality certification in terms of metric precision is not required.

Keywords: GIS; digital cartography; algorithms; spatial accuracy; analysis; OpenStreetMap

1. Introduction

OpenStreetMap (OSM) [1] is currently the largest collaborative and openly licensed collection of geospatial data, widely used in many projects as an alternative to or integrated with authoritative data. OSM was founded in 2004 by Steve Coast as one of the first widespread efforts to provide a mapping platform for volunteered data capture [2]. It started as a mapping exercise for the United Kingdom but it spread quickly to the entire world. Coast's idea was simple: combining worldwide local geographic data collected by a large number of widespread people who have local knowledge, makes it possible to build a geodatabase of the world [3].

In line with its collaborative mission, OSM data are available under the Open Database License [4]. Maps from OSM have a Creative Commons Attribution-ShareAlike 2.0 license (CC BY-SA) [5]. This license allows everybody to use, distribute, transmit and adapt the data, as long as OSM and its contributors are credited. The license is viral: anyone who alters or builds upon OSM data must distribute the result only under the same license.

Currently, there are almost 4.5 million registered OSM members [6], more than 1 million contributors [7], and the impressive geospatial dataset is supported by software systems and applications, tools and web-based information stores such as wikis [3].

Many authors and commentators have concerned about the rapid and sustained success of OSM. One factor is certainly Web 2.0 [8], which makes it to develop large scale collaborative projects easier where hundreds or thousands of people are able to contribute simultaneously. A second factor is the availability of low-cost, high-quality, and high-accuracy positioning systems as standalone dedicated GPS units or embedded in portable devices such as smartphones. A third factor is related to the rising of so-called citizen science, i.e., scientific activities in which non-professional scientists volunteer to participate in data collection, analysis, and dissemination of a scientific project [9]. The volunteer practice and the outcome of the activities of OSM contributors, which are part of what is called volunteered geographic information (VGI) [10], can also be considered to be a component of "geographic" citizen science [9], both because of the scientific tools the volunteers make use of (remotely sensed images, GPS receivers, and map editing software) and the final collaborative aim they share, which is, mapping the Earth.

In turn, the phenomenon of citizen science can be seen as part of the new attitude towards more general open access, as well as collaborative and sharing approaches to information resources, which was named collective intelligence [11]. On the one side, this results in the willingness of citizens to participate in the knowledge production of the world; on the other side, these collaborative projects, of which OSM can be considered to be one of the most relevant examples, are inclusive and welcome anyone to take part in as a contributor, proposing a role and activities to everybody: beginners, expert level geographers, or software developers.

While the low level for entering and contributing has been one of the keys to the success of the initiative, the counterpoint is that, following a survey made by Budhathoki and Haythornthwaite [12], only 25% of the participants have "professional GIS experience" and therefore we are expecting a lower quality of the database compared with other sources, like data of national, regional and local mapping agencies as well as that produced by professionals and companies.

Some web tools are provided for mitigating and reducing errors. Examples related to geometry or topology are: the OSM Map Compare tool [13], which allows visual comparison of OSM map layers with other popular mapping systems such as Google maps [14], Bing maps [15], HERE maps [16], ESRI maps [17], etc.; Ma Visionneuse [18], which allows OSM to be compared with IGN (French National Institute of Geographic and Forest Information) France layers, amongst others; OSM Inspector [19], which shows potential errors like long segments in polygons and polylines, called "ways" in OSM, self intersecting ways, polygons, or polylines which are represented by only one point, called "nodes" in OSM, and polygons or polylines containing duplicate nodes (for details about the topological model of OSM, see the following sections).

A useful application, named Taginfo [20] helps to check the tags and therefore the thematic quality. JOSM Validator [21] is a core feature of one of the most advanced editing tools of OSM. It checks and fixes a wide variety of problems, including topological errors, unclosed polygons and overlapping areas.

Osmose [22] and Keep Right [23] highlight errors in geometry/topology, tags, attribution, and other general OSM errors. MapRoulette [24] is a gamified application to fix errors in OSM. Each challenge proposes a set of tasks, which vary in difficulty, allowing contributors to choose the types of errors they feel more confident about fixing. DeepOSM [25] trains a neural network with OSM tags and aerial imagery, allowing the prediction of mis-registered roads in OSM. The Grass&Green project [26] is meant to correct tagging or classification of land use features involving grass or green areas.

Despite this huge number of applications for alleviating or correcting errors, some inaccuracies still remain and assessing the quality of the OSM spatial database is an issue on the agenda.

The ISO/TC 211 (technical committee) of the International Organization for Standardization (ISO) defines geographic information quality as the totality of characteristics of a product that bear on its ability to satisfy stated and implied needs. The quality measures that are considered for assessing VGI [27] are in general the most significant of the ISO 19157: 2013 Geographic Information—Data quality [28]: positional, thematic and temporal accuracy, completeness and consistency, even if there is still lack of formal standards for OSM.

"Accuracy" refers to the degree of closeness between a measurement of a quantity and the value which is accepted as true for that quantity. "Completeness" is assessed with respect to features, attributes, and model [29]. It can be evaluated in terms of commission errors, i.e., excess of data, and omission errors, i.e., absence of data. "Consistency" evaluates the coherence in the data structures; the errors resulting from the lack of it can be classified as referring to the conceptual model, domain, format and topology.

The assessment of OSM is a hot research topic and the majority of scholars have been contributing to this topic comparing the database against authoritative ones.

Hecht et al. [30], for instance, evaluated OSM building completeness in two regions of Germany applying different methods. The results highlighted a low degree of completeness, which was better in urban areas than rural areas, and also that the choice of method used for assessing the completeness has a high effect on the estimated value. Fan et al. [31], in comparing the building OSM dataset with the official one of Munich (Germany), found its high completeness over the city. However, with respect to the positional accuracy, the result was not so good as an average offset of about 4m exists between the two datasets. Conversely, they found that the footprints in the OSM dataset were highly similar to those in the reference dataset in terms of shape, the main difference being in fewer details (i.e., fewer points) in the polygons.

A different method, based on homologous point detection, was proposed by Brovelli et al. [32] on the city of Milan. The results seemed promising and the authors, considering also that there is not a consolidated and unique way for assessing spatial accuracy and completeness of the buildings dataset of OSM, decided to propose a new method and to test it on a significant case study covering the Italian region of Lombardy, which has an extent larger than the half of Switzerland and comprises both rural and highly populated urban areas. Moreover, as map matching approaches are more challenging in dense urban areas, where buildings are located close to each other and similar in shape and size [33], the authors also analyzed the outcomes focusing on the capitals of the provinces of the region.

The proposed methodology is partially derived from previous work of Brovelli and Zamboni [34,35] whose aim was authoritative map matching and warping. The method was based on the characteristics of the compared maps, specifically the existence of well-defined and rigorous prescriptions for their production. In this new approach, a different equality function was used because of the heterogeneity of the methods (and related accuracies) used by volunteers in collecting data.

The outcome of the paper is twofold: firstly, the presentation of the approach, which even though it still needs to be refined and compared to other methods, has good performance. Secondly, the evaluation of the positional accuracy and completeness of a significant dataset; the result on 940,000 buildings in the OSM map and about 2.8 million buildings in the Lombardy Regional Topographical Database (DBT) map shows that the quality of the OSM buildings is comparable to that of the regional technical authoritative map at the scale of 1:5000 everywhere, even in dense urban areas.

The remainder of this paper is structured as follows: Section 2 gives an overview of research related to this paper; Sections 3 and 4 respectively describe the method implemented for assessing the spatial accuracy and the completeness; Section 5 presents the results of the test areas, and Section 6 summarizes the outcomes of the whole work and draws some activities for the future.

2. Related Work

As said, in recent years many researchers have been working on the assessment of OSM data. While significant attention has been paid to OSM positional accuracy assessment and completeness, fewer authors have investigated semantic, temporal and thematic accuracy, and consistency [36] and none, to the best of the authors' knowledge, have assessed all the elements of data quality. As the method presented in the paper allows the assessment of positional accuracy and feature completeness, the analysis of the previous literature was mainly concentrated on these two aspects.

In the beginning, studies focused on the quality assessment of OSM road networks, which were the primary subjects of the OSM survey. The first studies date back to 2009. In their works, Ather [37], Haklay [38] and Koukoletsos et al. [39] assessed the positional accuracy of OSM streets in England (in an area corresponding to about 100 square kilometers) by a visual comparison of a limited number of roads with an authoritative dataset and a statistical approach based on a buffer technique developed by Goodchild and Hunter [40]. Moreover the completeness of the dataset over all England was estimated comparing the lengths of the roads in OSM with those of the Ordnance Survey vector datasets (the official dataset used for the assessment). Kounadi [41] did a similar experiment in Athens, where he considered around 300 roads and obtained results comparable to Hakley's, i.e., an average difference between OSM and official roads of about 6 m and an average overlap of nearly 80%.

Cipeluch et al. [42] visually analyzed roads of five case study cities and towns in Ireland, finding slightly different results case by case, but highlighting that the OSM dataset merited attention if compared with other data, like the imagery available in Google Maps and Bing Maps. Moreover, they concluded that there is a need to develop metrics that allow the measurement of both accuracy and coverage at neighbourhood, county, and country levels so that the quality of the dataset can be quantified.

In the following years, other research studies were based on the buffer zone methodology [43–47]. The results are not homogeneous if we compare the different areas investigated. In Europe generally the spatial accuracy and the completeness were good enough, while in South Africa for instance, the dataset did not meet the accuracy requirements for the integration with the authoritative database.

Al-Bakri and Fairbairn [48] assessed OSM in areas of England and Iraq by comparing reference survey data sets, and again the buffer method was used. They concluded that the integration of OSM data for large scale mapping applications was not viable.

Helbich et al. [49], in a case study of a city in Germany, used bi-dimensional regression analysis to evaluate the global geometries of the patterns and detected clusters of high and low precision by means of local autocorrelation statistics. They found that the OSM areas of high accuracy were primarily located in more populated parts of the city, leading to the conclusion that these areas were subject to more frequent validation, with consequent correction of errors, than rural areas.

Antoniou assessed the positional accuracy by evaluating, from geometry and semantics, the distance between corresponding intersections of the road network [50].

Girres and Touya [51] applied multiple methods to deal with the complexity of the analysis of the data, like the Euclidean distance for point features; the average Euclidean distance for linear features; the Hausdorff distance for linear features; and the surface distance, granularity, and compactness for area features. In their work, a limited number of homologous features, i.e., features representing the same object in the OSM and authoritative datasets, were selected and matched manually to avoid errors related to automatic processes. Differences in position were then computed on each pair of homologous objects. While the mean distance was acceptable, the standard deviation was definitely larger than the reference accuracy used for official datasets, showing that there was a huge heterogeneity in the quality of the data. Regarding completeness, they found that, using the number of objects as an indicator, OSM was far from being complete (around 10%); the completeness improved, however, when they considered the comparison between the total length/area of the objects, obtaining an average value of around 40%. This result clearly showed that shorter/smaller objects are more likely to be absent, reflecting the fact that volunteer contributors tend to map the most important elements in the road network.

In 2013, Canavosio-Zuzelski et al. [52] proposed a rigorous photogrammetric approach based on stereo imagery and a vector adjustment model for assessing the positional accuracy of several

OSM city streets. Forghani and Delavar [53] analyzed the data of one central urban zone of Tehran with a method combining different geometric elements—like road length, median center, minimum bounding geometry and directional distribution—concluding that the quality of OSM roads can be considered of medium level, mainly due to their heterogeneity. Brovelli et al. [54,55] introduced an automatic procedure based on geometrical similarity and a grid-based approach for the evaluation of road completeness and positional accuracy. The procedure was tested with good results on Paris, but it is completely flexible and can be reused by anybody because it is available as open source modules of the GIS GRASS [56].

The general comment about the assessment of OSM roads is that, even if many studies have been done, a general automatic solution does not yet exist and therefore there is still room in this field for investigating methods and developing new procedures.

Apart from the assessment of roads, in recent years researchers have also started assessing other objects in the OSM database, such as building footprints. To investigate the suitability of OSM data for the generation of 3D building models, Goetz and Zipf [57] provided a first quantitative analysis of OSM completeness, simply comparing the number of buildings mapped in OSM and the total number of buildings derived from the census data in Germany and showing that (in 2012) OSM covered around 30% of the total buildings. Hecht et al. [30] proposed four different methods for assessing the completeness with respect to the authoritative database: two respectively based on the comparison of building numbers and building areas calculated for reference unit zones (unit-based methods); and two respectively based on centroid and overlap for the detection of corresponding buildings in the two datasets (object-based methods). Their results indicated that the unit-based comparisons are highly sensitive to differences between the authoritative and the OSM data modeling, while object-based methods are more sensitive to positional mismatches of the OSM buildings. Anyhow, based on the analysis of many case studies, they concluded that object-based methods are preferable.

Fram et al. [58] assessed the quality of OSM buildings in different cities in the UK with the aim of investigating the potential of OSM data in applications of risk management solutions, such as natural catastrophe exposure models. The study was conducted applying the area unit-based method through a comparison against the authoritative datasets, and showed that OSM building completeness is very variable both within and between UK cities. Moreover, they tried to find a proxy variable for better computing the OSM completeness but they were not able to arrive at a satisfactory result.

Fan et al. [31] evaluated the quality of OSM in terms of completeness, semantic accuracy, positional accuracy, and shape accuracy by using building footprints of the official German dataset as reference data. Limiting the results to the indicators of interest for this paper, completeness was based on the area, identifying as corresponding objects those with an overlapping area that is larger than 30% of the smaller area of the two objects in OSM and the authoritative data. When one building of OSM matched only one building in the reference dataset (relation 1:1), the key points of the reference footprint were extracted using the Douglas-Peucker algorithm [59]. Next, the minimum bounding rectangles (MBRs) were calculated for the two polygons and their edges marked if they were located respectively on the edges of the corresponding MBR (OSM or reference). Again, the OSM MBR was shifted to the center of the reference MBR, in such a way that edges of these two MBRs could be matched if they were located (almost) on the same place. Finally, the edges of the footprints were matched if they were marked to the same edge of the MBRs. Regarding the positional accuracy, only buildings with a 1:1 relation were involved in the analysis and the accuracy was computed from the average distance between corresponding points in the pair of footprints in the two data sets. Finally, they also proposed computing the shape similarity between buildings, based on the turning function or tangent function introduced by Arkin et al. [60] for measuring the similarity of two polygons.

Törnros et al. [61] assessed OSM building completeness by comparison with a reference dataset, the official cadastre, and adopting the unit based method. However, they proposed a step forward with the computation of uncertainties given in terms of true positives, i.e., reference building areas that have been correctly mapped in OSM; false negatives, i.e., reference building areas not mapped in OSM;

and false Positives, i.e., OSM building areas not mapped in the reference data set. Their conclusion was that it is best to adopt the true positive rate (building areas overlapping in OSM and the reference data) as a method for estimating the building completeness.

In the same year, Müller et al. [62], compared the OSM buildings with the authoritative ones in Switzerland with a procedure based on the respective centroid distance and the comparison of the shape signature based on the Arkin algorithm [60]. In more detail, the threshold of the centroid distance for correspondent buildings in the two datasets was set to 20 meters and the average of the turning function for matching buildings corresponded to around 1.25.

Finally, Brovelli et al. [32] assessed the completeness and the positional accuracy in Milan (Italy) using, for the former, the area unit based approach with the computation of uncertainties already seen in other studies, and, for the latter, a new method based on the automatic matching of the points of the footprints. The work presented here is a step forward in the definition of this second approach, which aims at contributing to the debate about positional accuracy and completeness of OSM, a debate that is still open.

3. Methodology: Spatial Accuracy

The assessment of the spatial accuracy proposed in this paper is based on the evaluation of the distance between points representing the same features in two different maps (or layers) depicting the same area. The implemented algorithm works on vector layers considering the vertices of the map features as a set of coordinates. In detecting the homologous entity (in our case the building footprint), the algorithm emulates what a human operator would do: it compares the position, the shape and the semantics of the features on the two maps.

Obviously, to find such a correspondence, the two maps must have more or less the same scale and they must show more or less the same level of detail (LoD).

In cartography the scale is a well-defined concept ("ratio of the length of an object on the map by the length of the same object on the ground"); conversely, LoD is a vague notion which can be considered as the translation of map scale for use in geographic databases for which the scale is not fixed [63].

Speaking about OpenStreetMap, it is a community project and few guidelines have been established about the LoD in order to have rich datasets influenced by the diversity of the contributors. This is a tricky issue because on the one hand, this represents the fullness of the geodataset (and also one of the reasons for the success of the project); on the other hand, it can be a limitation because this diversity affects the resulting data quality [64]. This heterogeneity leads to LoD inconsistencies, i.e., some very detailed features and some less detailed features may coexist on the map. Generally, the people contributing to the map do not have a formal professional background as map-makers and do not use the same surveying tools (the use of professional tools is not required). This is the main reason for the diversified contributions, which can be very detailed or very poor, depending on the skill and scrupulousness of contributors and on the method used for mapping.

Data are collected in heterogeneous ways: in the field, doing what is called armchair mapping or as bulk import. In the first case, the OSM contributor walks around and records GPS points or tracks; generally, low cost receivers are used in this operation. In the second case, data are traced mainly by interpreting satellite imagery uploaded to the OSM platform or by integrating additional single or small free and open datasets from websites. Therefore, we can assume that majority of this data has more or less the same accuracy as the satellite images available for the area they refer to. The last method, considered as a supplement to data collected by individual volunteers, consists of importing free and open datasets, controlling the coherence with the existing data through a complex merging process. This collection method is done by expert users and generally data are authoritative, i.e., the LoD depends on the scale of the source dataset.

The heterogeneity of the collection affects the scale of the OSM geodatabase [65]. Moreover, the LoD varies not only from one theme layer to another (e.g., the buildings are detected with different detail than the road edges), but also from one feature to another in the same theme [51]. Moreover, it has been documented that the positional quality of features improves as more contributors add data or modify a feature [38].

On the other hand, if we consider authoritative maps, they are generally produced by (or under the control of) a national, regional or local mapping agency that conform to well-defined guidelines, i.e., the LoD is homogeneous in the whole map.

The maps to be considered for assessing the OSM spatial accuracy must be the most accurate possible, i.e., large scale maps at urban scale, 1:1000–1:2000 or, in the worst case, at regional scale 1:10,000. In the former case, the accuracy is such that we can consider our procedure as a validation. In the latter case, it is more like a comparison, even if the authoritativeness of the map makes it suitable as a reference.

Given that the map depicts similar details, we define homologous pairs as points that, considering the scale of the map and the consequent cartographic error, are at the same location in both geo-datasets and represent the same feature. An example can be the corner of a building or an isolated feature represented by a point, which, as already mentioned, is modelled with a node in OSM.

In our method, we decided to deal with the building footprint layer, detecting the homologous pairs representing the vertices of the buildings. At first glance, the problem seems to be trivial, but many factors can affect the differences of two layers representing buildings of the same area: different LoD, different update of the maps, errors, etc. If the building is a simple one, modelled as a rectangle in both geodatabases, finding the homologues is trivial. However, if the shape of the building is complex, with details of protruding parts such as terraces, balconies, stairs, etc.—the search becomes more challenging, specifically in zones where there is a high density of buildings close to each other.

The visual and manual detection of these homologous pairs is easy, but it is time consuming, especially if we are dealing with a big geodataset composed of millions of points. To avoid the time-consuming manual search for these corresponding points and the possible human errors in their detection, a strategy is needed to automate the procedure. The idea is to reproduce as much as possible what operators do when they try to mentally overlay the two maps. The first step consists in visually searching for the same features represented on the two different maps.

We can simplify this operation in three steps: the analysis of the position of the points that describe the vertices of the features (position comparison), the analysis of the segments (edges) joining the vertices and forming the polygon (shape comparison), and finally the content analysis, i.e., what the polygon represents (semantic comparison). Starting from the conceptual model that every cartographic entity is essentially defined by points (coordinates) and by the meaning of the points themselves (semantic attributes), the simplest way to search the homologous points can be summarized as follows: a point P_1 on a map m_1 is homologous to a point P_2 on a map m_2 if the two geographic shapes related to the two points correspond in geometry and semantics.

Referring to the semantic aspect and focusing only on the building footprints, the check can be easily done by extracting the layers corresponding to buildings from the two geodatabases. In the case of OSM, the features tagged as buildings are taken into account; in the case of the authoritative geodatabases, it depends on the conceptual schema and the adopted nomenclature. In the following, we assume that this step has already been carried out and we refer to the description of the case study presented in the next session for further details. Figure 1 shows an example of some homologous points that can be manually detected in the same area of two different maps.

Figure 1. Example of homologous points on two different maps.

Before starting the detecting operation, we perform a raw fitting of the two maps, consisting in an affine transformation, done by applying a least square estimate on (at least) five visually and manually selected points. In the case of a rectangular map, the best choice is to pick out four points at the corners and the fifth in the center of the map.

The affine transformation is widely applied in map conflation, especially when the coordinate systems of one or both of the datasets are unknown. In most cases, the coordinate systems of the maps are known and it should be possible to directly overlap the two datasets without a pre-alignment. Even if the affine transformation cannot decrease local deformations by applying the same geometric correction homogeneously over the entire dataset, it is always preliminarily applied by the algorithm in order to reduce any systematic misalignments eventually introduced by approximate geodetic transformation methods often used in the common Geographic Information System (GIS). Moreover, the pre-transformation allows the users to indistinctly apply the algorithm independently of the known or unknown coordinate reference systems, thus making it a generalized approach.

The algorithm allows the user to choose the type of affine transformation depending on the type of cartographic data taken into account:

- The general affine transformation, consisting of a roto-translation with anisotropic variation of scale and skew (six parameters), can be used when there is no information on the reference system of the map to be evaluated and/or the acquisition methods (e.g., digitization of scanned paper maps) may have been homogeneously distorted the map altering the corners (e.g., altering the scale only along the acquisition axis of the scanner);
- The conform transformation, consisting of a roto-translation with isotropic variation of scale (four parameters), can be used when you want to be sure that the transformation does not change the shape of the geometries (preserving the corners of the original map);
- The translation, consisting of a degeneration of the affine transformation in which only the two shifts along the Cartesian axes are estimated (two parameters), can be used when the two maps are not in the same reference system but these are known: it is therefore possible to apply a datum transformation, usually implemented by the most common GIS. In these cases, the translation can compensate for any slight misalignments due to the fact that in some cases these transformation formulas are not rigorous but derived from approximate estimates.

Finally, it is also possible to avoid the pre-alignment of the two maps when they are natively in the same reference system or the user is certain that the transformation between datums has been carried out with rigorous methods.

Depending on the choice of applying the affine transformation or not, the algorithm searches the homologous points in an iterative process or in one step.

When the affine transformation is used, starting from the five points manually selected by the user, an iterative process is executed where at each step the affine parameters are estimated and the set of homologous points are determined. The whole procedure is repeated until the number of detected points becomes stable (see Figure 2).

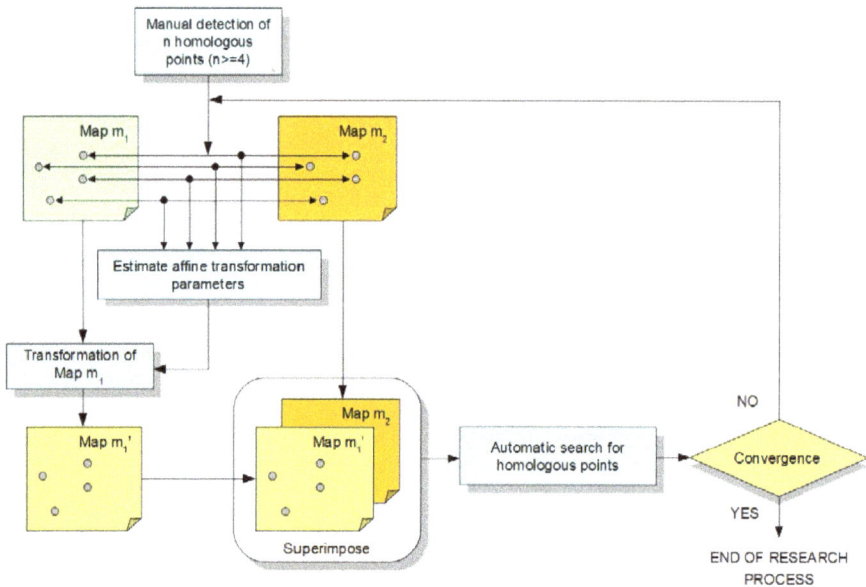

Figure 2. Homologous points detection algorithm.

When the transformation is not used, the set of homologous points are determined directly on the original maps and the schema of the algorithm represented in Figure 2 is reduced to the process box labeled with "Automatic search for homologous points".

The "Automatic search for homologous points" algorithm can be summarized as follows: if dist($P_1(i),P_2(k)$) is the distance from the point $P_1(i)$ on map m_1 to point $P_2(k)$ on map m_2, for each $P_1(i)$ ($i = 1, \dots , N$) on m_1 we search for the point $P_2(k)$ on m_2 which satisfies the condition of minimum distance from $P_1(i)$. If $P_1(i)$ and $P_2(k)$ are "geometrically compatible" (as described in detail below), $P_1(i)$ and $P_2(k)$ are set as homologous points.

The algorithm allows the choice of two different distances to measure the proximity of the candidate homologous points: a geometric distance and a statistic one. The former is the standard Euclidean distance while the latter is based on a Fisher test to establish if two candidate homologous points are compatible with the transformation model [34].

From the coordinates of the candidate homologous points and from the deterministic and stochastic model of the least squares approach used to estimate the transformation, we compute a variate F_0, which can be compared, with a fixed significance level α, with the critical value $F\alpha$ of a Fisher distribution of $(2, n - m)$ degrees of freedom. The first degree of freedom (the value 2) expresses the fact that we are considering a bi-dimensional problem. In the second degree of freedom, the number of observations used in the least square estimate—i.e., the n coordinates of the n/2 homologous points, and the m transformation parameters (6, 4, or 2, according to the chosen transformation) appear.

The test to accept the hypothesis H_0: {P_1 is homologous of P_2} can be formulated as follow: if H_0 is true then F_0 must be smaller than $F\alpha$ with probability $(1 - \alpha)$, otherwise H_0 is false.

Without detailing the test, we notice that moreover, to guarantee the uniqueness of the associations, for each point $P_1(i)$ on m_1 and the N points $P_2(k)$ ($k = 1, \ldots, N$) on m_2 which satisfy the hypothesis H_0, we selected the pair with smallest F_0.

The advantage of this approach, compared with the simple check of the standard geometric distance, is having a probability index that expresses the precision and the correctness of each homologous point association.

Beyond the distance, the "geometric compatibility" is based on the direction angles of the segments starting from the points and the inner corner of the edges of the polygon measured along the perimeter in a clockwise orientation.

Two points have an angle compatibility if both the direction angles of the common incoming segments and the corner angles are similar within a certain tolerance, hereafter indicated as α_{TOL} (see Figure 3).

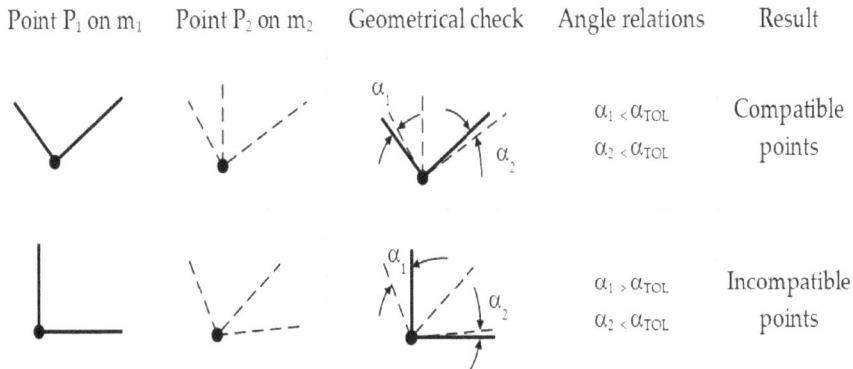

Figure 3. Examples of compatible/incompatible homologous points.

A step-by-step description of the algorithm with the use of the affine transformation can be summarized as follows:

- manual selection of five homologous points on the two maps m_1 and m_2;
- application of an affine transformation estimated using coordinates of the previous five points;
- repeat

  - for each point P_1 on the map m_1:

    - search the point P_2 on the map m_2 which satisfies the following conditions:

      - minimum distance from P_1 to P_2
      - the direction angles of all the incoming segments from the point P_1 are similar, within a certain tolerance α_{TOL}, to all the incoming segments from the point P_2
      - the inner corner of the edges P_1 measured along the perimeter of the polygon in a clockwise orientation is similar, within a certain tolerance α_{TOL}, to the inner corner of the edges P_2
      - if P_2 exists, set P_2 as the homologous point of P_1

- application of an affine transformation estimated using the new automatically detected homologous points

- until the count of homologous points converges

A step-by-step description of the algorithm without the use of the affine transformation can be summarized as follows:

- for each point P_1 on the map m_1:

  - search the point P_2 on the map m_2 which satisfies the following conditions:

    - minimum distance from P_1 to P_2
    - the direction angles of all the incoming segments from the point P_1 are similar, within a certain tolerance α_{TOL}, to all the incoming segments from the point P_2
    - the inner corner of the edges P_1 measured along the perimeter of the polygon in a clockwise orientation is similar, within a certain tolerance α_{TOL}, to the inner corner of the edges P_2

  - if P_2 exists, set P_2 as the homologous point of P_1

It is important to underline that the map transformation is used by the algorithm exclusively in the iterative search for the homologous pairs. The algorithm does not alter the coordinates of the input data and therefore the coordinates of the homologous points exactly match the coordinates of the original maps. In this way, the distance of the pairs can be used as a correct indicator for the assessment of the spatial accuracy of the maps.

A possible problem, also found in the test case shown in the next section, consists in not being able to find homologous points, in certain areas, due to the incompleteness of one or both maps. In fact, there may be situations in which in some areas there are data in the former map but not in the latter or vice versa. Generally, the reasons vary: maps may have been made or updated at different times and therefore what they depict is not exactly the same due to the evolution of the territory; they may depict different LoDs; or simply, and this is common enough in OpenStreetMap, some areas are less mapped because of a lack of volunteers in those zones who decided to map them.

Obviously, this is a factor that we have to consider in our analysis. As we are using an automatic detection of homologous pairs, some non-recognition of homologous points could be due to limitations inherent in the adopted method. Others, like the one mentioned, are unavoidable and even the most meticulous manual operator would not detect them.

4. Methodology: Completeness

For dealing with the incompleteness of the data, a parameter, $dist_{data}$, was defined. This can be calculated by comparing the number of detected points with the total number of vertices of all the buildings on the two maps (therefore we have two values, corresponding respectively to the former and latter map). Moreover, for dealing with the different levels of detail, to avoid considering missing buildings, a corrective calculation was introduced which does not consider the vertices of the buildings present in the first map that are not represented in the second one when counting the potential homologous pairs (i.e., the total number of vertices of the buildings).

Finally, we defined that a vertex on the first map does not have potential homologous in the second one (and must not be counted in the statistics) if there are no vertices in a significant neighborhood, where the significant neighborhood depends on the parameter $dist_{data}$. These points are called "isolated points".

The reasoning used for the completeness of the points can easily be extended to calculate the completeness of buildings. In fact, if all the vertices of a building are classified as isolate, then the building itself is considered isolated.

The percentage of isolated buildings can therefore be used, in addition to the quality of the data, to identify incomplete areas of a map (with respect to a more updated one) since the main and basic information usually depicted in a base map corresponds to streets and buildings.

The other factor that we mentioned and that has to be considered is the possible different levels of detail used to represent the same building on the two maps. According to the nominal scale of the maps (if this element was taken into account in its production; this is generally true for the authoritative maps) and to the different attention of the operators who digitized them, a building can be schematized for instance as a simple rectangle in a map or as a complex detailed polygon in the other one.

In this case, the more detailed map will have more points, many of which will not have a corresponding counterpart on the other map. These points, which we decided to call redundant points", should not be considered when counting potential homologous pairs, similarly to what we did for the isolated points.

Another set of points we did not consider is composed of all the vertices that do not represent significant variations in the shape of the buildings. These points are essentially intermediate vertices positioned along the effective edges and describe the geometry of not perfectly straight lines or that are used to model non-accentuated curves as a succession of segments with slight angular variations.

Also, in this case, the positioning and the number of these vertices depend on the subjectivity of the digitizer and on the scale of representation of the map: therefore, it is advisable not to take them into consideration. The parameter α_{tol}, already defined to compute the angular compatibility of the vertices, can be used to set the threshold under which not to consider an edge significant and therefore disregard it in the search algorithm.

5. The Lombardy Region Case Study

5.1. The Regional Topographical Database (DBT)

The methodology discussed in the previous section was applied to Lombardy, which is one of the northern regions of Italy, with an extent of about 23,900 km^2, a population of about 10 million and a population density of 420 people per km^2. This area was chosen because of its high level of urbanization and because of the availability of a good authoritative map to be used for checking the quality of the OpenStreetMap data. The official vector base map of the Lombardy region is named Regional Topographical Database (DBT). The DBT is the digital reference base for all planning tools made both by local authorities and the region, as defined in article 3 of the Regional Law 12/2005 for the Government of the Territory. It is a geographic database comprising various digital territorial information layers that represent and describe the topographic objects of the territory. Its main contents are: buildings, roads, railways, bridges, viaducts, tunnels, natural and artificial watercourses, lakes, dams, hydraulic works, electricity networks, waterfalls, altimetric information (contour lines and elevation points), quarries and landfills, plant covers, etc. Each object consists of a cartographic feature and an alphanumeric table, to which any other descriptive information is added according to the thematic layer: use and state of conservation of the building (residential, industrial, commercial, etc.), type of road surface (asphalted, starred, composite pavement, etc.), type of vegetation (divided into forests, pastures, agricultural crops, urban green, areas without vegetation), etc.

The survey scale is very detailed for urban areas (1:1000–1:2000) and at medium-scale for extra-urban areas (1:5000–1:10,000).

The DBT is carried out in collaboration with local authorities to have a unitary and homogeneous cartographic reference for all municipalities, provinces, the Lombardy region, other authorities and professionals. It is the main cartographic data used to build a regional Territorial Information System (SIT) in which all the thematic data and the plans of the various authorities converge. The DBT is the appropriate basis for municipal urban planning and other land planning tools. Moreover, it is the reference for all cartographic elaborations for anyone who wants to present a project to a public administration.

The technical specifications of the DBT are defined in official documents where the geometric accuracy is also prescribed. The standard deviation σ used as reference to define the accuracy of the map is defined for each cartographic scale. The tolerance for each DBT scale is defined equal to 2σ. The distribution of the residuals (the difference between the coordinate of the points stored in the DBT and their real coordinates, measured on randomly extracted samples) is always considered a normal one and therefore, in the quality control phase, only 5% of the absolute values of the differences can be higher than the tolerances. To further guarantee the quality of the data, the technical specifications prescribe that the residuals must in no case exceed twice this value; the maximum acceptable difference, in an absolute value, is therefore equal to 4σ.

Regarding the planimetric content of the DBT, the standard deviation for the various scales is as follows: for the scale 1:1000 σ = 0.30 m; for the scale 1:2000 σ = 0.60 m; for the scale 1:5000 σ = 1.50 m; and for the scale 1:10,000 σ = 3.00 m.

Similarly to the planimetric content, the altimetric accuracy is also prescribed. The standard deviation for the various scales is as follows: for the scale 1: 1000 σ = 0.30 m; for the scale 1:2000 σ = 0.40 m; for the scale 1:5000 σ = 1.00 m; and for the scale 1: 10,000 σ = 2.00 m.

Since, in the OSM, the altimetric information is not defined for buildings, in our tests the altimetric quality was assessed.

5.2. Zonal Positional Accuracies

The whole area of the Lombardy region was considered for the comparison of the OSM buildings with the homologous DBT building layer and, in the first instance, the whole area was divided into squares using a regular grid of 7 × 7 cells (see Figure 4). Splitting the region into cells was a result of the very large amount of data to be analyzed (as an order of magnitude, millions of buildings are involved). It is a common solution of breaking down a problem into more sub-problems of the same type, until these become simple enough to be directly solved [66]. The selection of the cell size was made based on two aspects: the type of possible misalignment of the two maps and the available hardware resources. When maps have a homogeneous misalignment, the cells can be wider than in the case of different localized misalignments, where a limited number of cells is preferred since the affine transformation is able to locally compensate for these deformations. Regarding the hardware resources, with smaller cells the computational performance is better (both in terms of required computation time and memory). A preliminary analysis of a sample dataset was performed and the optimal size of the cell was set to about 1000 km^2. Considering the total area of the case study, the minimum cell number of a regular grid that contains the whole region is equal to 49 (7 × 7). The irregular shape of the region leaves 11 cells without data. Hence the following results will refer to a sub-dataset of 38 cells instead of 49.

Since both the DBT and the OSM are dynamic maps constantly updated in a non-homogeneous way, it is not possible to define a unique date of realization of the whole dataset. It is therefore difficult to have a time alignment of the two maps since it would be necessary to compare the update dates zone by zone. Anyhow, in order to have a "time stamp" of the data used in the following tests, both maps were downloaded from the official repository at the same time (August 2017).

The parameters for the homologous points search algorithm were set according to the DBT accuracy. The maximum distance within which to search for a homologous point was set to 4σ, corresponding to the maximum acceptable tolerance of the DBT.

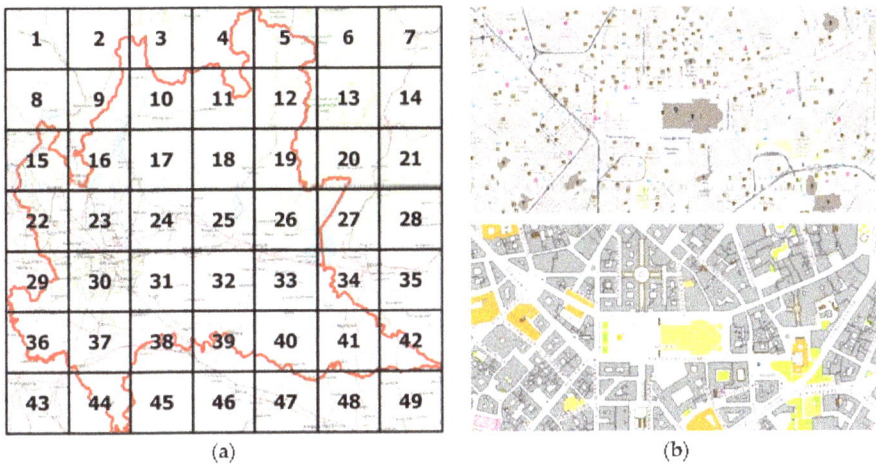

(a) (b)

Figure 4. (a) The regular grid used to analyze the Lombardy region delimited by red line; (b) one example of the OSM map (**up**) and the DBT map (**bottom**) of the same place.

Since the DBT Lombardy region is a mosaic of different local DBT at different scales (usually 1:1000 for historical centers of the city, 1:2000 for dense urbanized area, 1:5000 for peripheral sparse urbanized area and 1:10,000 for non-urbanized areas), the least restrictive tolerance among the urbanized areas was considered ($\sigma = 1.50$ m). The maximum distance was therefore set to 6.0 m.

Based on the same reasoning, the maximum distance used to consider a building to be isolated (isolated = without a corresponding homologous building on the other map), was set to twice the maximum tolerance and therefore set to 12.0 m.

Since only buildings were taken into account as geometric entities, the edges can define the shapes usually have corners of about 90 degrees and, more generally, greater than 45 degrees. With such a wide margin, it is therefore reasonable to not consider any data with vertices corresponding to angles of less than 10 degrees.

In the DBT, the building is defined as the whole of the volumetric unit that forms a body with a single building type; it may have several categories of use, it has a given state of conservation and it may have underground portions. Several alphanumeric attributes are available to describe it in detail. The main data are the building type (attribute: EDIFC_TY; values: generic house, terraced house, sports building, skyscraper, shed, monumental building, castle, etc.), the use (attribute EDIFC_USO; values: administrative, residential, public service, transportation services, commercial, industrial, etc.), and the state of conservation (attribute: EDIFC_STAT; values: in use, under construction, disused, etc.).

In order to make a direct comparison with the OSM buildings, only the geometries of the DBT buildings are taken into account, i.e., those being compatible with the OSM building model.

OpenStreetMap features are provided according to a topological data model [67]. The nodes describe points in the space by their latitude, longitude, and their identifier. The ways, which describe links, consisting of an identifier and an ordered list of between 2 and 2000 nodes. Relations allow the description of relationships between elements (which can be nodes, ways, and other relations). A feature is based on one of those three elements. Furthermore, it consists of a list of pairs (a key and a value) of called tags. Even if in principle arbitrary keys and values can be added to features, the OSM community agrees on certain key-value combinations for the most commonly used tags.

In the "standard" OSM model, buildings are features that have a tag with the key "building" [67]. The value of the tag may describe the type of accommodation (e.g., "apartments"), of commercial use (e.g., "warehouse"), of religious use (e.g., "cathedral"), or of civic/amenity use (e.g., "train station"), even if the most basic one is simply: "building = yes". For about 82% of all buildings [20], the value of the tag is purely "yes". For our purpose, we considered all typologies. There are currently about 282 million buildings in the OSM dataset [67], about 940 thousand in the Lombardy region. These buildings were used in our checking, subdividing them according to the grid we used for the DBT.

As explained in the previous theoretical section, the algorithm allows us to choose whether to apply a pre-alignment of the maps with an affine transformation (using an iterative approach) or to proceed directly to the search of the homologous points in one step.

In the following tests, all the solutions were investigated in order to analyze the differences, both in terms of homologous points found and in terms of reported statistical accuracy.

We expect that as a result of increasing the number of parameters of the transformation from 0 parameters (no transformation) up to six parameters (general affine transformation), the alignment of the two maps improves and therefore the number of homologous points, detected by the algorithm, increases. The greater the initial misalignment of the two maps, the greater the increase will be. Conversely, if the two maps are already well aligned, the number of homologous points is stable when the transformation changes.

Regardless of the number of points, if the accuracy of the OSM map is substantially homogeneous on the examined territory, the statistics on the distance between the homologous pairs should not change significantly. In fact it is important to remember that the transformation of the map is used exclusively in the iterative search process of the homologous pairs. At the end of the process the statistics on the distance of the pairs are calculated on the coordinates of the original maps (otherwise the results would no longer represent the accuracy of the original OSM map but the accuracy of the geometrically altered OSM).

The results for each cell obtained with no transformation in the search process are shown in Figure 5, while the results obtained using the general affine transformation in the search process are shown in Figure 6. Specifically, for each cell in Figures 5a and 6a, the number and percentage of homologous points detected by the algorithm are reported, while Figures 5b and 6b show the mean (M) and the standard deviation (S) of the distance of the points in the two maps. To get an indication of the correction to the statistics due to the highest number of homologous points detected using the affine transformation, the average differences are about 0.02 m for the mean and 0.03 m for the standard deviation. The results confirm that, in our case, the two datasets were already aligned since the pre-alignment does not introduce significant improvements on the number of points detected. We know that this is not true in general; for instance, in case of Munich an offset of about four meters on average in terms of positional accuracy was found [31]. Anyhow, the purpose of the proposed method was to be general and to allow global transformation for alleviating possible misalignments, which makes the automatic search of homologous pairs more difficult.

The percentages of homologous points detected with respect to the total potentially usable points (below indicated with P) can theoretically be computed both in the OSM map (P_{OSM}) and in the DBT map (P_{DBT}). By calculating these two values, it emerged that P_{OSM} was always significantly greater than P_{DBT} and it can be explained taking into consideration the redundant points. In the OSM map, the buildings are less detailed than in the DBT map and therefore there are more vertices in the DBT map that do not have a corresponding point in the OSM map. These points cannot be used by the algorithm and the result of the ratio between the number of homologous points detected and the total number of the points in the map inevitably decreases.

(a)

	2,375 (87%)	1,683 (86%)	7,799 (86%)	12,727 (85%)		
1,358 (80%)	27,991 (77%)	82,437 (84%)	53,282 (79%)	15,781 (82%)		
155,763 (82%)	208,201 (84%)	147,816 (94%)	17,550 (83%)	26,373 (81%)	2,672 (80%)	
436,840 (85%)	660,830 (80%)	242,231 (85%)	134,419 (80%)	134,349 (78%)	11,492 (76%)	
159,593 (92%)	355,844 (80%)	235,248 (86%)	34,548 (78%)	102,031 (88%)	6,582 (79%)	
2,363 (85%)	170,047 (80%)	39,517 (92%)	44,582 (87%)	5,985 (81%)	50,770 (83%)	66,131 (98%)
	116,995 (75%)	797 (78%)		245 (80%)	14,756 (99%)	

(b)

	M: 2.0 S: 1.2	M: 2.1 S: 1.2	M: 2.6 S: 1.3	M: 2.3 S: 1.3		
M: 1.7 S: 1.1	M: 2.3 S: 1.5	M: 1.7 S: 1.3	M: 2.1 S: 1.3	M: 2.2 S: 1.3		
M: 2.0 S: 1.2	M: 1.3 S: 1.4	M: 0.6 S: 1.2	M: 2.4 S: 1.3	M: 2.3 S: 1.3	M: 2.2 S: 1.2	
M: 1.4 S: 1.2	M: 1.5 S: 1.3	M: 1.8 S: 1.3	M: 2.3 S: 1.3	M: 2.3 S: 1.4	M: 2.3 S: 1.3	
M: 0.3 S: 0.8	M: 1.4 S: 1.1	M: 1.2 S: 1.3	M: 2.1 S: 1.4	M: 1.4 S: 1.6	M: 2.4 S: 1.3	
M: 1.7 S: 1.1	M: 1.4 S: 1.0	M: 0.7 S: 1.1	M: 0.7 S: 1.2	M: 2.1 S: 1.3	M: 0.7 S: 1.2	M: 0.0 S: 0.2
	M: 2.0 S: 1.2	M: 1.9 S: 1.0		M: 1.8 S: 1.0	M: 0.0 S: 0.1	

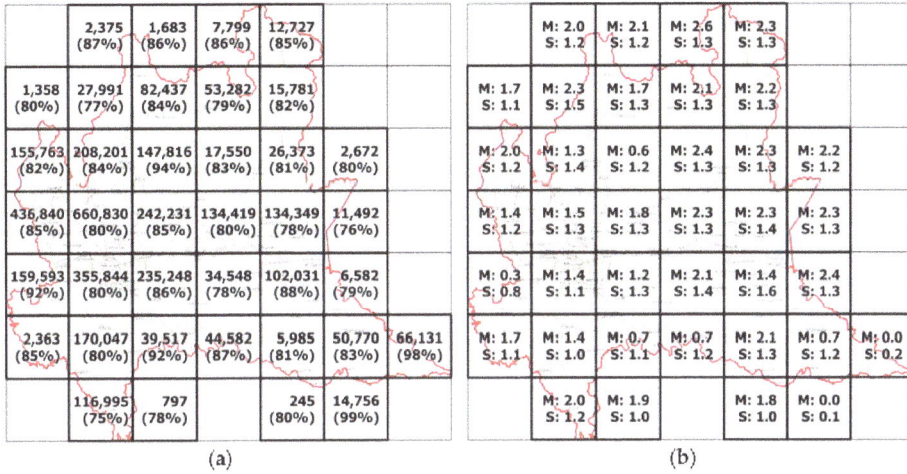

Figure 5. Homologous points detected without transformation in the search process. (**a**) Number and percentage of detected points with respect to the total number of OSM points; (**b**) mean (M) and standard deviation (S) in meters of the distance of the homologous points.

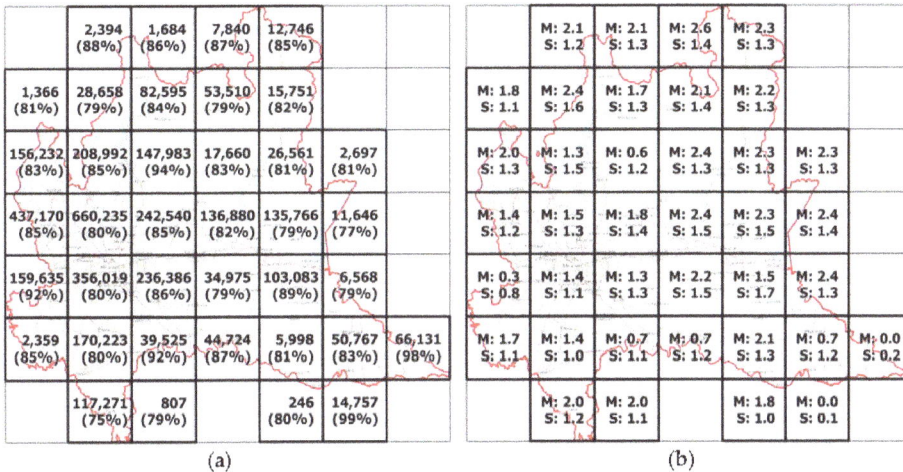

(a)

	2,394 (88%)	1,684 (86%)	7,840 (87%)	12,746 (85%)		
1,366 (81%)	28,658 (79%)	82,595 (84%)	53,510 (79%)	15,751 (82%)		
156,232 (83%)	208,992 (85%)	147,983 (94%)	17,660 (83%)	26,561 (81%)	2,697 (81%)	
437,170 (85%)	660,235 (80%)	242,540 (85%)	136,880 (82%)	135,766 (79%)	11,646 (77%)	
159,635 (92%)	356,019 (80%)	236,386 (86%)	34,975 (79%)	103,083 (89%)	6,568 (79%)	
2,359 (85%)	170,223 (80%)	39,525 (92%)	44,724 (87%)	5,998 (81%)	50,767 (83%)	66,131 (98%)
	117,271 (75%)	807 (79%)		246 (80%)	14,757 (99%)	

(b)

	M: 2.1 S: 1.2	M: 2.1 S: 1.3	M: 2.6 S: 1.4	M: 2.3 S: 1.3		
M: 1.8 S: 1.1	M: 2.4 S: 1.6	M: 1.7 S: 1.3	M: 2.1 S: 1.4	M: 2.2 S: 1.3		
M: 2.0 S: 1.3	M: 1.3 S: 1.5	M: 0.6 S: 1.2	M: 2.4 S: 1.3	M: 2.3 S: 1.3	M: 2.3 S: 1.3	
M: 1.4 S: 1.2	M: 1.5 S: 1.3	M: 1.8 S: 1.4	M: 2.4 S: 1.5	M: 2.3 S: 1.5	M: 2.4 S: 1.4	
M: 0.3 S: 0.8	M: 1.4 S: 1.1	M: 1.3 S: 1.3	M: 2.2 S: 1.5	M: 1.5 S: 1.7	M: 2.4 S: 1.3	
M: 1.7 S: 1.1	M: 1.4 S: 1.0	M: 0.7 S: 1.1	M: 0.7 S: 1.2	M: 2.1 S: 1.3	M: 0.7 S: 1.2	M: 0.0 S: 0.2
	M: 2.0 S: 1.2	M: 2.0 S: 1.1		M: 1.8 S: 1.0	M: 0.0 S: 0.1	

Figure 6. Homologous points detected using the general affine transformation in the search process. (**a**) Number and percentage of detected points with respect to the total number of OSM points; (**b**) Mean (M) and standard deviation (S) in meters of the distance of the homologous points.

For these reasons the value P_{OSM}, referred to the less detailed map, can be considered a reliability/quality index of the statistics reported in the figure as it represents the correct percentage of used data compared to potentially usable data.

The final results of the whole Lombardy region obtained, considering all the homologous points detected in each cell and differentiated by the type of transformation used in the search process, are reported in Table 1. With respect to the previous statistics, the percentiles are also reported in order to have more information about the compatibility of the maps.

Table 1. Points distance statistics for the whole Lombardy region.

Transformation	Number of Points	M (m)	S (m)	Percentile (m)							
				20%	40%	60%	80%	85%	90%	95%	100%
None	3,790,003	1.460	1.346	0.020	0.863	1.555	2.500	2.867	3.377	4.197	6.000
Translation	3,797,752	1.475	1.369	0.020	0.868	1.564	2.518	2.891	3.409	4.250	7.312
Conform	3,799,273	1.478	1.374	0.020	0.870	1.565	2.522	2.895	3.416	4.260	7.729
General Affine	3,800,380	1.480	1.378	0.020	0.870	1.567	2.525	2.899	3.421	4.269	9.533

Also in this case, the statistics confirm the expected results, since the number of homologous points varies only 0.3% from the use without transformation to the use of the general affine transformation and the mean and standard deviation are almost unchanged.

The important result to underline is that the positional accuracy of the OSM buildings of the analyzed area, statistically speaking, is at least compatible with the quality of a Regional Topographical Database at the scale of 1:5000, with an average deviation (with respect to the certificated points of the DBT) below the expected tolerance $2\sigma = 3.0$ m.

Furthermore, analyzing the individual results in detail, we found a higher quality for some areas with respect to the average quality of the whole dataset. In particular, the cells 17, 29, 38, 39, 41, 42, and 48 have an average deviation in the distance between homologous points of less than 1.0 m and are therefore even compatible with a DBT at the scale of 1:2000 (the average deviation is respectively equal to 1.2, 0.8, 1.1, 1.2, 1.2, 0.2, and 0.1 m).

Through a visual analysis of the data, it was possible to find that in these cells different buildings of the OSM map were exactly identical to the homologous buildings represented in the DBT map. For these portions of territory the OSM mappers have most probably preferred to directly import (bulk importing) buildings from the DBT database, which is published with an open license, rather than digitizing them from scratch using other reference sources. The equality of buildings introduces distances exactly equal to 0.0 m between the homologous points and the statistics improved significantly in these areas.

Specific studies focused on methods for identifying bulk imports exist [68,69]. For our purposes, that was to inspect the massive import of several buildings directly from another source (i.e., the DBT map), it was however sufficient to exploit the OSM changeset information.

A changeset consists of a group of changes made by a single user over a short period of time. One changeset may for example include the additions of new elements to OSM, the addition of new tags to existing elements, changes to tag values of elements, deletion of tags, and also deletion of elements [70]. Changesets can be directly accessed using the following URL schema: https://www.openstreetmap. org/changeset/\T1\textless{}Changesetnumber\T1\textgreater{}. Another option is to use the query feature and select a feature which shows the feature details and the last changeset for it.

By analyzing this information cell by cell for a sample of identical buildings results in the two maps, it was possible to note that the OSM buildings were loaded with the same changeset in a very short period of time (in a few seconds) and therefore using an automated bulk import procedure.

However, this information was not sufficient to confirm that the source of the imported data was exactly the same as the DBT map. In order to verify this and to quantify for each cell the number of identical buildings present in both maps, a specific comparison procedure was carried out: we decided to consider two buildings to be identical if all the vertices have a distance less than a threshold of 0.1 m. In Figure 7a, the spatial distribution of the OSM buildings identical to the DBT buildings are reported and in Figure 7b the percentage values with respect to the total OSM buildings in each cell are reported.

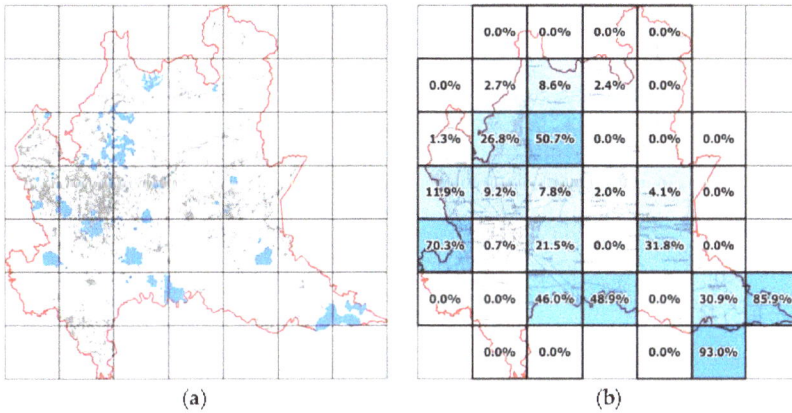

Figure 7. (**a**) OSM buildings identical to the DBT buildings; (**b**) percentage of identical buildings in each cell with respect to the total number of OSM buildings.

The percentage of identical buildings for cells 17, 29, 38, 39, 41, 42, and 48 analytically confirms the previous hypothesis about the improved statistics in these areas.

5.3. Positional Accuracies on the Province Capitals

To confirm the results about the accuracy of the OSM map, another analysis was carried out by using a different approach in the subdivision of the dataset. Instead of splitting the region using an abstract geometric criterion in which each regular cell contains a significantly heterogeneous territory (highly urbanized areas, expanses of agricultural areas, inhabited mountainous areas, etc.), administrative boundaries were considered and the provincial capitals were analyzed for detecting the positional quality of OSM in these cities.

In Figure 8, the analyzed areas are reported together with a table containing the number of homologous points and the relative statistics for each city. In the table, the percentages of identical buildings measured in the two maps are also reported, and the high values found in cities of Lecco, Lodi and Pavia explain the high precision of the OSM and DBT map alignment for these three cities.

As a general comment, these results confirm those obtained in the previous test and allow us to compare the quality of the OSM map to that of a DBT map at scale 1:5000.

It is finally worth underlining that the OSM map is not fully compliant with the metric requirements of a DBT map since it is not guaranteed that 95% of the points have a tolerance lower than 2σ. The different acquisition techniques of OSM and DBT have however to be considered in order understand the origin of some errors and to give a correct interpretation of the results. In the case of the DBT, for example, the perimeter of the building on the ground is outlined in the worst cases by using stereoscopic images and in the best cases by a field topographic survey; conversely, in OSM orthophotos (often with roofs partially hiding the perimeter of the buildings on the ground) are used. DBT is a product of professionals; conversely, OSM is contributed by citizens. If we consider the value for money, OSM can be considered to be a good product for integrating into authoritative data, especially in zones where DBT lacks information.

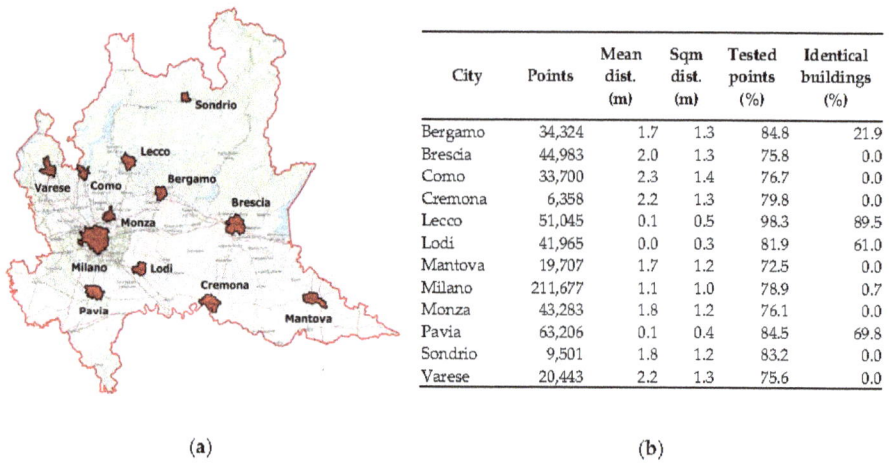

City	Points	Mean dist. (m)	Sqm dist. (m)	Tested points (%)	Identical buildings (%)
Bergamo	34,324	1.7	1.3	84.8	21.9
Brescia	44,983	2.0	1.3	75.8	0.0
Como	33,700	2.3	1.4	76.7	0.0
Cremona	6,358	2.2	1.3	79.8	0.0
Lecco	51,045	0.1	0.5	98.3	89.5
Lodi	41,965	0.0	0.3	81.9	61.0
Mantova	19,707	1.7	1.2	72.5	0.0
Milano	211,677	1.1	1.0	78.9	0.7
Monza	43,283	1.8	1.2	76.1	0.0
Pavia	63,206	0.1	0.4	84.5	69.8
Sondrio	9,501	1.8	1.2	83.2	0.0
Varese	20,443	2.2	1.3	75.6	0.0

(a)   (b)

Figure 8. (a) Provincial capitals of the Lombardy region; (b) homologous points statistics for each capitals.

5.4. Completeness

With respect to this completeness, our procedure can also be useful for evaluating the relative local and global territorial coverage of the two maps. In Figure 9, two examples of incomplete areas are shown. On the left, an area (labeled with the letter "A") is shown where the buildings are only present in the DBT map; on the right, another area (labeled with the letter "B") is shown where the buildings are only present in the OSM map. The results of the detection of isolated buildings on both maps are shown in Figure 10.

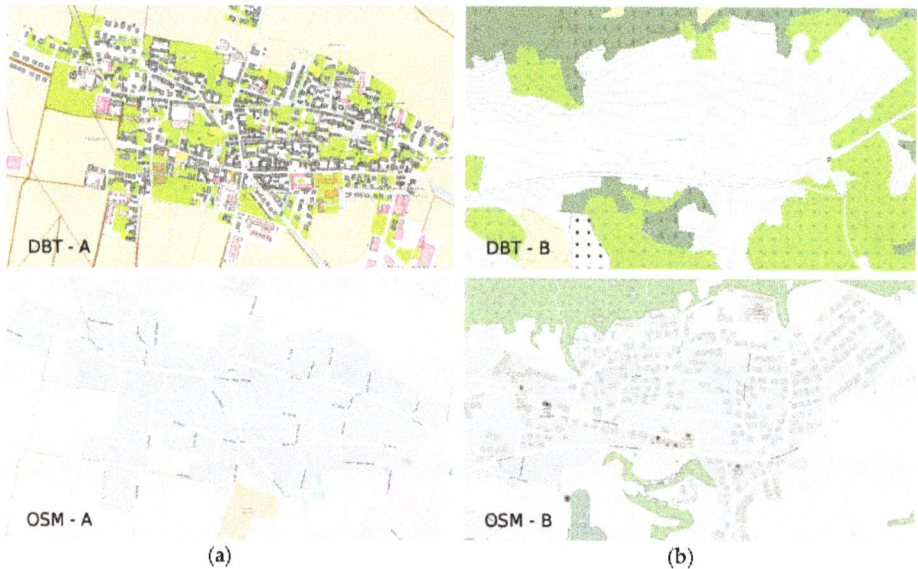

(a)   (b)

Figure 9. (a) Buildings only depicted in the DBT map; (b) buildings only depicted in the OSM map.

1.2%	1.9%	1.2%	2.2%		
0.3%	9.7%	1.8%	2.9%	13.7%	
1.7%	9.0%	9.1%	48.9%	30.1%	47.8%
1.3%	11.8%	14.9%	8.5%	8.7%	30.2%
1.0%	2.0%	3.7%	2.8%	13.0%	75.3%
80.6%	11.5%	8.6%	2.6%	45.5%	30.2%	11.6%
	2.2%	2.1%		93.0%	5.5%

(a)

	35.2%	73.0%	15.9%	50.5%	
49.4%	42.4%	55.9%	62.5%	76.8%	
48.2%	37.9%	47.5%	80.6%	68.4%	10.0%
46.2%	45.7%	62.9%	70.6%	63.5%	42.6%
59.4%	45.7%	53.9%	88.3%	76.4%	20.6%
94.6%	51.7%	79.6%	74.2%	93.5%	39.7%	0.4%
	5.3%	7.5%		81.2%	2.7%

(b)

Figure 10. (a) Percentages of OSM buildings without a corresponding DBT building; (b) percentages of DBT buildings without a corresponding OSM building.

The percentages of buildings in the OSM maps without a corresponding building on the DBT map are shown on the left, while the percentage of buildings in the DBT map without a corresponding building on the DBT map are shown on the right.

The analysis of completeness, given in terms of the count of buildings appearing in the authoritative dataset but that are not present in OSM, shows an average percentage in the whole region equal to 57%. Conversely, the count of buildings appearing in the OSM map that are not present in the authoritative dataset, shows an average percentage in the whole region equal to 9%. These results highlight that the DBT map is more complete than the OSM map, but that there is still a part of DBT area not yet mapped, or that does not contain updates already present in the OSM map. It is important to highlight that, as previously explained, the two maps are updated asynchronously and at irregular intervals and with this proposed method, it is not automatically possible to distinguish incompleteness due to non-mapping from incompleteness due to non-updating.

Finally, it must be taken into account that in the very low-urbanized areas and in the boundary cells, the completeness percentages are statistically less significant because the amount of data is much lower than in the other cells. This is also one of the reasons for the variability in the local results.

5.5. Performance of the Assessment of Spatial Accuracy and Completeness

Due to the huge number of homologous points detected by the automatic search algorithm (about 3.8 million homologous points), both a systematic and a statistical validation of the results using a manual approach are impracticable. Even if the geometric controls used by the algorithm can already be considered to be a good guarantee of the correctness of the results, an automatic validation system is currently being studied in order to provide a tool to cross-validate the accuracy and the completeness of the assessment.

In the meantime, a first manual check of the results was applied by analyzing 30 randomly selected buildings (adding up to a total of 30 buildings × 38 cells = 1140 buildings) for each cell and by visually verifying for each of them the correctness and completeness of the homologous points. In the same way, the method used to compute the completeness of the maps was also verified, selecting 30 buildings for each cell automatically highlighted by the algorithm as "isolated" in the OSM map and 30 buildings automatically highlighted by the algorithm as "isolated" in the DBT map. The results of the manual check are reported in Table 2. "Wrong points" is the number of homologous points incorrectly identified by the algorithm (correctness index), while the "missing points" is the number of homologous points not detected by the algorithm but that a human operator would have identified (completeness index). For checking the completeness of the maps, both the numbers of buildings classified by the algorithm only in the OSM map and only in the DBT map that was checked are reported, as well as the total number of wrong classifications. The small percentages found for the errors prove that the method is reliable.

Table 2. Validation of the assessment of spatial accuracy and completeness.

Homologous Points Check				Completeness Maps Check			
Number of buildings	Homologous points	Wrong points	Missing points	Number of building	Only in OSM	Only in DBT	Wrong classification
1140	5347	172 (3%)	269 (5%)	1140	570	570	0 (0%)

6. Conclusions

For several years, the base topographic maps have been used exclusively by administrative and governmental authorities and copyright protection was often justified by the high costs incurred for the creation and updating of the contents.

The advent of both VGI and the open-source philosophy has significantly contributed to the spread of a new type of geographic data, built with the voluntary contribution of the people and accessible for free without special licensing restrictions (e.g., ODbL [4] and CC BY-SA [5] license used by OSM).

For the official maps, there are well-defined production and updating techniques and this guarantees that the data has rigorous metric precision. For the data contributed by citizens and, in this specific case, for the OSM map, one of the problems still open is the control of its positional accuracy. This paper presents a possible external check, based on the comparison with certified authoritative data.

Test cases were carried out comparing the Topographic Data Base buildings of the Lombardy region with the OSM buildings, and some detailed statistics with respect to the assessment of spatial accuracy of the data were reported.

The work exploits an automatic search algorithm of homologous pairs between two different maps that allows us to significantly increase the volumes of compared data; this approach avoids human subjectivity in the selection of the features to compare and increases the data redundancy in the statistical estimation with respect to the limited number of control points that can (realistically) be manually checked.

Although the work requires further study and refinements, the first overall result on more than 3 million homologous points and about 940 thousand buildings in the OSM map and about 2.8 million buildings in the DBT map, allows us to say that the quality of the OSM buildings is comparable to that of the regional technical authoritative map at the scale of 1:5000. The OSM building map can therefore be considered to be a valid base map for both direct use (territorial frameworks, map navigation, urban analysis, etc.) and for derived use (background for the production of thematic maps) in all those cases in which a precision of 1:5000 is required. Moreover, it can be used for integration with the authoritative map at this scale (or greater) where it is not complete and rigorous quality certification in terms of metric precision is not required.

An improvement already planned for future work aims at solving a specific issue that the algorithm is not yet able to solve automatically. Since the two maps have inherent (and unavoidable) differences in their data creation and updating procedures for their features/buildings, the proposed methodology could incorrectly match already demolished buildings with new buildings roughly built at the same location. In order to avoid this false matching, a more sophisticated check must be introduced, taking into account not only the point compatibility, but also the compatibility of the whole building. However, it is important to highlight how these possible commission errors, at least for the examined case, will not affect the results presented in this paper. In fact, we automated the analysis of 2.8 million buildings and computed the statistics on 3.8 million homologous points, while in the Lombardy region, considering as an example the period of five years 2011–2015 [71], when about 21,000 buildings (both residential and non-residential) were built or expanded. As this value corresponds only to 0.8% of all buildings in Lombardy, even if all the vertices of these buildings were incorrectly classified by the algorithm as homologous points, the statistics would not be significantly changed.

Author Contributions: M.A.B. and G.Z. contributed to the design and implementation of the research, to the analysis of the results and to the writing of the manuscript. G.Z. wrote the code, while M.A.B. supervised the project.

Funding: This work was supported by URBAN-GEO BIG DATA, a Project of National Interest (PRIN) funded by the Italian Ministry of Education, University and Research (MIUR)—ID. 20159CNLW8.

Acknowledgments: The authors want to thank Lombardy Region and OpenStreetMap© contributors for making available as open their databases.

Conflicts of Interest: The authors declare no conflicts of interest.

References

1. OpenStreetMap. Available online: https://www.openstreetmap.org (accessed on 25 April 2015).
2. Haklay, M.; Weber, P. Openstreetmap: User-generated street maps. *IEEE Pervasive Comput.* **2008**, *7*, 12–18. [CrossRef]
3. Mooney, P.; Minghini, M. A Review of OpenStreetMap Data. In *Mapping and the Citizen Sensor*; Foody, G., See, L., Fritz, S., Mooney, P., Olteanu-Raimond, A.-M., Fonte, C.C., Antoniou, V., Eds.; Ubiquity Press: London, UK, 2017; pp. 37–59.
4. Open Data Commons Open Database License (OdbL). Available online: https://opendatacommons.org/licenses/odbl (accessed on 25 April 2018).
5. Creative Commons, Attribution-ShareAlike 2.0 Generic (CC BY-SA 2.0). Available online: https://creativecommons.org/licenses/by-sa/2.0 (accessed on 25 April 2018).
6. OSMstats, Statistics of the Free Wiki World Map. Available online: https://osmstats.neis-one.org (accessed on 25 April 2018).
7. Stats—OpenStreetMap Wiki. Available online: https://wiki.openstreetmap.org/wiki/Stats (accessed on 25 April 2018).
8. O'Reilly, T. What is Web 2.0: Design Patterns and Business Models for the Next Generation of Software. Available online: http://www.oreilly.com/pub/a/web2/archive/what-is-web-20.html (accessed on 25 April 2018).
9. Haklay, M. Citizen Science and Volunteered Geographic Information—Overview and Typology of Participation. In *Crowdsourcing Geographic Knowledge: Volunteered Geographic Information (VGI) in Theory and Practice*; Sui, D.Z., Elwood, S., Goodchild, M.F., Eds.; Springer: Berlin, Germany, 2013; pp. 105–122.
10. Goodchild, M.F. Citizens as sensors: The world of volunteered geography. *GeoJournal* **2007**, *69*, 211–221. [CrossRef]
11. Levy, P. *L'Intelligence Collective. Pour une Anthropologie du Cyberespace*; La Découverte: Paris, France, 1994; ISBN 10 2707126934.
12. Budhathoki, N.R.; Haythornthwaite, C. Motivation for Open Collaboration: Crowd and Community Models and the Case of OpenStreetMap. *Am. Behav. Sci.* **2012**, *57*, 548–575. [CrossRef]
13. Map Compare. Available online: http://mc.bbbike.org/mc (accessed on 25 April 2018).
14. Bing Maps. Available online: https://www.bing.com/maps (accessed on 25 April 2018).
15. Google Maps. Available online: https://www.google.com/maps (accessed on 25 April 2018).
16. HERE Maps. Available online: https://wego.here.com (accessed on 25 April 2018).
17. Esri Maps. Available online: https://livingatlas.arcgis.com/en/ (accessed on 25 April 2018).
18. Ma Visioneeuse. Available online: http://mavisionneuse.ign.fr/visio.html (accessed on 25 April 2018).
19. OSM Inspector. Available online: http://tools.geofabrik.de/osmi (accessed on 25 April 2018).
20. OpenStreetMap Taginfo. Available online: https://taginfo.openstreetmap.org/ (accessed on 25 April 2018).

21. JOSM Validator—OpenStreetMap Wiki. Available online: http://wiki.openstreetmap.org/wiki/JOSM/Validator (accessed on 25 April 2018).

22. Osmose—OpenStreetMap Wiki. Available online: https://wiki.openstreetmap.org/wiki/Osmose (accessed on 25 April 2018).

23. Keep Right—OpenStreetMap Wiki. Available online: https://wiki.openstreetmap.org/wiki/Keep_Right (accessed on 25 April 2018).

24. Map Roulette. Available online: http://maproulette.org (accessed on 25 April 2018).

25. DeepOSM. Available online: https://libraries.io/github/trailbehind/DeepOSM (accessed on 25 April 2018).

26. Ali, A.L.; Sirilertworakul, N.; Zipf, A.; Mobasheri, A. Guided Classification System for Conceptual Overlapping Classes in OpenStreetMap. *ISPRS Int. J. Geo-Inf.* **2016**, *5*. [CrossRef]

27. Fonte, C.C.; Antoniou, V.; Bastin, L.; Bayas, L.; See, L.; Vatseva, R. Assessing VGI data quality. In *Mapping and the Citizen Sensor*; Foody, G., See, L., Fritz, S., Mooney, P., Olteanu-Raimond, A.-M., Fonte, C.C., Antoniou, V., Eds.; Ubiquity Press: London, UK, 2017; pp. 137–164.

28. Geographic Information—Data Quality. 2013. Available online: https://www.iso.org/standard/32575.html (accessed on 25 April 2018).

29. Senaratne, H.; Mobasheri, A.; Loai Ali, A.; Capineri, C.; Haklay, M. A review of volunteered geographic information quality assessment methods. *Int. J. Geogr. Inf. Sci.* **2017**, *31*, 139–167. [CrossRef]

30. Hecht, R.; Kunze, C.; Hahmann, S. Measuring completeness of building footprints in OpenStreetMap over space and time. *ISPRS Int. J. Geo-Inf.* **2013**, *2*, 1066–1091. [CrossRef]

31. Fan, H.; Zipf, A.; Fu, Q.; Neis, P. Quality assessment for building footprints data on OpenStreetMap. *Int. J. Geogr. Inf. Sci.* **2014**, *28*, 700–719. [CrossRef]

32. Brovelli, M.A.; Minghini, M.; Molinari, M.E.; Zamboni, G. Positional accuracy assessment of the OpenStreetMap buildings layer through automatic homologous pairs detection: The method and a case study. *Int. Arch. Photogramm. Remote Sens. Spat. Inf. Sci.* **2016**, *41*, 615–620. [CrossRef]

33. Yong, H.; Sungchul, Y.; Chillo, G.; Kiyun, Y.; Wenzhong, S. Line segment confidence region-based string matching method for map conflation. *ISPRS J. Photogramm. Remote Sens.* **2013**, *78*, 69–84. [CrossRef]

34. Brovelli, M.A.; Zamboni, G. A step towards geographic interoperability: The automatic detection of maps homologous pairs. In Proceedings of the UDMS '04, Chioggia, Italy, 27–29 October 2004.

35. Brovelli, M.A.; Zamboni, G. Adaptive Transformation of Cartographic Bases by Means of Multiresolution Spline Interpolation. *ISPRS Int. Arch. Photogramm. Remote Sens. Spat. Inf. Sci.* **2004**, *35*, 206–211.

36. Antoniou, V.; Skopeliti, A. Measures and indicators of VGI quality: An overview. *ISPRS Ann. Photogramm. Remote Sens. Spat. Inf. Sci.* **2015**, *2*, 345–351. [CrossRef]

37. Ather, A. A Quality Analysis of Openstreetmap Data. Master's Thesis, University College of London, London, UK, 2009.

38. Haklay, M.; Basiouka, S.; Antoniou, V.; Ather, A. How many volunteers does it take to map an area well? The validity of Linus' Law to volunteered geographic information. *Cartogr. J.* **2010**, *47*, 315–322. [CrossRef]

39. Koukoletsos, T.; Haklay, M.; Ellul, C. An automated method to assess Data Completeness and Positional Accuracy of OpenStreetMap. *GeoComputation* **2011**, *3*, 236–241.

40. Goodchild, M.F.; Hunter, G.J. A simple positional accuracy measure for linear features. *Int. J. Geogr. Inf. Sci.* **1997**, *11*, 299–306. [CrossRef]

41. Kounadi, O. Assessing the Quality of OpenStreetMap Data. Master's Thesis, University College of London, London, UK, 2009.

42. Ciepluch, B.; Jacok, R.; Mooney, P.; Winstanley, A.C. Comparison of the accuracy of OpenStreetMap for Ireland with Google Maps and Bing Maps. In Proceedings of the Ninth International Symposium on Spatial Accuracy Assessment in Natural Resources and Environmental Sciences, Leicester, UK, 20–23 July 2010; pp. 337–340.

43. Koukoletsos, T.; Haklay, M.; Ellul, C. Assessing data completeness of VGI through an automated matching procedure for linear data. *Trans. GIS* **2012**, *16*, 477–498. [CrossRef]

44. Zielstra, D.; Zipf, A. A comparative study of proprietary geodata and volunteered geographic information for Germany. In Proceedings of the 13th AGILE International Conference on Geographic Information Science 2010, Guimaraes, Portugal, 11–14 May 2010.

45. Wang, M.; Li, Q.; Hu, Q.; Zhou, M. Quality Analysis of Open Street Map Data. *ISPRS Int. Arch. Photogramm. Remote Sens. Spat. Inf. Sci.* **2013**, *5*, 155–158. [CrossRef]

46. Siebritz, L.A.; Sithole, G. Assessing the Quality of OpenStreetMap Data in South Africa in Reference to National Mapping Standards. In Proceedings of the Second AfricaGEO Conference, Cape Town, South Africa, 1–3 July 2014.

47. Graser, A.; Straub, M.; Dragaschnig, M. Towards an open source analysis toolbox for street network comparison: Indicators, tools and results of a comparison of OSM and the official Austrian reference graph. *Trans. GIS* **2014**, *18*, 510–526. [CrossRef]

48. Al-Bakri, M.; Fairbairn, D. Assessing the accuracy of crowdsourced data and its integration with official spatial datasets. In Proceedings of the Ninth International Symposium on Spatial Accuracy Assessment in Natural Resources and Environmental Sciences, Leicester, UK, 20–23 July 2010; pp. 317–320.

49. Helbich, M.; Amelunxen, C.; Neis, P. Comparative Spatial Analysis of Positional Accuracy of OpenStreetMap and Proprietary Geodata. In Proceedings of the Geoinformatics Forum, Salzburg, Austria, 3–6 July 2012.

50. Antoniou, V. User Generated Spatial Content: An Analysis of the Phenomenon and Its Challenges for Mapping Agencies. Ph.D. Thesis, University College London (UCL), London, UK, 2011.

51. Girres, J.F.; Touya, G. Quality assessment of the French OpenStreetMap dataset. *Trans. GIS* **2010**, *14*, 435–459. [CrossRef]

52. Canavosio-Zuzelski, R.; Agouris, P.; Doucette, P. A Photogrammetric Approach for Assessing Positional Accuracy of OpenStreetMap© Roads. *ISPRS Int. J. Geo-Inf.* **2013**, *2*, 276–301. [CrossRef]

53. Forghani, M.; Delavar, M.R. A Quality Study of the Open Street Map Dataset for Tehran. *ISPRS Int. J. Geo-Inf.* **2014**, *3*, 750–763. [CrossRef]

54. Brovelli, M.A.; Minghini, M.; Molinari, M.E. An automated GRASS-based procedure to assess the geometrical accuracy of the OpenStreetMap Paris road network. *ISPRS Int. Arch. Photogramm. Remote Sens. Spat. Inf. Sci.* **2016**, *41*, 919–925. [CrossRef]

55. Brovelli, M.A.; Minghini, M.; Molinari, M.; Mooney, P. Towards an automated comparison of OpenStreetMap with authoritative road datasets. *Trans. GIS* **2017**, *21*, 191–206. [CrossRef]

56. OSM Roads Comparison. Available online: https://github.com/MoniaMolinari/OSM-roads-comparison/tree/master/GRASS-scripts (accessed on 25 April 2018).

57. Goetz, M.; Zipf, A. OpenStreetMap in 3D—Detailed Insights on the Current Situation in Germany. In Proceedings of the 15th AGILE International Conference on Geographic Information Science, Avignon, France, 24–27 April 2012.

58. Fram, C.; Chistopoulou, K.; Ellul, C. Assessing the quality of OpenStreetMap building data and searching for a proxy variable to estimate OSM building data completeness. In Proceedings of the 23rd GIS Research UK (GISRUK) Conference, Leeds, UK, 15–17 April 2015.

59. Wikipedia—Ramer-Douglas-Peucker Algorithm. Available online: https://en.wikipedia.org/wiki/Ramer-Douglas-Peucker_algorithm (accessed on 25 April 2018).

60. Arkin, E.M.; Chew, L.P.; Huttenlocher, D.P.; Kedem, K.; Mitchell, J.S.B. An Efficiently Computable Metric for Comparing Polygonal Shapes. *IEEE Trans. Pattern Anal. Mach. Intell.* **1991**, *13*, 209–216. [CrossRef]

61. Törnros, T.; Dorn, H.; Hahmann, S.; Zipf, A. Uncertainties of completeness measures in OpenStreetMap—A case study for buildings in a medium-sized German city. *ISPRS Ann. Photogramm. Remote Sens. Spat. Inf. Sci.* **2015**, *2*, 353–357. [CrossRef]

62. Müller, F.; Iosifescu Enescu, I.; Hurni, L. Assessment and Visualization of OSM Building Footprint Quality. In Proceedings of the 27th International Cartographic Conference (ICC 2015), Rio de Janeiro, Brazil, 23–28 August 2015.

63. Touya, G.; Antoniou, V.; Christophe, S.; Skopeliti, A. Production of Topographic Maps with VGI: Quality Management and Automation. In *Mapping and the Citizen Sensor*; Foody, G., See, L., Fritz, S., Mooney, P., Olteanu-Raimond, A.-M., Fonte, C.C., Antoniou, V., Eds.; Ubiquity Press: London, UK, 2017; pp. 137–164.

64. Coleman, D.J.; Georgiadou, Y.; Labonté, J. Volunteered geographic information: The nature and motivation of producers. *Int. J. Spat. Data Infrastruct. Res.* **2009**, *4*, 332–358. [CrossRef]

65. Touya, G.; Reimer, A. Inferring the scale of OpenStreetMap features. In *OpenStreetMap in GIScience*; Jokar Arsanjani, J., Zipf, A., Mooney, P., Helbich, M., Eds.; Springer International Publishing: Cham, Switzerland, 2015; pp. 81–99, ISBN 978-3-319-14280-7.

66. Wikipedia—Divide and Conquer. Available online: https://en.wikipedia.org/wiki/Divide_and_conquer (accessed on 25 April 2015).

67. OpenStreetMap Wiki. Available online: https://wiki.openstreetmap.org/wiki (accessed on 25 April 2018).

68. Zielstra, D.; Hochmair, H.H.; Neis, P. Assessing the effect of data imports on the completeness of OpenStreetMap—A United States case study. *Trans. GIS* **2013**, *17*, 315–334. [CrossRef]
69. Zielstra, D.; Hochmair, H.H.; Neis, P.; Tonini, F. Areal delineation of home regions from contribution and editing patterns in OpenStreetMap. *ISPRS Int. J. Geo-Inf.* **2014**, *3*, 1211–1233. [CrossRef]
70. Changeset—OpenStreetMap Wiki. Available online: https://wiki.openstreetmap.org/wiki/Changeset (accessed on 25 April 2018).
71. ASR Lombardia—Annuario Statistico Regionale. Available online: http://www.asr-lombardia.it/ASR/ regioni-italiane/costruzioni-opere-pubbliche-e-mercato-immobiliare/attivita-edilizia (accessed on 25 April 2018).

International Journal of
isprs *Geo-Information*

MDPI

Article

Increasing the Accuracy of Crowdsourced Information on Land Cover via a Voting Procedure Weighted by Information Inferred from the Contributed Data

Giles Foody [1,*]**, Linda See** [2]**, Steffen Fritz** [2]**, Inian Moorthy** [2]**, Christoph Perger** [2]**,
Christian Schill** [3] **and Doreen Boyd** [1]

[1] School of Geography, University of Nottingham, Nottingham NG7 2RD, UK;
    doreen.boyd@nottingham.ac.uk
[2] International Institute for Applied Systems Analysis (IIASA), Schlossplatz 1, A-2361 Laxenburg, Austria;
    see@iiasa.ac.at (L.S.); fritz@iiasa.ac.at (S.F.); moorthy@iiasa.ac.at (I.M.); pergerch@iiasa.ac.at (C.P.)
[3] Faculty of Environment and Natural Resources, Albert-Ludwig University, 79085 Freiburg, Germany;
    christian.schill@felis.uni-freiburg.de
*   Correspondence: giles.foody@nottingham.ac.uk; Tel.: +44-115-951-5430

Received: 22 January 2018; Accepted: 21 February 2018; Published: 25 February 2018

Abstract: Simple consensus methods are often used in crowdsourcing studies to label cases when data are provided by multiple contributors. A basic majority vote rule is often used. This approach weights the contributions from each contributor equally but the contributors may vary in the accuracy with which they can label cases. Here, the potential to increase the accuracy of crowdsourced data on land cover identified from satellite remote sensor images through the use of weighted voting strategies is explored. Critically, the information used to weight contributions based on the accuracy with which a contributor labels cases of a class and the relative abundance of class are inferred entirely from the contributed data only via a latent class analysis. The results show that consensus approaches do yield a classification that is more accurate than that achieved by any individual contributor. Here, the most accurate individual could classify the data with an accuracy of 73.91% while a basic consensus label derived from the data provided by all seven volunteers contributing data was 76.58%. More importantly, the results show that weighting contributions can lead to a statistically significant increase in the overall accuracy to 80.60% by ignoring the contributions from the volunteer adjudged to be the least accurate in labelling.

Keywords: crowdsourcing; volunteered geographic information (VGI); ensemble; classification accuracy; latent class analysis

---

1. Introduction

Members of the general public have for centuries made substantial contributions to science. The inputs range greatly and include the observations of environmental features by an individual and the processing of vast datasets by teams of citizens working in parallel in subjects ranging from astronomy to zoology. Technological developments such as the internet have greatly facilitated the recent strong rise in citizen science activity [1]. Additional technological advances, such as those that have allowed inexpensive and location-aware devices to become commonplace, have been associated with a substantial increase in citizen science activity within geography for which spatial data sets are important. This type of activity has been described in a variety of ways including neogeography, volunteered geographic information, user-generated content, and crowdsourcing [2]. The latter term will be used in this article. Crowdsourcing has become a popular means of acquiring geographic information. Indeed, the rise of the citizen sensor and growth of volunteered geographic information has revolutionised aspects of contemporary geoinformatics and mapping [3–6]. The power of the crowd has been harnessed in a wide range of mapping applications such as building damage mapping to aid post-disaster humanitarian aid [7,8] through scientific studies of the Earth [9] to the provision of complete open mapping at local to global scales such as OpenStreetMap [10]. Crowdsourcing has greatly changed mapping practice and also allows information that was otherwise impossible or at least impractical to obtain by other means to be acquired. One growth area in geoinformatics has been crowdsourcing as a source of ground reference data on land cover to inform analyses of satellite remote sensing imagery [11]. This is an important and growing application area, with citizens having the potential to provide the ground reference data that are needed to fully exploit the potential of remote sensing as a source of information on land cover.

A major problem with the volunteered geographic information (VGI) on land cover provided by the citizen community is that it can be of variable and typically unknown quality, resulting in concern over data accuracy and fitness for purpose [12–15]. The volunteers providing the data may, for example, vary greatly in their skill and ability to provide accurate class labels. Some contributors may simply be enthusiastic but unskilled while others, and quite commonly so [16], may actually have considerable relevant expertise [17,18]. Nonetheless, the power of the crowd is such that its combined wisdom helps generate a final high quality crowdsourced product.

The collective view of the crowd can be obtained in a variety of ways. Commonly, a simple democratic voting procedure is used to bring together the individual inputs from the volunteers and determine a single crowdsourced view. As such, it is common to find that a consensus or ensemble approach to labelling is used with crowdsourced data [14,15]. In these approaches, the contributions from each volunteer are often equally weighted. While ensemble approaches often appear to work well there are still concerns on the variation in quality of data acquired by citizens [19]. This is apparent in relation to performance relative to other citizens, but also within an individual's own set of contributions as performance might vary within given task. For example, in labelling-based tasks, a volunteer may be able to accurately label a sub-set of the classes but not the rest and so contribute quite differently to another volunteer with a different skill-set. A common concern is that a basic ensemble approach weights each contributor's inputs equally even though the volunteers may be of very different ability. This can give rise to a range of potential problems. For example, one volunteer, who may have considerable relevant expertise, may correctly label a case but this lone voice could be lost among the contradictory labelling provided by less informed members of the crowd who may be very numerous. As such, the composition of the crowd is important [17] and there may be a desire to weight contributions unequally to avoid problems of mob rule.

A variety of ways to facilitate effective use of VGI have been proposed. It is, for example, possible for trusted contributors to act as gatekeepers or to check the credibility of a contribution in relation to its known geographic context [20]. These various approaches to try and assure the quality of VGI are, however, not a panacea. It would, for example, be perfectly possible for a gatekeeper acting in good faith to be a barrier to the provision of accurate information from a new but presently untrusted contributor who actually has more skill and knowledge than the gatekeeper. Other means to try and enhance the quality of VGI have included the acquisition of information on the confidence of labelling. For example, volunteers may be asked when labelling cases to indicate for each one their confidence in the class allocation made [18]. This might then allow cases labelled with considerable uncertainty to be filtered out so that only cases labelled with high confidence are used. However, this type of approach has problems. Volunteers may have inflated views on their ability and in some instances, for example, ignorant people will still confidently label cases [21]. An enhancement of this basic method could be based on the surprisingly popular approach that focuses on labelling that is more popular than predicted [22]. Variations in volunteer performance would still be expected. If, however, this variation could be quantified then it may be possible to use this information to enhance analyses. For example, information on the performance of volunteers in terms of their ability to label cases obtained from the data may be used to enhance the accuracy of land cover maps [23]. Estimates of volunteer performance could also be used to weight simple voting procedures, perhaps acting to amplify the contributions from volunteers deemed skilled while down-weighting or even ignoring contributions from volunteers deemed to be inaccurate data sources. Thus, it would be possible to recognise that contributions vary in value and seek to weight them unequally within an ensemble approach. In previous work, it was shown that it is possible to characterise the quality of volunteers in terms of the accuracy of their labelling for each class using only the contributed data [24,25]. Here, the aim is to go beyond the characterisation of the quality of the volunteered data and show how this information, and other information inferred from the contributed data, may be used to enhance the final crowdsourced label that may be applied to VGI.

The key aim of this paper is to explore some simple scenarios for enhancing the accuracy of crowdsourced data on land cover obtained via visual interpretation of satellite sensor images provided via an internet based collaborative project. The paper seeks to show that useful information to inform an ensemble classification that employs a weighted voting strategy can be inferred from the volunteered data and this can be used to increase the overall accuracy of the ensemble classification.

2. Data

The data used comprised land cover class labels obtained from a group of volunteers for a set of 299 satellite sensor images of locations selected randomly over the global land mass. These data were acquired via an open call for data collection through the Geo-Wiki project [26,27] and were used in earlier research [24,25]. The data are available for downloading from the PANGAEA repository as documented in [28]. Each volunteer was invited to view the series of satellite sensor images and assign each a land cover label from a defined list of 10 classes: tree cover, shrub cover, herbaceous vegetation/grassland, cultivated and managed, mosaic of cultivated and managed/natural vegetation, regularly flooded/wetland, urban/built-up, snow and ice, barren, and open water. The volunteers were aided in this task by a brief on-line tutorial and no constraints were put upon contribution. An example of the interface used to collect the data is shown in Figure 1.

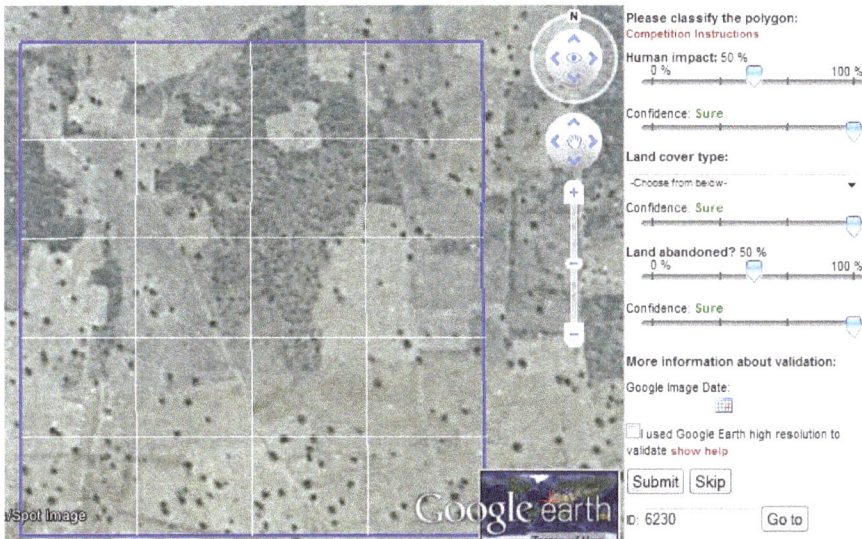

Figure 1. The Geo-Wiki interface used to collect information on land cover type among other features visible from the satellite sensor imagery.

In total, 65 volunteers contributed to the project but their contributions varied greatly in completeness. The amount of images labelled spanned the full spectrum possible, with one volunteer labelling a single image while a few labelled all 299; the average number of images labelled by a volunteer was approximately 110 images. Here, attention is focused on the labels provided by the 10 volunteers who labelled most if not all of the 299 images; these 10 volunteers labelled at least 289 images each. Consequently, this group of volunteers annotated broadly the same set of images reducing the potential for problems such as optimistic bias in their labelling that could occur by skipping the complex to label cases and focusing on only the easier images. The focus on a relatively small group of volunteers is also in keeping with suggestions in the literature [15,25] as well as a means of balancing the competing pressures of seeking multiple annotations but wishing to label many cases [29].

Although a key focus of this article is on information obtained directly from the crowdsourced data without any independent reference data set, a reference data set was formed to help demonstrate and confirm the approach used. Thus, a reference data set was generated simply to confirm the value of the approaches to be adopted, ensuring that the results and interpretations are credible. Three of the 10 selected contributors were experts who also revisited the entire set of 299 images to derive a ground reference data set after discussion amongst themselves informed by their own set of labellings. Although this reference data set is unlikely to be perfect and represent a true gold standard reference which can lead to misestimation [30] it is, however, of a type that is common in major mapping programmes (e.g., [31]). These ground reference data were used to assess the accuracy of the labelling generated from the data contributed by the remaining seven volunteers. This approach reduced the potential for complications caused by missing data and meant that for most of the 299 images, a set of seven class labels were defined. Each label was treated here as a vote for the relevant class and used in simple ensemble methods to obtain a single crowdsourced land cover class label for each image. To maintain anonymity these seven volunteers were labelled A–G.

3. Methods

The work focuses on four scenarios. The first scenario is a benchmark test of the value of the crowd. In this, the accuracy with which individual volunteers classified the images is compared to the accuracy of the classification obtained from the volunteers as a whole using a basic majority voting approach to label each image from the set of labels generated for it by the seven volunteers. Here, accuracy was measured relative to the reference data set generated from the three expert contributors and expressed as the percentage of correctly allocated cases.

All additional analyses sought to use information inferred from the data contributed by the volunteers to weight the voting procedure. Here, the weighting focused on the skill of the volunteers in terms of their ability to label each class and on the relative abundance of the classes in the data set. Information on both of the latter variables was inferred from the results of a latent class analysis of the volunteered data.

The latent class analysis uses the observed data contributed by the volunteers to provide information on an unobserved or latent variable which in this case is the actual land cover. A standard latent class model to describe the relationship between the observed and latent variables was used and can be written as

$$f(y_i) = \sum_{x=1}^{C} P(x) \prod_{v=1}^{V} f(y_{iv}|x)$$

where $f(y_i)$ is a vector representing the complete set of responses obtained from the V volunteers $(1 < v < V)$ contributing data for the case i, C is the number of classes, and x the latent variable [32,33]. Assuming that the model is found to fit with the observed data, the parameters of this model provide the information to inform weighted voting approaches. Specifically, the $f(y_{it}|x)$ parameters of the model represent the conditional probabilities of class membership. Thus, for example, these model parameters indicate the conditional probability that a case allocated a class label by a volunteer is actually a member of that class; in the geoinformatics community, this probability is often referred to as the producer's accuracy for the specified class. Critically, for each volunteer, it is possible to obtain a conditional probability of class membership for each class, indicating the volunteer's skill in labelling each class. The average conditional probability calculated over all classes was also used as a measure of the volunteer's overall skill. In addition, the other latent class model parameter, $P(x)$, indicates the prevalence or abundance of the classes. A feature to note here is that the information on both volunteer skill and class abundance is inferred from only the contributed data.

The information on per-class producer's accuracy for each class and each volunteer could be used to weight the contributions from the volunteers. Of the many ways to approach this task, in Scenario 2 any label (i.e., vote) for a class from a volunteer whose accuracy in labelling of that class was estimated to be substantially less than the maximum accuracy observed for that class was deleted. Here, the focus was on instances for which there was a very large difference in the accuracy relative to that observed for the most accurate volunteer. The approach was implemented here by ignoring the label provided by a volunteer if that volunteer's estimated accuracy for that specific class, rounded to a whole number, was more than 30% less than the highest estimated accuracy for that class associated with another volunteer. This, in effect, was seeking to determine if removing votes from volunteers known to be inaccurate on a specific class would help the overall labelling task. Note that while the labels for a class may be ignored, the other class labels provided by a volunteer would still be used, it is only the labels for class(es) on which the volunteer's performance was viewed as insufficiently high that are removed.

In Scenario 3, the entire contribution from a volunteer with low overall accuracy, expressed as the mean of the producer's accuracy estimated over all classes, were down-weighted to zero by their removal. In essence this was seeking to explore the effect of 'silencing' an inaccurate contributor. Here, this was undertaken twice: the contributions from the volunteer deemed least accurate were removed (Scenario 3a) and the contributions from the two volunteers deemed least accurate were removed (Scenario 3b).

The measure of overall accuracy used in Scenario 3 weights each class equally but accounts for variations in class abundance could further enhance the analysis. This approach would, for example, reduce the effect of poor performance on classes that are rare and so have little impact on the overall proportion of cases correctly classified. Given this context, Scenario 4 sought to extend the analysis one step further and weight the per-class producer's accuracy values estimated for the volunteers by class abundance information estimated from the latent class model. Here, the contributions from the most inaccurate contributor were again removed. In addition, the research sought to explore the effect of magnifying the input of the most accurate contributor, here achieved by duplicating their contributions, effectively making a vote count twice. This weighting is relatively arbitrary and different results could be expected at other settings. In total three different approaches were explored: the magnification of the contributions of most accurate contributor (Scenario 4a), the magnification of the contributions from the most accurate contributor and the removal of the data from the least accurate contributor (Scenario 4b) and the removal of the contributions from least accurate contributor (Scenario 4c).

The overall accuracy of a crowdsourced set of class labels was expressed as the percentage of cases whose labelling agreed with that in the reference data set. The statistical significance of differences in overall accuracy was calculated using the McNemar test. The latter focuses on the discordant cases, the cases which were allocated correctly in only one of the pair of classifications compared. The test is based on the normal curve deviate, z, and the null hypothesis of no significant difference is rejected if the value of z obtained is greater than the critical value of $|1.96|$; the sign is important for a hypothesis with a directional component for which the critical value of z at the 95% level of confidence is 1.645.

4. Results and Discussion

A reference data set, to be used purely for illustrative purposes and ensure credibility of the results, was obtained from the three expert contributors who allocated labels after reaching a consensus.

The labelling from these contributors showed moderate levels of pairwise agreement (with 66.6–69.9% pairwise agreement; kappa coefficients varied from 0.55–0.61) and final class allocations were made after discussion amongst the experts informed by their own initial labelling. It was apparent that the classes varied greatly in abundance. Two classes (regularly flooded/wetland and snow and ice) were determined to be absent in the reference data set, although some cases were sometimes incorrectly labelled as belonging to these classes.

The accuracy with which each volunteer classified the set of satellite sensor images is highlighted in Table 1. The accuracy of the classifications from each and every volunteer was less than that obtained by combining their contributions with a simple majority vote approach. The most accurate individual, for example, provided a set of labels with an overall accuracy of 73.91% while the ensemble classification obtained via the use of the majority vote procedure applied to the volunteered data had an accuracy of 76.58%. This result confirms the oft-stated view that the crowd can be more accurate than the individuals in it.

Table 1. Per-class and overall classification accuracies (%) for the seven volunteers obtained from the latent class model

Class	A	B	C	D	E	F	G
Tree cover (T)	100	86.27	74.73	62.60	73.23	67.43	66.51
Shrub cover (S)	64.44	74.54	83.47	71.13	50.81	69.65	60.61
Herbaceous vegetation/Grassland (H)	69.54	71.22	73.27	45.03	64.65	47.52	24.79
Cultivated and managed (C)	94.16	92.66	100	70.31	87.14	20.17	91.82
Mosaic (M)	54.87	73.8	95.34	97.75	67.9	64.74	67.5
Regularly flooded/wetland (R)	0	0	0	0	0	0	0
Urban/built-up (U)	50	25	50	50	50	50	25
Snow and ice (I)	0	0	0	0	0	0	0
Barren (B)	38.7	0	11.99	0	50.9	30.25	0
Open water (O)	25	25	25	25	25	25	25
Overall (mean)	49.67	44.85	51.38	42.18	46.96	37.48	36.12
Overall (mean weighted by class size)	59.1	59.27	64.86	51.98	55.12	34.93	50.57

Although the simple majority voting approach provided a basic ensemble approach to classification that was more accurate than its component parts, the testing of the three other scenarios sought to explore the possibility to raise the accuracy of the crowdsourced labelling further by weighting the contributions from the volunteers, notably by their skill or accuracy inferred from the latent class analysis.

The estimates of producer's accuracy obtained from the latent class analysis for each volunteer with regard to each class (Table 1) highlight that volunteers vary greatly in their skill and ability to label the imagery. In addition to the variation between volunteers there was variation in the accuracy of classes within the set of data contributed by the volunteers. For example, it was evident that an individual could be very highly accurate with regard to one class but inaccurate with another. For example, Volunteer A had estimated accuracy values of 100% and 54.87% for the tree cover and mosaic classes. In relation to the latter, note also that Volunteer D's estimated accuracy values were almost the direct opposite with 62.60% and 97.75% for the tree cover and mosaic classes, respectively. In addition, it was evident that a volunteer with generally low accuracy could still be highly accurate on a specific class. This was evident for Volunteer G who was only highly accurate on one class: cultivated and managed, which also was a relatively abundant class.

In Scenario 2, the vote for a class by a volunteer was removed if that volunteer was highly inaccurate in the labelling of that specific class in comparison to the other volunteers. The effect of removing the votes for a class from a volunteer deemed to be unskilled for the labelling of that class increased the accuracy of the overall ensemble approach using the majority voting procedure to 78.26%.

An alternative approach to using the estimated information on volunteer labelling accuracy is to remove all contributions from volunteers adjudged to provide labels of low or insufficient accuracy. This was explored in Scenario 3. It was evident in Scenario 3a that by dropping the entire set of contributions of the least accurate volunteer (Volunteer G, with a mean producer's accuracy of 36.12%)

the accuracy of the ensemble classification could increase to 77.92%. Moreover, the largest ensemble accuracy observed in Scenario 3, 79.59%, was obtained in Scenario 3b when the contributions from the two least accurate volunteers (Volunteers F and G) were ignored. It was also evident that the accuracy of the contributions by these two volunteers were noticeably less accurate than from the other volunteers (Table 1).

In addition to information on the accuracy with which each volunteer can classify the classes, the latent class model also indicates the prevalence or abundance of the classes. This information on class abundance inferred from the analysis was used to adjust the estimates of overall volunteer accuracy, here expressed as the average producer's accuracy. The weighted overall accuracy values (Table 1) revealed that one volunteer (Volunteer C) was noticeably more accurate and one noticeably less accurate (Volunteer F) than the remaining set; note that after weighting for class abundance Volunteer F rather than G is associated with the lowest labelling accuracy. Increasing the weight of the accurate volunteer by duplicating their contributions (i.e., giving each vote a weight of two) increased accuracy. For example, increasing the vote for the most accurate volunteer in Scenario 4a raised the accuracy of the ensemble from the benchmark value of 76.58% to 78.59%. Furthermore, ignoring the labels from the least accurate volunteer in addition further increased accuracy to 79.59% in Scenario 4b. However, it was also apparent that a more accurate ensemble could be achieved in Scenario 4c by solely removing the contributions of the least accurate volunteer, which yielded an ensemble classification with an accuracy of 80.60%. It should be noted that at the 95% level of confidence, this latter ensemble classification was also significantly more accurate than that achieved by increasing the weighting for the most accurate volunteer ($z = 4.31$) and by additionally ignoring the least accurate volunteer's data ($z = 3.90$). The ensemble classification arising through the removal of the contributions from the least accurate volunteer (Scenario 4c) was also the most accurate of all classifications reported in the study and significantly different at the 95% level of confidence to the benchmark classification based on the standard majority voting rule ($z = 5.54$).

The ability to increase the accuracy of the crowdsourced labelling by weighting the voting process is highlighted in Figure 2 which shows the overall accuracy of classifications relative to the reference data for individuals and from each of the four scenarios for ensemble classification discussed. Additional summary data for each of the classifications arising from the scenarios reported is provided in Table 2 and the full confusion matrix provided for the classifications arising from the basic ensemble (Table 3) and Scenario 4c (Table 4).

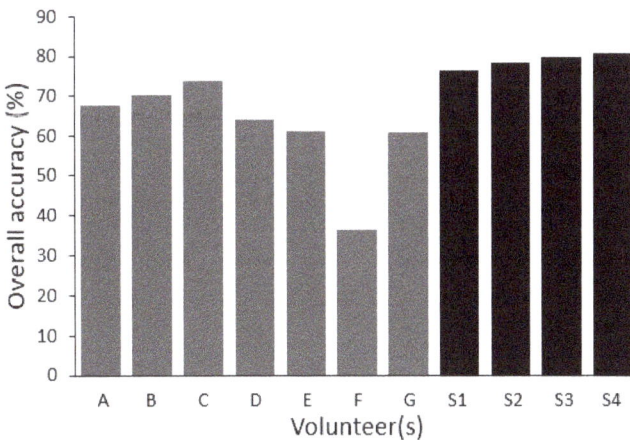

Figure 2. Overall classification accuracy determined relative to the reference data set for each of the individual volunteers (grey bars) and the highest accuracy from each of the four scenarios for an ensemble classification (S1–S4, black bars).

Table 2. Summary of the correct allocations observed in each scenario (S1–S4c) reported for each class and the class size (n) in the reference data set; complete confusion matrices for two key classifications are given in Tables 3 and 4.

Class	n	S1	S2	S3a	S3b	S4a	S4b	S4c
T	47	35	32	34	36	34	35	35
S	20	15	15	15	15	16	16	16
H	24	17	19	16	17	16	16	17
C	119	98	108	99	101	104	106	106
M	85	60	55	65	65	61	61	63
R	0	0	0	0	0	0	0	0
U	1	1	1	1	1	1	1	1
I	0	0	0	0	0	0	0	0
B	2	2	2	2	2	2	2	2
O	1	1	1	1	1	1	1	1
Total	299	229	233	233	238	235	238	241

Table 3. Confusion matrix for the benchmark classification of Scenario 1; columns show the class label in the reference data set and rows the label determined in the scenario.

Class	T	S	H	C	M	R	U	I	B	O	Total
T	35	1	1	0	10	0	0	0	0	0	47
S	8	15	3	0	2	0	0	0	0	0	28
H	2	3	17	0	6	0	0	0	0	0	28
C	0	0	0	98	5	0	0	0	0	0	103
M	2	0	1	20	60	0	0	0	0	0	83
R	0	0	0	0	0	0	0	0	0	0	0
U	0	0	0	0	1	0	1	0	0	0	2
I	0	0	0	0	0	0	0	0	0	0	0
B	0	1	2	1	1	0	0	0	2	0	7
O	0	0	0	0	0	0	0	0	0	1	1
Total	47	20	24	119	85	0	1	0	2	1	299

Table 4. Confusion matrix for the classification of Scenario 4c; columns show the class label in the reference data set and rows the label determined in the scenario.

Class	T	S	H	C	M	R	U	I	B	O	Total
T	35	1	1	0	7	0	0	0	0	0	44
S	8	16	3	0	4	0	0	0	0	0	31
H	2	3	17	1	3	0	0	0	0	0	26
C	1	0	0	106	7	0	0	0	0	0	114
M	1	0	1	12	63	0	0	0	0	0	77
R	0	0	1	0	0	0	0	0	0	0	1
U	0	0	0	0	1	0	1	0	0	0	2
I	0	0	0	0	0	0	0	0	0	0	0
B	0	0	1	0	0	0	0	0	2	0	3
O	0	0	0	0	0	0	0	0	0	1	1
Total	47	20	24	119	85	0	1	0	2	1	299

Figure 2 highlights that each ensemble approach yielded a classification that was more accurate than that arising from the individual contributors alone. It also highlights that the relative accuracy of the classifications weighted by class abundance obtained from the individuals inferred from the latent class analysis (Table 1) corresponds with the actual accuracy assessed relative to the reference data (Figure 2). In particular, the relatively low accuracy of the labelling provided by Volunteer F is evident and it is the removal of these data in Scenario 4c that resulted in the largest, and statistically significant, increase in accuracy over the benchmark classification of Scenario 1. From earlier research [25], the accuracy with which the data contributed by each of the volunteers may be characterised could increase if the number of volunteers also increased, paving the way for further refinement of the analysis.

The results, especially from Scenario 4c, show that, for the data set used, the removal of inaccurate data is of more value than the enhancement or amplification of more accurate data sources. It should be noted that this latter issue may reflect the composition of the volunteers used in this study. Given that all seven volunteers had contributed labels for virtually all of the images, it may be that these people have a high level of motivation which could be used as a proxy variable to indicate high skill sets so it was the removal of the occasional anomalously poor inputs that was important rather than efforts to amplify good quality contributions. Had the set of volunteers been of more mixed ability, notably if made up of a large number of true amateurs, then less expertise might be present and different trends may have been observed. Similarly, it should be noted that the results may, of course, be specific to the data set used and the information inferred could be used in other ways (e.g., to inform labelling in tie-break situations by allocating to the class indicated by the relatively more accurate labellers).

Finally, it should be stressed that the information on volunteer skill and class abundance to weight the voting procedure were all inferred from the contributed data alone. In many applications there may be little or no reference data available to allow a standard assessment of the accuracy of labelling and comparison of classifications such as that provided by Figure 2. Critically, however, all of the information contained in Table 1 was obtained from the set of contributed crowdsourced labels only; this includes the information on class size which was obtained from the latent class model. Thus, the information on per-class and overall classification accuracy needed to enhance the voting method is inferred entirely from just the contributed data; the reference data were only used in this study to provide supporting evidence that the approaches discussed actually did impact on accuracy. The quality of the crowd-sourced estimates may also increase if data from additional volunteers are available [25]. As well as providing an intrinsic approach to the assessment of contributed data quality, the approach has additional advantages. Since only the contributed data are required, there is, therefore, no need to use a proportion of the crowdsourced data to measure the variables directly, perhaps via some dedicated ground based research or use of additional experts. There is also no use of external auxiliary information. Further enhancements could be made by expressing skill in different ways; the measure of accuracy used may not always be ideal and other approaches could be used to focus more directly on the objectives of a specific study (e.g., weighting by unequal costs of errors). Critically, however, this article has gone beyond earlier work to show that the quality of contributed data can be estimated from the data alone to demonstrate how crowdsourced labelling can be enhanced via simple weighted voting methods without any additional data.

5. Conclusions

The results have highlighted that the wisdom of the crowd can be used to generate a single crowdsourced set of land cover annotations that are more accurate than those achieved by any individual in the crowd. More importantly, estimates of the skill of each individual in terms of classifying classes and on the abundance of the classes that were inferred from the contributed data may be used to increase the accuracy of the crowdsourced labels; reference data were used here to confirm the validity of the approach, but are not required for its implementation. In this study, a significant increase in the accuracy of labelling of land cover from satellite sensor imagery was obtained by down-weighting the contributions adjudged to be of relatively low overall accuracy for the task. This was most apparent when the estimation of the volunteer's skill, expressed here as the average producer's accuracy calculated over all classes, was weighted by class abundance. It is evident that very simple methods may be used to increase the quality of crowdsourced data which should hopefully further facilitate the use of crowdsourcing of geographic data.

Acknowledgments: This work benefitted from funding from the EU COST Action TD1202 and the EU Horizon 2020 funded project LandSense (No. 689812) as well as founding work funded by the British Academy (reference SG112788) and EPSRC (reference EP/J0020230/1). We are also grateful to the editor and the two referees who provided constructive comments to enhance this article.

Author Contributions: G.F., L.S., S.F., and I.M. discussed the idea; L.S., S.F., C.P., C.S., and D.S. contributed underpinning research; G.F. undertook the analyses and wrote the paper with inputs from all authors.

Conflicts of Interest: The authors declare no conflict of interest. The founding sponsors had no role in the design of the study; in the collection, analyses, or interpretation of data; in the writing of the manuscript, or in the decision to publish the results.

References

1. Bonney, R.; Shirk, J.L.; Phillips, T.B.; Wiggins, A.; Ballard, H.L.; Miller-Rushing, A.J.; Parrish, J.K. Next steps for citizen science. *Science* **2014**, *343*, 1436–1437. Available online: http://science.sciencemag.org/content/343/6178/1436 (accessed on 12 February 2018). [CrossRef] [PubMed]

2. See, L.; Mooney, P.; Foody, G.; Bastin, L.; Comber, A.; Estima, J.; Fritz, S.; Kerle, N.; Jiang, B.; Laakso, M.; et al. Crowdsourcing, citizen science or Volunteered Geographic Information? The current state of crowdsourced geographic information. *ISPRS Int. J. Geo-Inf.* **2016**, *5*, 55. Available online: http://www.mdpi.com/2220-9964/5/5/55 (accessed on 12 February 2018). [CrossRef]

3. Goodchild, M.F. Citizens as sensors: The world of volunteered geography. *GeoJournal* **2007**, *69*, 211–221. Available online: https://link.springer.com/article/10.1007/s10708-007-9111-y (accessed on 11 December 2017). [CrossRef]

4. Pullar, D.; Hayes, S. Will the future maps for Australia be published by 'nobodies'. *J. Spat. Sci.* **2017**, 1–8. Available online: http://www.tandfonline.com/doi/abs/10.1080/14498596.2017.1361873 (accessed on 11 December 2017). [CrossRef]

5. Capineri, C.; Haklay, M.; Huang, H.; Antoniou, V.; Kettunen, J.; Ostermann, F.; Purves, R. *European Handbook of Crowdsourced Geographic Information*; Ubiquity Press: London, UK, 2016. Available online: https://doi.org/10.5334/bax (accessed on 12 February 2018).

6. Foody, G.; See, L.; Fritz, S.; Mooney, P.; Olteanu-Raimond, A.M.; Fonte, C.C.; Antoniou, V. *Mapping and the Citizen Sensor*; Ubiquity Press: London, UK, 2017. Available online: https://doi.org/10.5334/bbf (accessed on 12 February 2018).

7. Goodchild, M.F.; Glennon, J.A. Crowdsourcing geographic information for disaster response: A research frontier. *Int. J. Dig. Earth* **2010**, *3*, 231–241. Available online: http://www.tandfonline.com/doi/full/10.1080/17538941003759255 (accessed on 11 December 2017). [CrossRef]

8. Kerle, N.; Hoffman, R.R. Collaborative damage mapping for emergency response: The role of Cognitive Systems Engineering. *Nat. Hazards Earth Syst. Sci.* **2013**, *13*, 97–113. [CrossRef]

9. Goodchild, M.F.; Guo, H.; Annoni, A.; Bian, L.; de Bie, K.; Campbell, F.; Craglia, M.; Ehlers, M.; van Genderen, J.; Jackson, D.; et al. Next-generation digital earth. *Proc. Natl. Acad. Sci. USA* **2012**, *109*, 11088–11094. Available online: http://www.pnas.org/content/109/28/11088.full (accessed on 11 December 2017). [CrossRef] [PubMed]

10. Mooney, P.; Minghini, M. A review of OpenStreetMap data. In *Mapping and the Citizen Sensor*; Foody, G., See, L., Fritz, S., Mooney, P., Olteanu-Raimond, A.-M., Fonte, C.C., Antoniou, V., Eds.; Ubiquity Press: London, UK, 2017; pp. 37–59, ISBN 978-1-911529-17-0. Available online: https://doi.org/10.5334/bbf (accessed on 11 December 2017).

11. Foody, G.M. Citizen science in support of remote sensing research. In *International Geoscience and Remote Sensing Symposium*; IEEE: Piscataway, NJ, USA, 2015; pp. 5387–5390. Available online: http://ieeexplore.ieee.org/stamp/stamp.jsp?tp=&arnumber=7326952 (accessed 24 February 2018).

12. Flanagin, A.J.; Metzger, M.J. The credibility of volunteered geographic information. *GeoJournal* **2008**, *72*, 137–148. Available online: https://link.springer.com/article/10.1007/s10708-008-9188-y (accessed on 11 December 2017). [CrossRef]

13. Koswatte, S.; McDougall, K.; Liu, X. VGI and crowdsourced data credibility analysis using spam email detection techniques. *Int. J. Dig. Earth* **2017**, 1–13. Available online: http://www.tandfonline.com/doi/abs/10.1080/17538947.2017.1341558 (accessed on 11 December 2017). [CrossRef]

14. Fonte, C.C.; Antoniou, V.; Bastin, L.; Bayas, L.; See, L.; Vatseva, R. Assessing VGI data quality. In *Mapping and the Citizen Sensor*; Foody, G., See, L., Fritz, S., Mooney, P., Olteanu-Raimond, A.-M., Fonte, C.C., Antoniou, V., Eds.; Ubiquity Press: London, UK, 2017; pp. 137–163, ISBN 978-1-911529-17-0. Available online: https://doi.org/10.5334/bbf (accessed on 11 December 2017).

15. Haklay, M.; Basiouka, S.; Antoniou, V.; Ather, A. How many volunteers does it take to map an area well? The validity of Linus' Law to volunteered geographic information. *Cartogr. J.* **2010**, *47*, 315–322. Available online: http://www.tandfonline.com/doi/abs/10.1179/000870410X12911304958827 (accessed on 7 December 2017). [CrossRef]

16. Brabham, D.C. The myth of amateur crowds: A critical discourse analysis of crowdsourcing coverage. *Inf. Commun. Soc.* **2012**, *15*, 394–410. Available online: http://www.tandfonline.com/doi/abs/10.1080/1369118X.2011.641991 (accessed on 11 December 2017). [CrossRef]

17. Comber, A.; Mooney, P.; Purves, R.S.; Rocchini, D.; Walz, A. Crowdsourcing: It matters who the crowd are. The impacts of between group variations in recording land cover. *PLoS ONE* **2016**, *11*, e0158329. Available online: https://doi.org/10.1371/journal.pone.0158329 (accessed on 5 December 2017). [CrossRef] [PubMed]

18. See, L.; Comber, A.; Salk, C.; Fritz, S.; van der Velde, M.; Perger, C.; Schill, C.; McCallum, I.; Kraxner, F.; Obersteiner, M. Comparing the quality of crowdsourced data contributed by expert and non-experts. *PLoS ONE* **2013**, *8*, e69958. Available online: https://doi.org/10.1371/journal.pone.0069958 (accessed on 5 December 2017). [CrossRef] [PubMed]

19. Salk, C.F.; Sturn, T.; See, L.; Fritz, S. Limitations of majority agreement in crowdsourced image interpretation. *Trans. GIS* **2017**, *21*, 207–223. [CrossRef]

20. Goodchild, M.F.; Li, L. Assuring the quality of volunteered geographic information. *Spat. Stat.* **2012**, *1*, 110–120. Available online: http://www.sciencedirect.com/science/article/pii/S2211675312000097 (accessed on 7 December 2017). [CrossRef]

21. Kruger, J.; Dunning, D. Unskilled and unaware of it: How difficulties in recognizing one's own incompetence lead to inflated self-assessments. *J. Pers. Soc. Psychol.* **1999**, *77*, 1121–1134. Available online: http://dx.doi.org/10.1037/0022-3514.77.6.1121 (accessed on 5 December 2017). [CrossRef] [PubMed]

22. Prelec, D.; Seung, H.S.; McCoy, J. A solution to the single-question crowd wisdom problem. *Nature* **2017**, *541*, 532–535. Available online: https://www.nature.com/articles/nature21054 (accessed on 7 December 2017). [CrossRef] [PubMed]

23. Gengler, S.; Bogaert, P. Integrating crowdsourced data with a land cover product: A Bayesian data fusion approach. *Remote Sens.* **2016**, *8*, 545. Available online: http://www.mdpi.com/2072-4292/8/7/545/htm (accessed on 12 February 2018). [CrossRef]

24. Foody, G.M.; See, L.; Fritz, S.; Van der Velde, M.; Perger, C.; Schill, C.; Boyd, D.S. Assessing the accuracy of volunteered geographic information arising from multiple contributors to an internet based collaborative project. *Trans. GIS* **2013**, *17*, 847–860. Available online: http://onlinelibrary.wiley.com/doi/10.1111/tgis.12033/full (accessed on 7 December 2017). [CrossRef]

25. Foody, G.M.; See, L.; Fritz, S.; Van der Velde, M.; Perger, C.; Schill, C.; Boyd, D.S.; Comber, A. Accurate attribute mapping from volunteered geographic information: Issues of volunteer quantity and quality. *Cartogr. J.* **2015**, *52*, 336–344. Available online: http://www.tandfonline.com/doi/abs/10.1080/00087041.2015.1108658 (accessed on 7 December 2017). [CrossRef]

26. Fritz, S.; McCallum, I.; Schill, C.; Perger, C.; See, L.; Schepaschenko, D.; van der Velde, M.; Kraxner, F.; Obersteiner, M. Geo-Wiki: An online platform for improving global land cover. *Environ. Modell. Softw.* **2012**, *31*, 110–123. Available online: http://www.sciencedirect.com/science/article/pii/S1364815211002787, (accessed on 11 December 2017). [CrossRef]

27. See, L.; Fritz, S.; Perger, C.; Schill, C.; McCallum, I.; Schepaschenko, D.; Duerauer, M.; Sturn, T.; Karner, M.; Kraxner, F.; et al. Harnessing the power of volunteers, the internet and Google Earth to collect and validate global spatial information using Geo-Wiki. *Technol. Forecast. Soc. Chang.* **2015**, *98*, 324–335. [CrossRef]

28. Fritz, S.; See, L.; Perger, C.; McCallum, I.; Schill, C.; Schepaschenko, D.; Duerauer, M.; Karner, M.; Dresel, C.; Laso-Bayas, J.-C.; et al. A global dataset of crowdsourced land cover and land use reference data. *Sci. Data* **2017**, *4*, 170075. [CrossRef] [PubMed]

29. Boyd, D.; Jackson, B.; Wardlaw, J.; Foody, G.; Marsh, S.; Bales, K. Slavery from space: Demonstrating the role for satellite remote sensing to inform evidence-based action related to UN SDG number 8. *ISPRS J. Photogramm. Remote Sens.* **2018**, in press.

30. Foody, G.M. Assessing the accuracy of land cover change with imperfect ground reference data. *Remote Sens. Environ.* **2010**, *114*, 2271–2285. Available online: https://www.sciencedirect.com/science/article/pii/S0034425710001434 (accessed on 12 February 2018). [CrossRef]

31. Scepan, J.; Menz, G.; Hansen, M.C. The DISCover validation image interpretation process. *Photogramm. Eng. Remote Sens.* **1999**, *65*, 1075–1081. Available online: https://www.asprs.org/wp-content/uploads/pers/1999journal/sep/1999_sept_1075-1081.pdf (accessed on 11 December 2017).

32. Vermunt, J.K.; Magidson, J. Latent class analysis. In *The Sage Encyclopedia of Social Science Research Methods*; Lewis-Beck, M., Bryman, A.E., Liao, T.F., Eds.; Sage Publications: Thousand Oaks, CA, USA, 2003; Volume 2, pp. 549–553.

33. Vermunt, J.K.; Magidson, J. Latent class models for classification. *Comput. Stat. Data Anal.* **2003**, *41*, 531–537. Available online: http://www.sciencedirect.com/science/article/pii/S0167947302001792 (accessed on 7 December 2017). [CrossRef]

isprs International Journal of
Geo-Information

MDPI

Article

OSM Data Import as an Outreach Tool to Trigger Community Growth? A Case Study in Miami

Levente Juhász * and Hartwig H. Hochmair

Fort Lauderdale Research and Education Center, University of Florida, Davie, FL 33314, USA;
hhhochmair@ufl.edu
* Correspondence: levente.juhasz@ufl.edu; Tel.: +1-954-577-6392

Received: 1 January 2018; Accepted: 12 March 2018; Published: 15 March 2018

Abstract: This paper presents the results of a study that explored if and how an OpenStreetMap (OSM) data import task can contribute to OSM community growth. Different outreach techniques were used to introduce a building import task to three targeted OSM user groups. First, existing OSM members were contacted and asked to join the data import project. Second, several local community events were organized with Maptime Miami to engage local mappers in OSM contribution activities. Third, the import task was introduced as an extra credit assignment in two GIS courses at the University of Florida. The paper analyzes spatio-temporal user contributions of these target groups to assess the effectiveness of the different outreach techniques for recruitment and retention of OSM contributors. Results suggest that the type of prospective users that were contacted through our outreach efforts, and their different motivations play a major role in their editing activity. Results also revealed differences in editing patterns between newly recruited users and already established mappers. More specifically, long-term engagement of newly registered OSM mappers did not succeed, whereas already established contributors continued to import and improve data. In general, we found that an OSM data import project can add valuable data to the map, but also that encouraging long-term engagement of new users, whether it be within the academic environment or outside, proved to be challenging.

Keywords: OpenStreetMap; VGI; community mapping; data analysis; GIS education; data import

---

1. Introduction

OpenStreetMap (OSM) is one of the most prominent Volunteered Geographic Information (VGI) [1] projects to date that implements a collaborative workflow and aims to create a freely available map database of the entire world. VGI users in general, and in the case of OSM specifically, use a set of tools, such as field surveys, on-screen digitizing from aerial imagery, and software to create verifiable information on the ground [2]. The success of OSM is based on a large and active user base that interacts with other contributors, and validates and corrects errors made by them [3]. OSM data is released under the Open Database License (ODbL) (https://opendatacommons.org/licenses/odbl/), which allows to freely copy, distribute, transmit and adapt the data as long as its source is credited. Derivative work needs to be released under the same license. ODbL prohibits the use of copyrighted material (e.g., commercial maps) without explicit permission.

As OSM is a collaborative project, local contributors often organize social events (so-called mapping parties) all over the world. These mapping parties are effective ways for local community building and social collaboration within OSM [4], facilitating face to face meetings among online data contributors. A prime goal of mapping parties is to introduce OSM to new members through hands-on mapping sessions. These sessions can include joint field surveys (e.g., to record house numbers) and data editing tutorials (e.g., to teach how to trace roads from imagery). The effect of mapping parties on user and data growth has been analyzed in various studies. For example, it was observed that during a mapping party, participants tend to edit more than usual [5]. This increased activity is more pronounced for light and medium contributors than for heavy users. This fact could be due to the leading role of heavy users in organizing the mapping parties. Similar behavior was also observed for another collaborative project, Wikipedia, where the most committed users took up organizational roles [6]. Another study describes the organizational and planning aspects of a mapping party held in connection with a geospatial conference [7]. The organizers concluded that, although contributed data was of very high quality, on a wider scale the mapping party had not contributed a very large amount of data. Not all of the data collected during the field survey was uploaded to OSM due to lack of time during the mapping party, incomplete training, and users' lack of confidence in using OSM tools. Analyzing contribution patterns after a mapping party held in London, 50% of new OSM members were found to stop contributions in the week after the event [8]. Another study estimated that only 64% of new OSM contributors "survive" their first day, after which the estimated survival rate decreases [9], suggesting that the 50% withdrawal rate observed in [8] is not specific to mapping parties. Apart from mapping parties, OSM shows other characteristics of a social project. For example, after the Haiti Earthquake in 2010, a new project called the Humanitarian OpenStreetMap Team (HOT) emerged to generate freely available geographic data in areas affected by natural disasters [10]. As a response to that event, 600 remotely located volunteer mappers built a base layer map for Haiti nearly from scratch. This map was then used in the field by response teams to support residents and save lives. In 2013, HOT evolved into a registered US non-profit organization (https://www.hotosm.org) that aims to create and provide free, up-to-date maps for relief organizations responding to natural and man-made crises. Their mapping efforts primarily use an online tool called the Tasking Manager (TM).

Besides field surveys and on-screen digitizing from remote sensing imagery, OSM also allows the integration of other datasets available under licenses compatible with ODbL (e.g., CC0 (https://creativecommons.org/share-your-work/public-domain/cc0/)). This usually triggers subsequent user contributions and edits of imported data. Permissible datasets include public domain data that is often published by government agencies. Importing data through automatic means is one of the most controversial topics within the OSM community as this method is different from the core approach of OSM, which is to manually add verifiable data to the map [11]. However, the general consensus is that imports, if carefully executed, add value to OSM. The OSM community discusses import related issues in a dedicated channel (https://lists.openstreetmap.org/listinfo/imports). Numerous OSM data import tasks have been executed so far [12], and some studies have evaluated the effect of data imports on OSM data quality and user participation. For example, one study described the effects of US Census TIGER/Line import on data completeness [13]. Challenges associated with the matching of tags between imported data sources and the OSM tagging structure are discussed in [14]. In [15], the authors discuss inconsistencies in the level of detail within VGI data and found examples of OSM data imports that cause this problem. For example, buildings imported from the French cadastral sources can overlap with land parcels imported from the European CORINE Land Cover dataset because of the different scale of those data sources. OSM data imports can be beneficial for the data donor as well. For example, the Department of National Resources of Canada allowed the OSM community to import their national dataset with the hope that the community would further improve it. These improvements (upon approval) could then be fed back into the national dataset [16].

The research presented in this study describes experiences with a local OSM data import in Miami-Dade County, Florida. More specifically, it evaluates how effective a building import task is at engaging different targeted community groups in OSM participation. The import task of this study integrates a public domain dataset (building footprints) in Miami-Dade County. Targeted communities were asked to join the import project and to manually edit OSM data in the hope that this mapping experience would trigger community growth. These targeted groups were (1) existing members of the OSM community contacted through the OSM site; (2) users reached by Maptime Miami, a local chapter of an initiative built around open knowledge and geospatial technologies; and (3) students enrolled in two courses of the Geomatics Program at the University of Florida who were introduced to the import task as part of their course work. The success of VGI projects generally depends on participants, therefore understanding their motivation is key for the success of such projects [17]. The motivation of VGI users can be divided into two distinct categories, namely extrinsic motivation, that is, related to outside factors (e.g., receiving compensation) and intrinsic motivation that originates directly from the user (e.g., gaining new knowledge or recreation) [18]. Our different outreach techniques were expected to reach users with both motivation categories.

OSM building imports are useful to improve the quality of the OSM building layer, which is generally still low compared to that of road network data. The completeness of buildings mapped in OSM relative to official data from national mapping and cadastral agencies has been examined in several studies. For example, it was found that completeness levels vary widely between different cities, e.g., between 12% and 48% for selected parts of Germany [19], and between 30% and 75% for three cities in the United Kingdom [20]. Completeness evaluation and positional accuracy assessment was also performed for Milan, Italy [21], which revealed a decreasing trend in completeness from the city center towards the outskirts. Positional accuracy was found to be similar across the city, probably due the constant accuracy of the underlying imagery from which buildings were traced. It is worth mentioning, however, that calculated completeness values differ strongly between applied methods. One study measured building completeness for a medium-sized German city with two common unit-based methods and found that the count ratio method underestimates building completeness, whereas the area ratio method overestimates it [22]. Another study examined completeness, semantic accuracy, position accuracy, and shape accuracy of OSM building footprints for Munich, Germany, revealing a high completeness and semantic accuracy, whereas in terms of shape some architectural details are missing [23]. In [24], the authors proposed an intrinsic approach for OSM quality assessment. As part of OSM data analysis for Madrid (Spain), San Francisco (USA) and Yaoundé (Cameroon), they found that buildings imported through a bulk upload lack attribute completeness when compared to areas without bulk upload.

2. Materials and Methods

2.1. Miami-Dade Large Building Import

On 16 May 2016, Maptime Miami (https://www.meetup.com/Maptime-Miami/) proposed an OSM data import of Large Building Footprints (http://gis.mdc.opendata.arcgis.com/datasets/1e87b925717747c7b59979caa7779039_1) from Miami-Dade County's Open Data repository to kick-start Miami's OSM, which lags behind other major cities in the United States both in terms of contributor numbers and data completeness. To ensure that a data import is not harmful for OSM, such projects need to adhere to strict guidelines, which include local community buy-in, announcements on different OSM channels (Wiki, mailing lists) and the ability for the community to review and test both the data to be imported and the methods used during the import process. These guidelines were followed for this project and the import was discussed within the US OSM community.

Other building imports, such as the ones in Los Angeles and New York City rely exclusively on the OSM community and therefore require many active contributors to manually review and import buildings. Due to the low number of OSM contributors in South Florida and the lack of existing buildings in OSM, the first author of this paper, as part of Maptime Miami, implemented a hybrid approach that consisted of an automated bulk upload of buildings and a manual community review of remaining buildings where needed. For this purpose, a software tool (https://github.com/jlevente/MiamiOSM-buildings) was developed and open sourced to pre-process the building dataset, to perform quality checks and to separate the dataset into two parts. Hence, one part of the dataset was uploaded automatically, and the other one was set aside for the community to review. The latter set contained buildings with detected conflicts (overlap with existing OSM buildings, road or railroad features, geometry errors, etc.) whereas the rest of the dataset (i.e., buildings with no geometry issues and no overlaps) was uploaded automatically from a dedicated import account (https://www.openstreetmap.org/user/MiamiBuildingsImport) with upload scripts. A description of the workflow and the software tool is available at https://github.com/jlevente/MiamiOSM-buildings. In Figure 1a, the features in green represent buildings in the automatic bucket (i.e., no conflicts) whereas those in red represent buildings for the community review process. The dataset consists of 95,536 large buildings that are defined as structures in commercial, industrial or other non-residential areas. Additionally, structures larger than approximately 750 m^2 (e.g., townhomes, condominiums) are also classified as large buildings. The dataset was derived from high resolution aerial imagery by both automatic photogrammetric methods and manual digitization and is available for download unprojected (in geographic coordinates). After visual inspection, the building layer was found to be of high quality. Buildings also contained building height information in feet, which was converted to meters and also imported along with the geometry. Additionally, an address dataset was spatially merged with the buildings to provide accurate street level information along with the buildings. A total of 84,348 buildings were uploaded automatically (green features in Figure 1a), which left nearly 11,000 buildings for the manual review process (red features in Figure 1a). All import buildings were tagged with the "ref:miabldg" (http://wiki.openstreetmap.org/wiki/Key:ref:miabldg) key for easy identification.

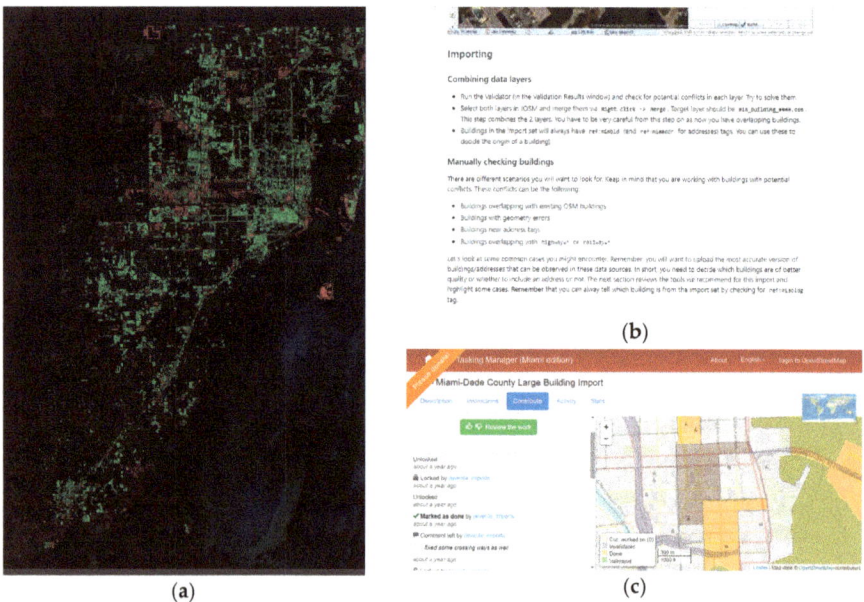

Figure 1. Import buildings (automatically uploaded—green, for manual review—red) in Miami-Dade County (**a**); excerpt from the import tutorial (**b**); and interface of the Tasking Manager (TM) instance (**c**).

The remaining 11,000 buildings were split by US Census Block Groups to provide a manageable number of buildings for manual review. A custom workflow was developed and explained in a detailed tutorial (Figure 1b). The workflow uses the JOSM editor since it can load multiple datasets and since it provides superior tools for data editing compared to the Web based iD editor [25]. The tutorial used screenshots, explanations and specific instructions detailing how to execute the import steps. The tutorial was tested by multiple members of Maptime Miami before releasing it. To administer the progress and to provide a central interface for users, a dedicated TM instance was set up at http://tasks.osm.jlevente.com. A screenshot of the interface is shown in Figure 1c. On this site, contributors can log in with their OSM username and then select a block group within Miami-Dade County to work on. Once a user selects an area, it will be locked for an hour to avoid concurrent edits. This lock is visible on the website for other users currently browsing the site. The TM instance contains hyperlinks to the tutorial and provides an easy way to load data into JOSM. For example, by pressing a button in TM, JOSM on the user's computer loads data from the selected area and zooms to the extent of the import area. The general steps of the import workflow are as follows:

1.   User logs on to TM
2.   User selects and locks an area to work on
3.   In TM, user loads current OSM data coverage into JOSM
4.   In TM, user loads import building dataset into JOSM
5.   In JOSM, user merges the import and OSM datasets into one single layer
6.   In JOSM, user works on resolving conflicts and refers to the tutorial if needed
7.   In JOSM, user runs the validation tool to ensure all data is correct and ready for upload
8.   User uploads data to OSM
9.   User marks TM task as "done" or unlocks it if task is unfinished

2.2. Outreach Techniques and Target Audiences

To reach a sufficient number of contributors for the project, different user groups were targeted and introduced to the import. This also allows to explore the willingness of different user groups to participate in this import and provides a better understanding of the impact different user groups have on OSM.

2.2.1. Students

The import project was introduced at two courses in the Geomatics Program at the University of Florida, which are GIS Programming (Fall 2016, graduate level) and GIS Analysis (Spring 2017, undergraduate and graduate level). Participation in this study was voluntary and students received extra credit to complete this task. Both courses are offered in an online format and therefore students were located in different parts of Florida. In both courses a lecture was dedicated to the import where students were given an overview of the import task and received information about the available resources (tutorial, TM, etc.). A hands-on editing session that illustrated the import process in detail was demonstrated live and also recorded to make it available for review later on. To earn full extra credit, students were asked to import at least 50 buildings.

Before the submission deadline, seven students needed assistance and troubleshooting associated with the assignment. Encountered problems included technical issues with JOSM and fixing errors in the submitted OSM edits. The early edits of a few students contained some building outlines traced from aerial imagery, but without the "building = yes" tags. These students did not initially realize the importance of tags and were asked to fix their edits so that the added buildings would be recognized as buildings in OSM. In one instance, some changesets that only contained overlapping (hence incorrect) buildings needed to be manually reverted. This was due to skipped steps 6 and 7 described in Section 2.1 from a student's side.

2.2.2. Existing OSM Community and Local Community

On 1 August 2016, the 50 most active mappers in Miami-Dade County between March 2015 and August 2016 were contacted through the OSM messaging system. A contributor's activity was measured by the total number of edits (including geometry changes, feature additions, modifications and deletions) observed in all changesets of the user whose centroids were located in Miami. Since messaging involved navigating to these user's profiles, automatic filtering of bots was not necessary. Two of the original top 50 accounts were removed from the list as one was found to be a bot, and the other one was the first author's OSM profile. These users were replaced with

OSM users originally in the 51st and 52nd place. Figure 2a shows the spatial distribution of OSM changesets (cyan transparent rectangles) around Miami-Dade County within this period, where the most active areas correspond well to the Miami metropolitan area. An introductory message was sent to the top 50 users, informing them about the import and listing all the resources (chat room, code repository, tutorials, meetups). It was assumed that the most active local mappers could be reached with this method.

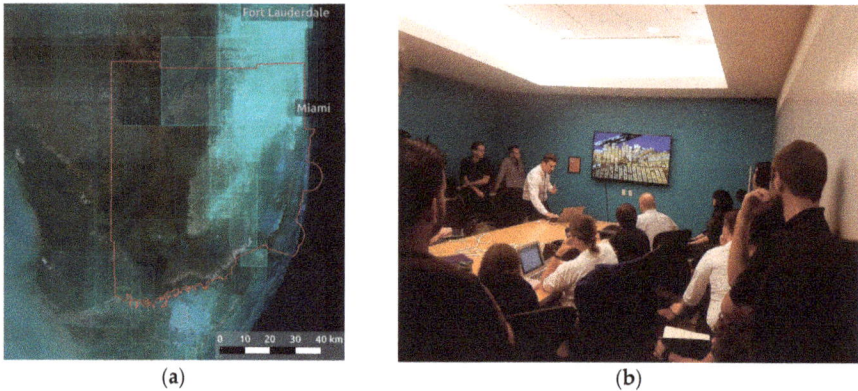

(a)   (b)

Figure 2. OSM changesets between March 2016 and August 2016 in Miami-Dade County (red outline) (**a**) and a Maptime Miami Meetup held on 26 September 2016 (**b**).

In addition to messaging OSM users, Maptime Miami organized meetings on the Meetup platform almost every month. These meetups were announced in different social media platforms (Facebook, Twitter, Meetup) and promoted by Maptime Miami and other Miami based community organizations, such as Code for Miami and Venture Café Miami. These meetups were organized around the import process. Most of the meetups included an interactive session where organizers helped new users getting started with OSM and importing buildings (Figure 2b). Between 1 August and 31 December 2016, a total of four meetups were dedicated to the import project.

Since OSM implements a free tagging system there is no control over how users indicate that their edit is related to the Miami-Dade Large Building Import project (if they indicate it at all). To identify users who directly interacted with this import process, we gathered usernames from three different sources. First, all users were extracted from the TM instance. Since this is the interface where users can download the import dataset, the users who contribute to this project according to the provided tutorial, will show up in this list. Another possible use case is when a user, instead of going through the import process, finds out about the project through editing OSM and chooses to improve the buildings that have been imported so far. These users tend to be more experienced and can be identified by analyzing a history dump and extracting all new features that match the "ref:miabldg" tag. There is another way for users to contribute to the import task without showing up in the TM or in the history dump. Namely, users could indicate the import process on the changeset level without marking individual features [25,26]. Our TM instance was configured so that the JOSM editor automatically populated the changeset comment field with the #miabuildings hashtag, which makes it possible to query these edits later.

For the remainder of the paper, users described in this section (contacted through direct messages, gathered from TM or history and changeset dumps) are referred to as community users. Besides the two targeted groups (students and community users), there will also be other OSM members that are not directly involved with the outreach activities described before, but who instead edit already imported buildings, e.g., by adding more attributes or refining building outlines.

3. Results

3.1. Participation Numbers

3.1.1. Students

Overall, 16 student submissions were received (Fall 2016, GIS Programming: 1/15 graduate students; Spring 2017, GIS Analysis: 14/26 graduate students, 1/3 undergraduate students). 15 of these 16 students received extra credit offered for the assignment. The difference in student activity between both courses might be due to the different nature of these classes as a programming class is rather technical and the focus is not on data sources and data analysis. The participation rate in the GIS Analysis class of 51% is a little higher than participation rates in extra credit activities in other studies with participation rates below 40% [27,28]. This might be due to the online nature of the extra credit opportunity, which did not involve commuting to campus. The grade distribution for GIS Analysis suggests that students from the whole grade spectrum participated in the bonus assignment. More specifically, 7/14 (50%) A-students, 7/13 (54%) B-students, 1/1 D-student, and 0/1 F-students participated, showing that the motivation across top (A) students, good (B) students, and poor (D, F) students to participate in the extra credit assignment is similar. This is somewhat different from earlier studies that showed that significantly more students who earned below the average and average elected not to participate in extra credit tasks [28]. A grade improvement due to the completed bonus assignment can be observed for seven out of the 15 participating students in this class.

The impact students had on OSM through participation in this extra credit assignment can be measured by the number of edits they made. On average, each student added 104 buildings (median: 87), although the assignment asked for a minimum of 50 buildings only. This resulted in a total of 1554 buildings in OSM through students. The median and mean number of buildings edited did not vary significantly between student performance (i.e., A through F letter grades considered before extra credit), which indicates that the work performance and motivation among all students who participated is comparable, independent of their overall course performance.

Contributions to the OSM mapping platform are in general, predominantly made by male users [29,30]. As opposed to this, the extra credit assignment did not reflect this usual gender bias. More specifically, the GIS Analysis course had a total enrollment of 29 students (31.0% female), with 15 students participating in the extra credit assignment. Among these 15 participants, 40.0% were female, which is higher than the percent of female enrollment in the course (31.0%). However, the difference is not statistically significant, suggesting that, if a reward by grade is involved, male and female students are similarly motivated to participate in OSM contributions. This is in-line with previous findings from an earlier study about volunteer research participation among 193 undergraduate students [28], which suggests that the difference between participation rates of women and men may not be meaningful.

3.1.2. Community Users

The 50 most active OSM contributors between March 2015 and August 2016 in Miami-Dade County who were contacted via direct messages submitted between 1 and 594 changesets (mean: 92, median: 27) in Miami-Dade County during this period. The number of map edits per user ranged between 518 and 62,555 (mean: 4052, median: 1324). The OSM sign up date of these 50 users was extracted from the main API. A histogram shows that the majority of these top 50 users are long standing OSM members who registered to the project between December 2006 and April 2015 (Figure 3b). Only seven out to the 50 users responded to the initial query. Four mappers provided supportive feedback but were not able to help out due to busy schedules or unfamiliarity with the area. The three remaining users did contribute to the project, although their user names did not show in the TM. This means that their contributions lean towards quality checks and follow up fixes. In fact, these users opened several OSM notes, provided changeset discussions and fixed several data issues in the proximity of import buildings. These contributions are also valuable parts of data imports.

Thirteen of the "top 50" mappers are Mapbox (https://www.mapbox.com/) employees working for the Data team, which operates worldwide on creating new data, improving existing features and fixing errors reported by OSM users. The fact that these users appear in the "top contributors in Miami" list indicates a small and generally inactive OSM user base in Miami-Dade County.

The TM had 30 users listed, though not all of them contributed to the project through data edits. By analyzing the history dump after 1 August 2016, 34 users were identified to add original import buildings to OSM. 18 of these users also submitted changesets with the #miabuildings import hashtag and showed up in either the TM or in the list of users extracted from the history dump. After combining users that used the #miabuildings import hashtag with those that did not, and excluding student accounts and the official import account that was used to automatically upload buildings, 32 unique users were left that were considered community users as their interaction with the import process was first-hand. These 32 community users are responsible for around the same number of buildings (1547) as the student group (1554). However, 9 of the community users (identified through TM) did not add any import buildings to OSM, but rather ran some other edits. This shows that the initial interest in an import project (expressed by signing up for the TM with their OSM credentials) does not always result in actual contributions. The remaining 23 contributors added 67 import buildings to the project on average, which implies a smaller import rate than for students (see Table 1). A two-sample t-test showed that there was a significant difference in the log transformed number of imported buildings between community users (M = 2.6, SD = 1.9) and students (M = 4.5, SD = 0.6): $t(27.24) = 4.41, p < 0.001$. These results suggest that different user engagement techniques have a different effect on user activity. In this case, the higher activity of students could have been driven by their desire for a higher grade. As opposed to this, community users would not experience any short-term gain (e.g., monetary or prestige) from the import task. This means that although in the short run students handled more imports per user, in the longer run it can be expected that community users provide more data than non-community users, since social mappers were previously found to contribute continuously [5]. Although that latter study analyzed mapping parties, we consider them the same as our categorization of community mappers as they are working towards a defined goal (import buildings) and also meet face to face at social events occasionally.

The import task became an organic part of OSM where data were further edited by the community. Such edits include further refinements of building geometries and tag additions (e.g., the name of a hotel). A total of 177 OSM users that were otherwise not related to the import process have interacted with import buildings so far. This is similar to OSM users interacting with the pedestrian network imported as part of the TIGER dataset [13] or excessively editing ways after a local import [31]. Such observed follow-up edits demonstrate the additional benefits of data imports.

3.1.3. New and Existing Users

Besides our user distinction that is based on recruitment efforts (students, community), users can also be classified across these categories into new and established OSM users. Accordingly, it is possible to analyze if data imports engage new and existing users differently. Analyzing the OSM editing history of users contributing to data imports or edits, it was found that 23 users came for the first time in contact with OSM during the import task and could therefore be classified as new users. This includes all 15 participating students and eight newly registered users through community outreach (Section 2.2.2). All of the remaining users created their accounts at least three months before the actual import task began. The first two columns of Table 1 show that students were significantly more active in the import task than those new users who were recruited through community events, which is supported by a two-sample *t*-test on the log transformed number of buildings: $t(8.45) = 6.5; p < 0.001$. These different levels of activity can likely be attributed to different motivations between those two groups of newly engaged OSM users (see Section 3.1.2). To refine the activity analysis of community users, and specifically to identify the effect of the import task on the new OSM members we compare the activity of new community users to existing community users. A two-sample t-test conducted on the log transformed number of imported buildings between existing users (N = 15, M = 3.3, SD = 2.0) and new users gained through community outreach (N = 8, M = 1.4, SD = 1.3) shows that existing members add significantly more buildings (t(20.01) = −2.73, p = 0.01) than new community users. Furthermore, the effect of motivation on import activity remains significant (t(8.45) = 6.5, p < 0.001) when compared between students (extrinsic) and existing community members (intrinsic).

Table 1. Descriptive statistics of imported buildings by user groups.

	Students(New)	Community Users		
		(New)	(Existing)	Total
N	15	8	15	23
Total # of buildings	1554	69	1478	1547
Average # of buildings per user	103.6	8.7	98.5	67.2
Median # of buildings per user	87.0	3.5	24.0	16.0
SD of # of buildings per user	59.3	11.8	159.9	135.0

These results show that data imports, at least in the short run, benefit most from (a) existing community members and (b) highly motivated users who gain some economic benefit (such as extra credit which can lead to better job placement chances through better grades). As opposed to this, new community users without an obvious economic benefit tend to generate less data. A stable base of OSM community contributors is, however, necessary to keep OSM data up-to-date in the long run. Therefore, although only a small number of building imports were observed for new community members, a data import task like the one analyzed in this study, will help to retain the critical mass of OSM community users that is needed to sustain data quality in the long run.

3.2. Mapping Behavior

3.2.1. Temporal Aspects

Figure 3 shows the histograms of OSM sign up dates which were extracted from the main API for students (Figure 3a), for the top 50 users contacted via direct messages (Figure 3b) and for community users (engaged through Maptime Miami; Figure 3c). Student sign up dates follow closely specific academic events during the semester, such as the introduction of the extra credit assignment in a lecture (15 February 2017, shown with a vertical dashed line) or assignment deadlines. The due dates (solid vertical lines) for GIS Programming were 2 December 2016 and 29 March 2017 for GIS Analysis, respectively. Most of the contacted users from the "top 50-editing list" have prior mapping experience, which is reflected by the fact that the majority of these users signed up more than a year before the import project. Community users who interacted with the import dataset first-handed consist of both new and experienced mappers. 40% of the community user group signed up to the OSM platform after the first discussions in May 2016 and 35% of them after August 2016, when the tasks were made available to the public, resembling the group of new mappers. This suggests that increased social media activity and local outreach can be an effective method in recruiting new contributors.

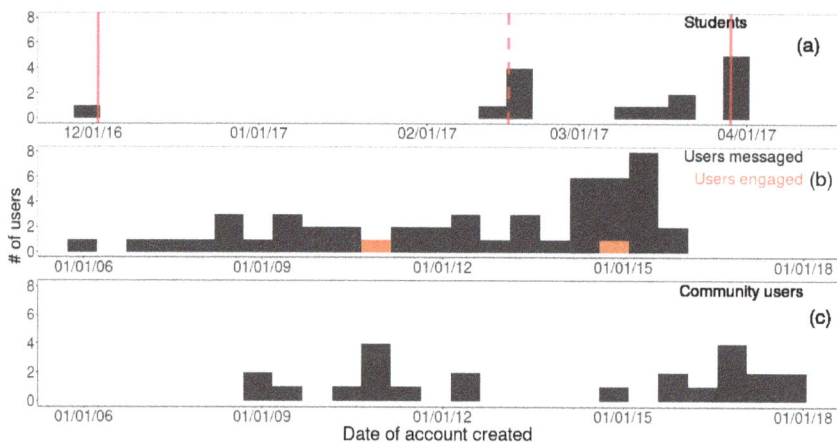

Figure 3. Histograms of sign up dates for different user groups. For students, assignment due dates (solid vertical lines) and first introduction to the project (dashed vertical line) are shown. Note that the horizontal time axes cover different date ranges for the user groups in (**a**–**c**).

To explore how different users interacted with the import task over time, their activities were plotted based on interactions with import buildings (addition, edits) between August 2016 and October 2017. A time-series visualization has been used in other studies to assess trends, seasonal and random components involved in OSM contribution activities [32]. Figure 4 shows the overall import related activity and the activity of different user groups over time. The activity of community users and students is directly associated with the import, as these groups were involved in the addition of original buildings. On the other hand, the group "other" is only indirectly associated with the import. Their activities include tag additions and follow up edits. The overall activity, which is the sum of the group activities, shows distinct peaks. This suggests that the import did not happen at a constant pace but that different events triggered increased activity over shorter periods of time. More specifically, dashed vertical lines in Figure 4 represent community related events (meetups), while solid vertical lines show due dates of home assignments for students. These events are listed in a chronological order in Table 2.

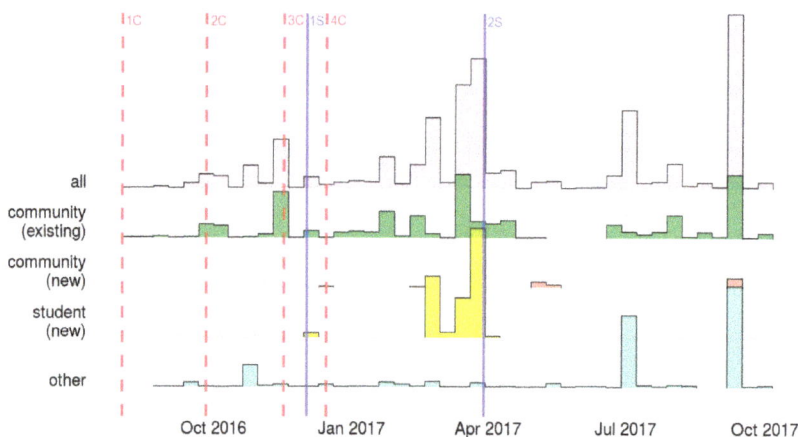

Figure 4. Import related activity levels over time for different user groups. Dashed vertical lines show community events, while solid vertical lines represent assignment deadlines for students.

Table 2. Description of events related to the import project.

Event	Event type	Event Description	Date
1C	Community	Technical discussion of software tools, general information on the import, intro messages sent out	1 August 2016
2C	Community	Presentation of automatic import results, hands-on mapping session	26 September 2016
3C	Community	Hands on mapping session dedicated to the import	17 November 2016
1S	Student	GIS Programming course bonus assignment due	2 December 2016
4C	Community	Hands on mapping session dedicated to the import	15 December 2016
2S	Student	GIS Analysis course bonus assignment due	9 March 2017

Community users (23 individuals) learned about the import project through various channels, such as meetups, message reach out, social media, OSM Wiki pages, and mailing list communications. Fifteen of these users are existing members and can therefore be classified as experienced mappers. There is an association between the amount of early contributions of this group (Fall 2016) and the community events held at that time (Figure 4). Event 1C did not trigger any significant activity, as it was a technical presentation along with general information about the proposed import process. As opposed to this, discernable import activities before the first hands on mapping session (2C) can be attributed to organizers testing the import process and to a few early users who followed the online conversations in the chat group. An even stronger increase in community activity can be observed during and after hands on mapping sessions 2C and 3C. A similar event (4C), however, did not have such an effect due to low participation before the holidays. It is also evident from the plot that the community user group remained active even when no more community events were organized. The motivation of these users can be classified as intrinsic as they were offered no monetary or other benefits or gains, yet they participated in the import and contributed to its success. This group mainly consists of locals. The continuing interest of these users in the building import can be explained by the pride of place concept [33], which describes the desire of the mapper to see one's own home town or region (Miami-Dade County in this context) on the map. Their behavior is also similar to previous findings about loyal OSM users who regularly check and update their "pet locations", which is the area where they edit most frequently [34].

Newly recruited community users show a different activity pattern. Surprisingly, no editing activity for these users was recorded until event 4C, even though three of the eight new users in this category signed up before that date. We attribute this to the fact that OSM, especially a data import task, may seem challenging and overwhelming at first. Our community events with high participation numbers did not seem to provide a good platform for engaging new contributors. In contrast, event 4C was not well attended, which provided an opportunity to dedicate more time and attention to newcomers who were present. Three users with no prior OSM experience attended this event, out of which one user (a local) successfully imported several buildings and added even more at later dates during that month. The remaining two users at this meetup were not interested in the import, but rather in general discussions about mapping. Figure 4 also shows that the long-term engagement of new community members is only sporadic. Unlike the existing community group, their contributions are ad hoc and can be traced back to social media posts or other events (e.g., HOT mapping), but then quickly vanish. This is similar to what has been revealed for mapping parties through user interviews, where users cannot be engaged for longer periods [34].

Students in the GIS Programming (one student) and GIS Analysis (15 students) classes focused their activities around the due dates of their home assignments (1S and 2S on Figure 4). Even though students were introduced to this extra credit task months before the due date (on 28 October 2016 for 1S and 15 February 2017 for 2S), their activity peaked right before the deadline and then quickly declined. This suggests that students were highly active before the deadlines but otherwise spent very little time on the task. This is in line with common practice of college students postponing assigned tasks until the day or night before due dates [35,36]. The figure also reveals that none of the students remained active after submitting their assignments, which indicates that our import task was not successful in attracting students to become permanent OSM contributors. Only two out of 16 students who completed the extra credit assignment have some ties to the project area in Miami-Dade, either by working or having grown up in Miami, based on class introductions posted by students. This general lack of ties to the study area for almost all students may explain the absence of motivation for students to voluntarily continue with OSM mapping activities after completing their assignment. Instead, students appear to be motivated primarily by the prospect of improved course grades, which can be classified as an extrinsic motivator [18].

The activities of the remaining user group ("other") in Figure 4 are not associated with either community events or student assignment deadlines. These contributions tend to follow a random pattern and could be a result of people spending their vacation in Miami and editing the map in the meantime or a regular OSM editor making some edits. The first distinct activity peak of other users in July 2017 is caused by one user (again, otherwise not related to the data import) adding building level information ("building:levels") to 342 of the import buildings. The other peak in September 2017 is related to HOT Hurricane Irma relief, which drew a large amount of editing activity to Miami. Interestingly, this humanitarian mapping project increased the building import activity of the community user group (both new and existing members) as well (September 2017 activity of the community group in Figure 4). This effect can be attributed to those local members of the community group that contribute to HOT projects as well. The fact that the Hurricane Irma mapping event overlapped with the import area gave these users an opportunity to further map their home region.

3.2.2. Spatial Aspects

The Tasking Manager logs user activity by storing when users accessed (i.e., locked) individual tasks. This information allowed us to explore which areas users prioritized for data imports and editing. Tasks were spatially subdivided into US Census block groups which contain approximately the same number of residents. Figure 5 provides an overview of the cumulative number of times a task was locked over time. There is no limit as to how many times a task can be locked by users. Even when a task is marked as "done", another user can still interact with it, for example to validate edits.

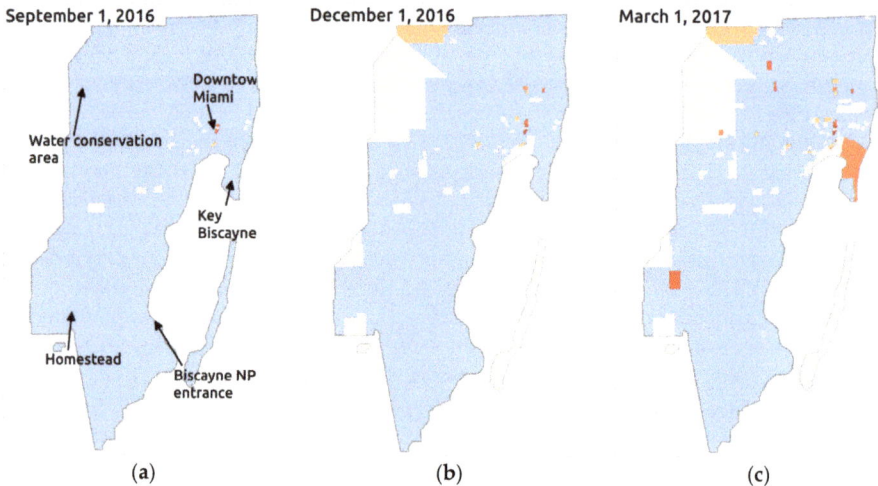

(a) (b) (c)

Figure 5. *Cont.*

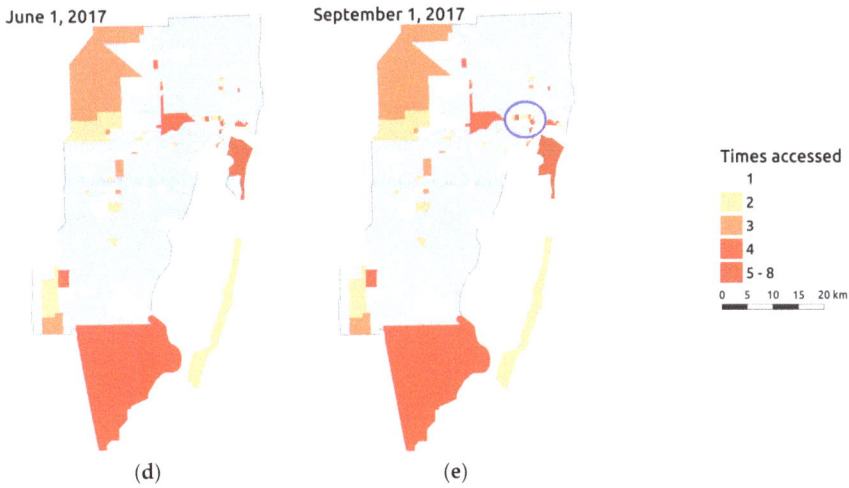

Figure 5. Cumulative number of times individual tasks from the Tasking Manager were accessed throughout September 2016 (**a**); December 2016 (**b**); March 2017 (**c**); June 2017 (**d**); and September 2017 (**e**).

The spatial distribution of activities shows a few distinct patterns. Since US Census block groups contain approximately the same number of residents, they tend to be larger around the edge of the project area, which has a low population density. Therefore, these census block groups appear more prominently on the map, which makes them more likely to be chosen by contributors, especially by those who lack local knowledge about the spatial layout of the study area. These larger census blocks show agricultural (Homestead), natural (Water conservation area), or industrial (e.g., quarries) characteristics and were often locked by users. Also, centrally located areas, where tasks are smaller in size, tend to be very popular among mappers, probably because of some mappers' interest in learning more about the city center regions. Accordingly, frequently locked tasks can be found in Downtown Miami (blue circle in Figure 5e), the financial district (Brickell) south of Downtown Miami, or Key Biscayne, which is a scenic and touristic island. Users locked 201 tasks, out of which 91 were marked as "done". These marked areas are highlighted in Figure 6 (green polygons) among other areas that showed only some or no activity. Approximately 45% of the areas end up being finished once a user locks them. A large number of tasks marked as done are found in natural areas, which require only a few or no building imports. Also, downtown areas showed a similarly high number of finished tasks, probably due to higher user interest in these areas. It has to be noted that a task is not automatically marked as "done". Therefore, in reality the number of tasks that are already finished could be higher. There was no evidence of users erroneously marking tasks as finished. Also, 41% of the total task areas (1591) contain no buildings to be imported. These tasks require no work, and therefore could easily increase the completion rate if set to "done". However, only 27 of the 650 tasks that involved no buildings have been marked as "done" so far.

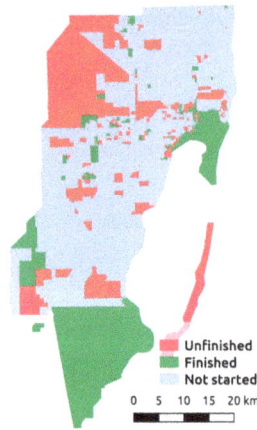

Figure 6. Finished tasks (green), tasks that have been worked on (red), and tasks that have not been worked on (blue).

Figure 7a shows that most tasks are locked only once (143), while only a few are locked more than 5 times. This can be expected because once an area has been correctly imported with all building conflicts removed, there is no need to work on it anymore. Accordingly, most tasks (174), whether finished or unfinished, were only worked on by one mapper (Figure 7b). A few popular tasks show that some areas remain interesting for OSM users, even though they are marked as "done". The most popular tasks were locked by three different users. These two distributions closely follow a power law function with an exponent value of 2.39 and an adjusted R^2 of 0.94 in the case of task locks (Figure 7a), and an exponent of 2.95 and an adjusted R^2 of 0.99 in the case of the number of users working on a task (Figure 7b). These heavy-tailed distributions follow a similar pattern observed many times in user-generated data. However, the level of information, especially in Figure 7b, is somewhat limited since only three data points were used for the regression.

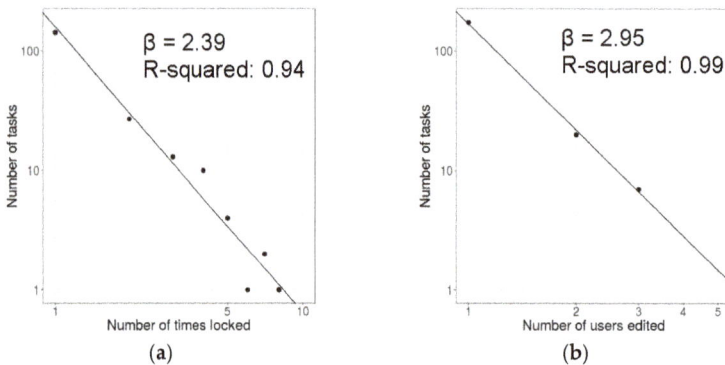

Figure 7. Fitted power law functions on task lock (**a**) and user (**b**) distributions (log-log plots).

3.2.3. Other OSM User Activities

Since OSM users were not limited to the import task analyzing user contributions outside Miami-Dade County provides further information about user characteristics and typical contribution behavior. Figure 8 shows the worldwide spatial distribution of changesets submitted by community users (blue areas) and graduate students (red triangles indicating centroids of changesets) between May 2016 and October 2017. The commitment of existing community members from the OSM community is reflected in Figure 8 through their changesets (blue areas) that cover significantly larger areas than the changesets of students (red triangles) and new community members (not visible due to small size and low volume). The majority of changesets submitted by existing community users can be found in the US, suggesting that these users are mostly US residents. Their interest in the entire country is also reflected through mapping activities in Hawaii and Puerto Rico, which are not part of the contiguous United States. A clear divide in contribution patterns along the US-Canadian border suggests that the interest of a user group can be influenced by administrative boundaries and cultural aspects. This can be explained by the pride of place concept [33,34]. National borders were also found to shape the spatial extent of mapping activities of individual users in other studies, for example when editing OSM based on Mapillary street level photos [26]. The extensive mapping of Columbia in South America by existing community users is a result of one user who divides his or her mapping efforts between Columbia and South Florida. The spatial distribution of student OSM activities is concentrated in Miami-Dade County, with only one student contributing outside the county. This student added several buildings in Bangladesh from aerial imagery and added also building names to existing features in that region. According to the class introduction, this student is originally from Bangladesh and most probably has personal knowledge and ties to the mapped area. Similarly, contributions of new community members are mainly found in South Florida (even outside Miami-Dade County) and in the Caribbean, suggesting that new users focus their editing activity on smaller areas than already established users.

Figure 8. Spatial distribution of OSM changesets submitted by participating users between May 2016 and October 2017.

Besides geographic coverage another distinct pattern between students and community users is the higher mapping activity associated with humanitarian projects (HOT) of the latter group. Since May 2016, 15 community users (11 existing and 4 new) submitted over 6200 changesets with #hotosm and #missingmaps hashtags. These changesets were mainly located in Haiti after Hurricane Matthew in 2016, in Africa (as part of Missing Maps (http://www.missingmaps.org/) projects in Tanzania, Nigeria, or Congo) and in Sri Lanka. An OSM data import project is similar to a humanitarian project in the sense that users work towards a specific goal while following centralized instructions. This explains that some community users contributed both to HOT and data import activities, in some instances within the project area (see also Section 3.2.1).

4. Summary and Conclusions

This study analyzed if and how a local building import task can help to engage students and targeted community groups in OSM participation and retention. One of our observations related to the organizational aspects of such data imports, is that the amount of information provided to participants can overwhelm prospective new users who are otherwise unfamiliar with collaborative mapping, and that only individual or small group tutoring on related tasks leads to the continuous engagement of a new community member. We also found that checking the early edits of new contributors and providing feedback are effective methods for ensuring high quality contributions later on.

Our results show that the type of engagement technique used to recruit users has a significant effect on the import activity. Students who were recruited as part of an extra credit assignment imported more buildings on average than users who were recruited at community meetings or through social media activity. The level of contribution activity is ultimately related to different motivations of users. In that regard, our results suggest that extrinsic motivation (i.e., students receiving extra credit) triggers more activity than the intrinsic motivation of community users, at least in the short run.

However, our experiment proved to be unsuccessful in retaining new users in the long run, regardless of their motivation. 23 new OSM members started mapping through the import project. The activity of students (15 individuals) closely followed academic deadlines, but no continuation of long-term activities could be observed. As previous research has already shown [5], mapping parties fail to retain newcomers almost completely, with no retention in the long term. This study expanded related research to the academic environment and found the same problem with extra credit activities among GIS students. Although participating students did contribute more data than required for the assignment, the import exercise did not retain the students as permanent OSM contributors. One possible approach to mitigate the latter problem could be to request continued OSM editing as part of project assignments so that students get used to a regular (e.g., weekly) OSM editing schedule. The remaining eight new users were recruited at community events, or through online outreach. The activity pattern of these users also corresponds to specific events, such as community events, social media posts or other related mapping activities in the study area. However, similarly to students, none of these new users became long-time contributors. Our experiments did not include follow up surveys or conversations with new users about their reasons for not continuing to contribute to the import task, which could be included in future import projects.

Though the study suggests that no long-term contributions can be expected from newly recruited mappers motivated by short-term extrinsic factors, it also shows the dedication of existing community members who demonstrated sustained editing activity levels throughout the study time frame. In addition, imports reach beyond those edits stemming from the direct import since these data also trigger long-term editing activities (e.g., adding attributes, fixing geometry errors) submitted by other, already engaged OSM user contributors. The study showed also that already established mappers do not change their contribution behavior through community events. Instead, they are active before and after the event, contributing to OSM on a regular basis. This is also in line with previous findings from OSM mapping parties [5]. The presented study also provides evidence of other users, who are otherwise unrelated to the project, to interact with the building import. More specifically, two distinct instances of this activity were one individual mapper adding more information to buildings (i.e., building levels) and another one with an increased editing activity due to an organized hurricane relief event by HOT.

To keep an active user base in OSM that will also ensure regular data updates and quality enhancement in the future, new ways of user recruitment and retention are necessary. While the presented study showed that extra credit assignments increase short-term engagement of students, highlighting fun aspects could be another potential component to retain new mappers in OSM on the longer term, as this was also found to be a major driver for other collaborative projects, such as Wikipedia [37]. For example, geo-gaming and gamification has been shown to be an attractive and incentivizing way of engaging a different audience in land cover validation [38] and the collection of crowd-sourced Points of Interest [39]. While there is no best recipe for how to integrate fun components into OSM mapping and import tasks, considering ideas from other platforms could provide some guidance and ideas, including the provision of reward diversity [40], or the interaction between participants, e.g., by responses to video recordings [41]. Future work will aim to integrate such components into mapping events and recruitment efforts and evaluate their efficacy. Based on experiences gained in the presented experiment, we recommend that similar projects put extra effort in interacting with prospective users who lack prior OSM experience. Providing a welcoming, personalized experience that addresses the special needs of these users might be a promising way to engage new users more effectively.

Acknowledgments: The authors would like to thank past and present organizers of Maptime Miami (Matthew Toro, Daniela Waltersdorfer, Nohely Alvarez, Adam Old and Ernie Hsiung) for keeping the small OSM community in South Florida alive. We are also thankful for OSM members who provided feedback on various channels, and finally, we thank all OSM users, community members and students who participated in the import. Publication of this article was supported by the University of Florida Open Access Publishing Fund.

Author Contributions: L.J. was the technical lead on the OSM import project and is a co-organizer of Maptime Miami. H.H.H. teaches both courses mentioned in the paper at the University of Florida. L.J. designed the experiments and analyzed the data. L.J. and H.H.H. both contributed to the writing of this manuscript.

Conflicts of Interest: L.J. is one of the current co-organizers of Maptime Miami and also contributed to this project by importing building data.

References

1. Goodchild, M.F. Citizens as Voluntary Sensors: Spatial Data Infrastructure in the World of Web 2.0 (Editorial). *Int. J. Spat. Data Infrastruct. Res.* **2007**, *2*, 24–32.
2. Haklay, M. Citizen Science and Volunteered Geographic Information: Overview and Typology of Participation. In *Crowdsourcing Geographic Knowledge*; Sui, D., Elwood, S., Goodchild, M., Eds.; Springer: Berlin, Germany, 2013; pp. 105–122.
3. Goodchild, M.F.; Li, L. Assuring the quality of volunteered geographic information. *Spat. Stat.* **2012**, *1*, 110–120. [CrossRef]
4. Haklay, M.; Weber, P. OpenStreetMap: User-Generated Street Maps. *IEEE Pervas. Comput.* **2008**, *7*, 12–18. [CrossRef]
5. Hristova, D.; Quattrone, G.; Mashhadi, A.J.; Capra, L. The Life of the Party: Impact of Social Mapping in OpenStreetMap. In Proceedings of the Seventh International AAAI Conference on Weblogs and Social Media, Cambridge, MA, USA, 8–11 July 2013; pp. 234–243.
6. Bryant, S.L.; Forte, A.; Bruckman, A. Becoming Wikipedian: Transformation of Participation in a Collaborative Online Encyclopedia. In Proceedings of the GROUP: International Conference on Supporting Group Work, Sanibel Island, FL, USA, 6–9 November 2005; pp. 1–10.
7. Mooney, P.; Minghini, M.; Stanley-Jones, F. Observations on an OpenStreetMap mapping party organised as a social event during an open source GIS conference. *Int. J. Spat. Data Infrastruct. Res.* **2015**, *10*, 138–150.
8. Mashhadi, A.; Quattrone, G.; Capra, L. The Impact of Society on Volunteered Geographic Information: The Case of OpenStreetMap. In *OpenStreetMap in GIScience (Lecture Notes in Geoinformation and Cartography)*; Jokar Arsanjani, J., Zipf, A., Mooney, P., Helbich, M., Eds.; Springer: Berlin, Germany, 2015; pp. 125–141.
9. Bégin, D.; Devillers, R.; Roche, S. Contributors' Withdrawal from Online Collaborative Communities: The Case of OpenStreetMap. *ISPRS Int. J. Geo-Inf.* **2017**, *6*, 340. [CrossRef]
10. Soden, R.; Palen, L. From crowdsourced mapping to community mapping: The post-earthquake work of OpenStreetMap Haiti. In Proceedings of the COOP 2014-Proceedings of the 11th International Conference on the Design of Cooperative Systems, Nice, France, 27–30 May 2014; pp. 311–326.
11. Mooney, P.; Minghini, M. A Review of OpenStreetMap Data. In *Mapping and the Citizen Sensor*; Foody, G., See, L., Fritz, S., Mooney, P., Olteanu-Raimond, A.-M., Fonte, C.C., Antoniou, V., Eds.; Ubiquity Press: London, UK, 2017; pp. 37–59.
12. OSM Wiki: Import Catalog. Available online: http://wiki.openstreetmap.org/wiki/Import/Catalogue (accessed on 20 January 2018).

13.   Zielstra, D.; Hochmair, H.H.; Neis, P. Assessing the effect of data imports on the completeness of OpenStreetMap—A United States case study. *Trans. GIS* **2013**, *17*, 315–334. [CrossRef]

14.   Mooney, P.; Corcoran, P. The annotation process in OpenStreetMap. *Trans. GIS* **2012**, *16*, 561–579. [CrossRef]

15.   Touya, G.; Brando-Escobar, C. Detecting level-of-detail inconsistencies in volunteered geographic information data sets. *Cartographica* **2013**, *48*, 134–143. [CrossRef]

16.   Beaulieu, A.; Bégin, D.; Genest, D. Community mapping and government mapping: Potential collaboration? In Proceedings of the Symposium of ISPRS Commission I, Calgary, AB, Canada, 15–18 June 2010; pp. 16–18.

17.   Fritz, S.; Linda, S.; Brovelli, M. Motivating and Sustaining Participation in VGI. In *Mapping and the Citizen Sensor*; Foody, G., See, L., Fritz, S., Mooney, P., Olteanu-Raimond, A.-M., Fonte, C.C., Antoniou, V., Eds.; Ubiquity Press: London, UK, 2017; pp. 93–117.

18.   Budhathoki, N.R.; Haythornthwaite, C. Motivation for open collaboration crowd and community models and the case of OpenStreetMap. *Am. Behav. Sci.* **2013**, *57*, 548–575. [CrossRef]

19.   Hecht, R.; Kunze, C.; Hahmann, S. Measuring completeness of building footprints in OpenStreetMap over space and time. *ISPRS Int. J. Geo-Inf.* **2013**, *2*, 1066–1091. [CrossRef]

20.   Fram, C.; Chistopoulou, K.; Ellul, C. Assessing the quality of OpenStreetMap building data and searching for a proxy variable to estimate OSM building data completeness. In Proceedings of the GIS Research UK (GISRUK) 2015 Proceedings, Leeds, UK, 15–17 April 2015; pp. 195–205.

21.   Brovelli, M.; Minghini, M.; Molinari, M.; Zamboni, G. Positional accuracy assessment of the OpenStreetMap buildings layer through automatic homologous pairs detection: The method and a case study. *Int. Arch. Photogramm. Remote Sens. Spat. Inf. Sci.* **2016**, *41*, 615. [CrossRef]

22.   Törnros, T.; Dorn, H.; Hahmann, S.; Zipf, A. Uncertainties of completeness measures in OpenStreetMap—A case study for buildings in a medium-sized German city. *ISPRS Ann. Photogramm. Remote Sens. Spat. Inf. Sci.* **2015**, *2*, 353–357. [CrossRef]

23.   Fan, H.; Zipf, A.; Fu, Q.; Neis, P. Quality assessment for building footprints data on OpenStreetMap. *Int. J. Geogr. Inf. Sci.* **2014**, *28*, 700–719. [CrossRef]

24.   Barron, C.; Neis, P.; Zipf, A. A comprehensive framework for intrinsic OpenStreetMap quality analysis. *Trans. GIS* **2014**, *18*, 877–895. [CrossRef]

25.   Juhász, L.; Hochmair, H.H. How do volunteer mappers use crowdsourced Mapillary street level images to enrich OpenStreetMap? In Proceedings of the 20th AGILE Conference on Geo-Information Science, Wageningen, The Netherlands, 18–21 Septermber 2017.

26.   Juhász, L.; Hochmair, H.H. Cross-Linkage Between Mapillary Street Level Photos and OSM Edits. In *Geospatial Data in a Changing World: Selected papers of the 19th AGILE Conference on Geographic Information Science (Lecture Notes in Geoinformation and Cartography)*; Sarjakoski, T., Santos, M.Y., Sarjakoski, L.T., Eds.; Springer: Berlin, Germany, 2016; pp. 141–156.

27.   Elicker, J.D.; McConnell, N.L.; Hall, R.J. Research Participation for Course Credit in Introduction to Psychology: Why Don't People Participate? *Teach. Psychol.* **2010**, *37*, 183–185. [CrossRef]

28.   Padilla-Walker, L.M.; Thompson, R.A.; Zamboanga, B.L.; Schmersal, L.A. Extra credit as incentive for voluntary research participation. *Teach. Psychol.* **2005**, *32*, 150–153. [CrossRef]

29.   Stephens, M. Gender and the GeoWeb: Divisions in the production of user-generated cartographic information. *GeoJournal* **2013**, *78*, 981–996. [CrossRef]

30.   Schmidt, M.; Klettner, S. Gender and experience-related motivators for contributing to openstreetmap. In Proceedings of the Action and Interaction in Volunteered Geographic Information (ACTIVITY) Workshop at AGILE 2013, Leuven, Belgium, 5 may 2013; pp. 13–18.

31.   Mooney, P.; Corcoran, P. Understanding the Roles of Communities in Volunteered Geographic Information Projects. In *Progress in Location-Based Services (Lecture Notes in Geoinformation and Cartography)*; Krisp, J., Ed.; Springer: Berlin, Germany, 2013; pp. 357–371.

32.   Bégin, D.; Devillers, R.; Roche, S. Contributors' enrollment in collaborative online communities: The case of OpenStreetMap. *Geo-Spat. Inf. Sci.* **2017**, *20*, 282–295. [CrossRef]

33.   Coleman, D.J.; Georgiadou, Y.; Labonte, J. Volunteered Geographic Information: The nature and motivation of produsers. *Int. J. Spat. Data Infrastruct. Res.* **2009**, *4*, 332–358.

34.   Napolitano, M.; Mooney, P. MVP OSM: A tool to identify areas of high quality contributor activity in OpenStreetMap. *Bull. Soc. Cartogr.* **2012**, *45*, 10–18.

35. Burchfield, C.M.; Sappington, J. Compliance with required reading assignments. *Teach. Psychol.* **2000**, *27*, 59–60.

36. Fernald, P.S. The Monte Carlo quiz: Encouraging punctual completion and deep processing of assigned readings. *Coll. Teach.* **2004**, *52*, 95–99.

37. Nov, O. What motivates wikipedians? *Commun. ACM* **2007**, *50*, 60–64. [CrossRef]

38. See, L.; Fritz, S.; Perger, C.; Schill, C.; McCallum, I.; Schepaschenko, D.; Duerauer, M.; Sturn, T.; Karner, M.; Kraxner, F. Harnessing the power of volunteers, the internet and Google Earth to collect and validate global spatial information using Geo-Wiki. *Technol. Forecast. Soc. Chang.* **2015**, *98*, 324–335. [CrossRef]

39. Juhász, L.; Hochmair, H.H. Where to catch 'em all?—A geographic analysis of Pokémon Go locations. *Geo-Spat. Inf. Sci.* **2017**, *30*, 241–251. [CrossRef]

40. Choi, J.; Choi, H.; So, W.; Lee, J.; You, J. A Study about Designing Reward for Gamified Crowdsourcing System. In *Design, User Experience, and Usability. User Experience Design for Diverse Interaction Platforms and Environments. DUXU 2014 (Lecture Notes in Computer Science, Vol. 8518)*; Marcus, A., Ed.; Springer: Cham, Switzerlands, 2014; pp. 678–687.

41. Spiro, I. Motion chain: A webcam game for crowdsourcing gesture collection. In *CHI'12 Extended Abstracts on Human Factors in Computing Systems*; ACM: Austin, TX, USA, 2012; pp. 1345–1350.

International Journal of
Geo-Information

MDPI

Article

Experiences with Citizen-Sourced VGI in Challenging Circumstances

Mustafa Hameed [1],*, David Fairbairn [1] and Suzanne Speak [2]

[1]   School of Engineering, Newcastle University, Newcastle NE1 7RU, UK; david.fairbairn@newcastle.ac.uk
[2]   School of Architecture, Planning and Landscape, Newcastle University, Newcastle NE1 7RU, UK;
     s.e.speak@newcastle.ac.uk
*   Correspondence: m.r.hameed@newcastle.ac.uk; Tel.: +44-75-7042-1032

Received: 20 October 2017; Accepted: 22 November 2017; Published: 26 November 2017

Abstract: The article explores the process of Volunteered Geographic Information (VGI) collection by assessing the relative usability and accuracy of a range of different methods (smartphone GPS, tablet, and analogue maps) for data collection among different demographic and educational groups, and in different geographical contexts within a study area. Assessments are made of positional accuracy, completeness, and the experiences of citizen data collectors with reference to the official cadastral data and the land administration system. Ownership data were validated by crowd agreement. The outcomes of this research show the varying effects of volunteers, data collection method, geographical area, and application field, on geospatial data handling in the VGI arena. An overview of the many issues affecting the development and implementation of VGI projects is included. These are focused on the specific example of VGI data handling presented here: a case study area where instability and lack of resources are found alongside strong communities and a pressing need for more robust and effective official structures. The chosen example relates to the administration of land in an area of Iraq.

Keywords: volunteer geographic information; positional accuracy; land administration systems

1. Introduction

The term 'Volunteer Geographic Information' (VGI) was introduced by Goodchild [1], to describe the widespread participation of the private citizen in creating geographic information, a function that, for centuries, had been reserved to official agencies [2]. The majority of VGI projects have concentrated on building web applications that allow citizens, using the Internet and contemporary technology, to access and edit or create features with reference to maps, satellite images, and ground-based methods. However, such projects may not always be successful in developing countries, such as Iraq, where Internet connections may not be good, the majority of citizens have no knowledge of using web map applications, and there are problems in engaging with communities, especially in rural areas [3].

Only 25% of nations (mostly industrial countries, 35–50 in total) have a complete land registration system: lack of finance, limited institutional capacity, and ineffective political will, mean that 75% of the world's land parcels have not yet been registered. The majority of their occupants are the most vulnerable and poorest groups in society, and they live under threat of expulsion due to lack of security of tenure [4]. Clearly, improved systems and practices are required, and it is suggested that cooperation with the local community could accelerate the creation of land administration systems which are appropriate, realistic, sustainable, manageable, and effective, i.e., fit-for-purpose.

The priority of a 'fit-for-purpose' system is not necessarily high spatial accuracy, but rather the effective recording of ownership and provision of security of tenure for underprivileged communities [5]. Building such a system could incorporate efforts of local citizens, with different levels of education and background, and utilizing varying technologies. The majority of previous research in VGI concentrated on the general drive, nature, and applications of the crowdsourced data, i.e., data collated from contributions of a number (often with redundancy) of 'amateurs', non-officials, or volunteers. It mainly focused on the nature of projects such as OpenStreetMap (OSM) and Wikimapia, but with little research on the fitness for use of VGI in official domains, such as land administration. Keenja et al. [6] reported that "to date, limited empirical work has been undertaken in this domain: there remain many unanswered questions regarding the accuracy, authority, assuredness, availability, and ambiguity of crowdsourced data. Meanwhile, the potential for crowdsourcing to provide a low cost and high-speed solution in areas where cadastral coverage is lacking, is eagerly anticipated". Basiouka, Potsiou, and Bakogiannis [7] used volunteers to assess the possibility of using OSM for official, cadastral purposes, but the target group was college-educated surveying practitioners, rather than real members of the community. Other researchers, such as Grus and Hogerwerf [8], have reported on experiences of crowdsourcing in the Netherlands' Kadaster, concentrating on change detection, whilst de Almeida et al. [9] explored the role of VGI in capturing and utilizing 3D data for property cadastres.

The study presented here examines VGI data collection, involving a wide range of community citizens, and several digital and analogue collection methods acceptable to them, to investigate contributions to official land administration systems (LAS). The paper is based on research and fieldwork in Iraq undertaken in 2016. The next section covers the issues which affect the role of VGI in land administration projects. Section 3 introduces the research project undertaken in Iraq, and its results are reported in Section 4. Following discussion, it is suggested that, in areas of conflict or when official systems are under extreme stress, VGI may be the only realistic method of collecting usable data.

2. Contextual Issues in VGI for Land Administration

There are many issues which influence and directly affect the operation of VGI projects for the purposes of land management. Land administration is a crucial governmental function, and its effective delivery in a dysfunctional environment, such as modern-day Iraq, is difficult. The concept of public participation in such activity can be problematic for both authorities and citizens. Further, the nature and quality of data collected by volunteer citizens must be addressed. It may also be possible to learn from other investigations into the potential of VGI, and into the governance and methods of land administration. The most significant issues for the case study presented here are outlined in this section.

2.1. Approaches to the Public Participatory Collection of Geographic Data

Mass contributions of VGI are referred to as 'crowdsourcing' [10], generally taking advantage of contemporary mobile devices and the 'geo-web'. The terms 'collaborative mapping' and 'participatory GIS' can be used to describe the application of VGI to land administration; further VGI use is directed towards the concept of 'citizen science' [11]. In the handling of data related to land administration, the subject of the study outlined here, de Vries, Bennett, and Zevenbergen [12] refer to 'neo-cadastres', meaning land-based records built and maintained by citizens. Seeger [13] considers three specific aspects important to the use of VGI in cadastre- and land-based projects: the motivation for volunteer engagement with land administration, the quality of the data, and methods of validation and verification. The case study described here exposed an uncertain group of volunteers to concepts of participatory geographic data collection and its subsequent handling and processing.

2.2. Citizens' Motivation

A range of issues may affect or direct the work of a volunteer who collects and handles VGI [14]. Some citizens speak of altruism, professional and personal interest, intellectual stimulation, protection of possible personal investment in the locality, social reward, personal reputation, self-expression, opportunity, and 'pride of place', including improvement in public services [15], as positive reasons for their engagement in VGI collection and management. Basiouka and Potsiou [16] suggest that the main motivation for public volunteering might be to overcome bureaucracy and assist in opening land up to more development. More negative factors can also act as drivers, including the promotion of mischief, support of a contrary social, economic, or political agenda, or malice intent (similar to hacking or seeking criminal access to data). Participation can often be seen as recreational, with Cotfas and Diosteanu [17] suggesting that the public does not even need to be particularly aware or motivated for their participation. However, Tulloch [18] suggests that communities and individuals which engage in VGI achieve a higher level of 'empowerment', and this certainly did become a public motivation in this study.

2.3. VGI as a Contributor to Official Activity

Haklay et al. [19] present direct uses of VGI in governmental activities. Although some of these are not directly related to cadastral systems, they inform this project with experiences and difficulties that may face the application of VGI to such tasks. In Kibera, Nairobi's biggest informal settlement, VGI was used to create a basic topographic map, enhancing the surveying activities of the national mapping agency [20]. In a more integrated fashion, the Canadian government has developed a project for correcting and updating topographic maps using VGI [21]. VGI can also be used for improving public services, although there is evidence of resistance to its adoption as a regular information source for some local government tasks [22].

Official governmental activity, local, regional, national, and international, typically involves significant amounts of spatial data handling, but it is also characterized by shortcomings in resources which can lessen effectiveness, or may even dissuade communities from engaging with the economic, political, and social management of society. The advantages and experiences of VGI should be communicated to formal structures, including those engaged in administration of a major societal resource, land, and this was a major goal of this study.

2.4. Accuracy and Completeness Considerations for VGI

Since VGI is provided, in many cases, by people with little or no knowledge of the mapping process [23], it is necessary to verify the quality of this data and the potential benefits. Further, there is possible divergence between the quality of original cadastral map production and the accuracy of volunteer information, which indicates the need for verification of the latter. The quality of data can comprise several factors [24,25]:

- Positional accuracy is the 'nearness' of coordinate values of a VGI feature (e.g., a captured point) to a corresponding authoritative equivalent feature.
- Thematic or attribute accuracy refers to the reliable and reasonable correctness of semantic information attached to the point, line, and polygon features of the spatial database.
- Completeness refers to the comparison between different datasets for the same area of interest to find which features are included or excluded from a dataset.
- Temporal accuracy refers to the agreement between encoded and 'actual' temporal coordinates [26].
- Logical consistency refers to the identification and resolution of contradictions, relationships, and connections within a dataset [27].

The evaluation of the positional accuracy of VGI can utilize traditional statistical methods, such as root mean square error (RMSE) to describe the spatial error of point features. Fairbairn and Al Bakri [28] reviewed the spatial correspondence between VGI and official government data, finding that the RMSE for OpenStreetMap data against official topographic mapping data was consistently higher than established tolerances, with errors attributed to the low-precision devices, for example, personal GPS units and commercial imagery services, commonly used in VGI data collection. Such measures can vary within a dataset which covers different areas: Zielstra and Zipf [29] found that the quality of VGI became worse the further away it was collected from the urban core.

Completeness can be assessed by considering the difference between what is recorded (e.g., in the context of a VGI project such as OSM, number of houses or length of roads) and what is actually found in the real-world [30]. Haklay [31] suggested that it is possible to rely on such a numerical assessment, by simply comparing the total length of streets in OpenStreetMap (OSM) with Ordnance Survey (OS) data. Jackson et al. [32], undertook a study counting numbers of identified schools in an area and showing correspondence across four datasets. However, when they repeated the study basing it on specific attributes—e.g., names, addresses—they noted that, although the numbers were similar, the schools themselves were often not identifiably the same. Summarizing completeness based on quantities alone was insufficient to assess this factor.

2.5. Land Administration Systems in Developing Countries

Adlington and Tonchovska [33] argue that one of the main reasons for inefficiency of official land administration systems is a lack of funds. Characteristic of many changing economies in less developed countries, public projects in revising, updating or maintaining any official system may fail due to political change and economic transition from central to free markets: 49% of World Bank supported projects suffer from budget deficiency, exemplified by a study of land administration by Basiouka and Potsiou [16] in Bulgaria. McLaren [4] noted that lack of trained staff also makes official systems inefficient, with Enemark et al. [5] citing Rwanda as an example of this situation.

In practice, governmental systems can be inaccessible for most of the people, especially in developing countries. The lengthy and costly procedures in handling land data mean that poorer people may not register their ownership, and may buy or sell their land without reference to the formal system. Official systems usually record only legally registered land, leaving millions of people whose tenures are predominantly social, rather than legal, as unprotected occupiers [34]. A further issue is the difficulty faced by official systems after change of governmental regime, or following civilian or military conflict [35]. Even where usable systems exist, Al-Bakri and Fairbairn [36] noted shortcomings in completeness and currency as authorities struggle to record increasing numbers of plots and subdivisions, and changes of use. Supplementary or alternative collection, recording, and management procedures, including VGI-based techniques, can appear attractive.

2.6. Fit-For-Purpose Land Administration

Where the official cadastral system is weak or does not exist at all, 'fit-for-purpose' systems [5] can be considered. Their main aim is to provide security of tenure for underprivileged communities, using several key principles: general boundaries are used rather than fixed/monumented boundaries, meaning that the accuracy of the delineation process is not necessarily high, especially in rural and peri-urban areas where land values may be low; the use of cheaper satellite or aerial imagery as base mapping is promoted as suitable for land administration; the approach is participatory and inclusive, covering all tenure types, including both legal and de facto occupied; and the resultant system flexibility allows for easy data update and system upgrade. 'Fit-For-Purpose' approaches have been tested in this research study.

2.7. Evaluating the Use of VGI in the Land Administration System

De Vries, Bennett, and Zevenbergen [12] argue that the use of VGI in cadastral systems is faster, cheaper, and more fit-for-purpose than the traditional method of official survey and registration. VGI can also act as an interim cadastral solution for securing land rights with different levels of tenure security. However, along with Lanier [37] and Keen [38], they do acknowledge possible conflict between VGI methods and data, and the procedures of official organizations and experts.

Goodchild and Li [39] have concerns about the quality of VGI and the fact that only a small number of people can validate it. In land administration, similar concerns have been expressed by Navratil and Frank [40] who argued that it would be difficult to depend on VGI alone as an alternative for an official cadastral system. The testing of VGI validity, and assessment of conflicts between it and official records, are important in determining in its worth. Methods of verifying VGI may include the matching of data against ground truth or accepted values, quality control of the data flowline, the acquired reputation of the volunteer (both as a source of original data and as a checker of others' [41]), or confirmation by multiple data collection methods or individuals [42,43].

2.8. An Example from Iraq

In Iraq, the official land administration system has faced many problems since the US-led Occupation in 2003. Large-scale forgery of title deed documents dates from that period. The dysfunctional nature of official systems has led to the seizure of public buildings by people occupying them as their living space, with others illegally squatting on public land and building their own houses on it, bypassing the formal land registration system. Internal migration and displacement of large groups of the Iraqi population have exacerbated property ownership uncertainty and occupation disputes, and rebel political and military groups have established alternative governance in many areas. Current political and economic circumstances do not signify any improved situation, and the sub-optimal nature of the current land administration system in Iraq suggests that an alternative method based on informal data sources is the better approach to improvement.

Each of the contextual aspects presented here in Section 2 has been incorporated into this research, and several are explored in depth later. This wide-ranging set of issues has an effect on the investigation of possible impact of VGI, derived in difficult situations, on dysfunctional organisations charged with official geospatial data handling.

3. Establishing a VGI Project

A project was set up to test the applicability and value of VGI in enhancing the land administration system in a province in central Iraq, and the practical procedures employed to handle the VGI. The issues presented in Section 2 were major drivers in establishing a methodology for this project. The governorate of Babil was chosen, and field work was conducted in the region of its capital city, Al-Hillah, 100 km due south of Baghdad. Here, three types of locality—rural, peri-urban, and urban—were identified. For each type, several specific locations were chosen and the local communities were contacted through gatekeepers identified in collaboration with the local land administration office. The locations varied not only in topography, but in land utilization dynamics, demographic profile, and socio-cultural community. In addition, the coverage of formal land records and maps obtained from the official LAS professionals was different for each site. The variability in methodology extended to the testing of differing data collection methods in each of these areas.

3.1. Community Sampling

VGI data collection was undertaken in nine communities: four urban, three peri-urban, and two rural. Preparation involved ensuring engagement with those local leaders who were positive about the research program, and were able to introduce the project to residents. A total of 10–15 volunteer citizens per community were recruited, varying in gender, age, and educational level (Table 1). In a formal workshop environment, volunteers were given full training in the project requirements, in terms of security, anonymity in data handling and instrumentation, and procedures for data collection. Volunteers could choose which technology to use to gather the data. In addition, the implications of giving consent to participation were explained, and a brief overview of the formal land administration procedures was given. Training was delivered, with detailed instructions on both low- and high-tech methods, and on specific issues, such as identifying a plot's 'center point' and being precise in attribute recording.

Table 1. Characteristics of volunteers who mapped case study areas.

	Gender		Age			Education level		
	Male	Female	<30	30–50	>50	Uneducated	School	University
Urban Communities	23	18	15	19	7	6	18	17
(Four sample areas)	(56%)	(44%)	(37%)	(46%)	(17%)	(15%)	(44%)	(41%)
Peri-urban Communities	29	8	11	14	12	10	17	10
(Three sample areas)	(78%)	(22%)	(30%)	(38%)	(32%)	(27%)	(46%)	(27%)
Rural Communities	27	0	4	10	13	14	9	4
(Two sample areas)	(100%)	(0%)	(15%)	(37%)	(48%)	(52%)	(33%)	(15%)

3.2. VGI Collection

The methods of data collection are central to the testing of geometric and semantic/attribute data, as well as considering the varying nature of the environment visited, and the personal abilities of the individual volunteers. Three methods were developed and used: (i) smartphone with a GPS app uploaded for locating land parcel corners and attributing the resultant polygon; (ii) portable iPad tablet PCs with the official cadastral map uploaded, and overwriting and annotating capability provided through QGIS; and (iii) paper-printed aerial or satellite image, with clipboard and pencil for demarcation and annotation.

Initial training, interviews, and practical data collection took place with reference to the land parcel presently occupied by the volunteer, to establish familiarity with the method chosen—gathering GPS coordinates of plot boundaries, identifying parcels and annotating maps on a portable tablet, or demarcating and tracing plots on paper images of satellite scenes (Figure 1). It was recognised that scale is an important issue in VGI data collection [44]: for urban and peri-urban areas, tablet mapping and satellite imagery should render the base map data at a large scale (e.g., 1:1500), although smaller scales may be appropriate for rural areas.

Each volunteer surveyed a number of land parcels.

(i) (ii) (iii)

Figure 1. Data collection methods applied in different geographical contexts: **(i)** smartphone GPS in an urban centre; **(ii)** tablet computer in an urban centre; and **(iii)** paper aerial image in a rural area.

3.3. Practical Fieldwork

The volunteer groups themselves demarcated their precise data collection site (community extent) using standard paper mapping or imagery. The size of such sites was set at 250 m × 250 m in rural areas (maximum 100 agricultural plots), 150 m × 150 m in peri-urban zones (maximum 100 land parcels), and 150 m × 150 m in central urban districts (maximum six city blocks, perhaps including high-rise buildings). Each site was divided to identify specific sub zones for each volunteer to capture the land parcels' geometry. In addition, attribute data for each land parcel (e.g., owner; type of tenure; date of last transaction; land use; date of last land use change (e.g., agriculture to residential)) was captured across the whole site in duplicate by all volunteers. Multiple capture of such attribute data is necessary to provide validation data. The geometric and positional data captured by VGI could be tested against reference surveys, existing mapping or imagery, but the attribute data were validated by such crowdsourcing because: (a) official records of ownership are confidential and inaccessible; (b) subdivision has complicated ownership; and (c) contemporary information is required. Further attribute data reflecting environmental change was also recorded by the volunteers—change of river course, building demolition or construction, notification of heritage status, or change in land use. Such information allowed for tests of completeness and currency to be undertaken later, and comparisons made between the official records and the VGI.

4. VGI Activities and Outputs in Al-Hillah

A series of field data collection activities were undertaken in each case study area (Figure 2): communities were consulted, citizen volunteers identified, officials interviewed, data collected, and existing official mapping and records were collated. Much of the assessment of VGI methods and outputs involved comparisons with these official datasets. The land administration agency in Al-Hillah retains official records in both textual and in map form. Unfortunately, there are very few records which offer up-to-date information, the vast majority of maps having been produced before 2003 and typically hand-drawn on paper. These show parcel boundaries but little other topographic detail, generally limited to drainage and highways. The scale of such maps is typically 1:1000 for urban areas, 1:2500 for rural areas, and the name of the surveyor and date of survey is shown. One of the nine areas chosen had mapping produced in 1951, with others dating from the 1970s. Such documents are accurate in terms of survey, although the attribute information (e.g., owner's name, land use, etc.) can be vastly out-of-date, and some environments have also changed significantly: for example, in one rural area a new irrigation scheme has radically transformed the layout, and quantity, of land parcels. Unfortunately, most of the maps have no visible coordinate reference system or grid associated with them, but the availability of accurate and contemporary aerial orthophotos has allowed for georeferencing of the maps to take place, and for comparisons to be made between coordinates collected by volunteers on the ground and the records held on official mapping.

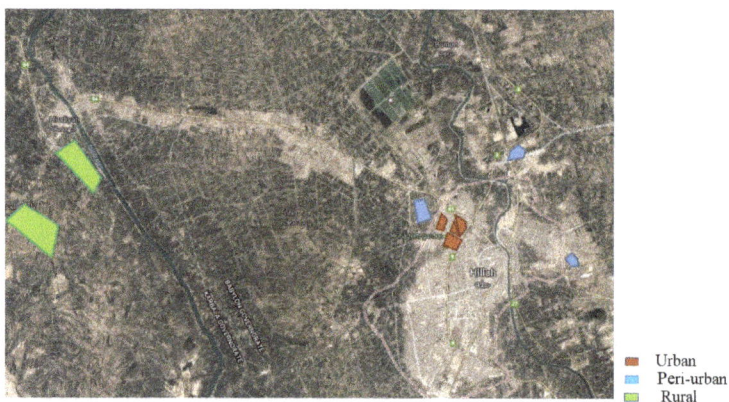

Figure 2. Study site locations, Al-Hillah, Iraq (Google Maps, 2016).

In urban areas, information about the coordinate system and an overlaid grid appear on some maps; but in peri-urban and rural areas all the maps needed to be spatially matched with georeferenced aerial orthophotos, using well-chosen reference points on the imagery.

4.1. Positional Accuracy Results from VGI

Significant outputs from this project include the testing of VGI datasets against the official records. Such accuracy testing is done by matching land parcel corners and boundaries from the VGI with the formal LAS documentation, which may be a map or a list of coordinates. Each of the three technologies was used to capture a dataset of land parcel corners in each of the nine areas (although the rural areas only used two technologies—smartphone GPS and analogue paper photo). Therefore, a significant number of points captured by each volunteer using all technologies was made available and accuracy comparisons could be calculated (examples of land parcel measurement for an urban and a rural area are shown in Figures 3 and 4). Positional accuracies for datasets created using each data collection method were calculated for each of the nine sites. This involved a RMSE analysis of coordinated corner points of land parcels, assessing the discrepancy between the positions captured using the three methods utilized and the coordinates of those points as shown in the official map records. A customized dashboard tool, developed in MATLAB for calculating and visualizing the RMSE and other measures [28], was used to quantify the discrepancies between VGI and official data (Figure 5).

Figure 3. Moharbeen urban area, Al-Hillah. VGI-defined land parcel boundaries using three methods of delineation.

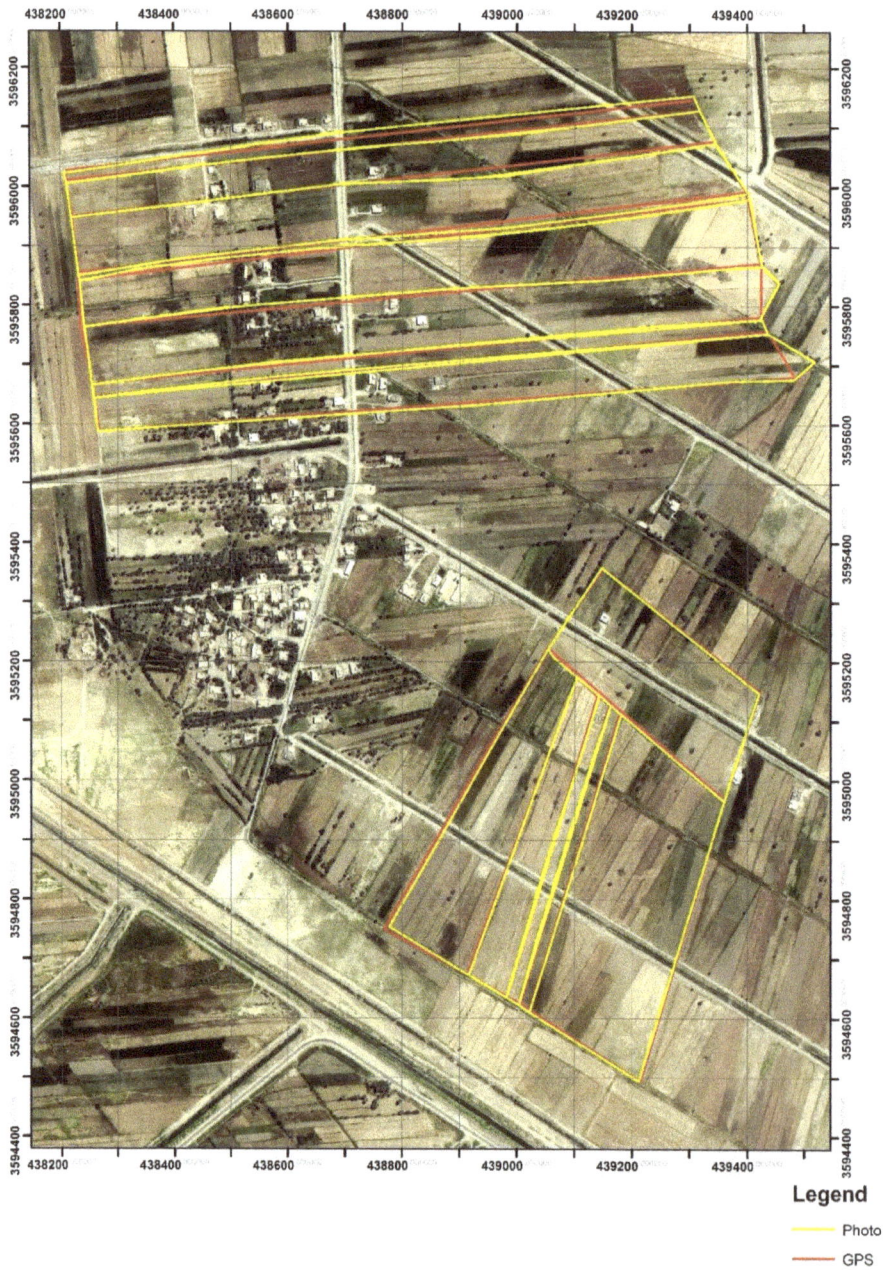

Figure 4. Aries rural area, near Al-Hillah. VGI-defined land parcel boundaries using two methods of delineation.

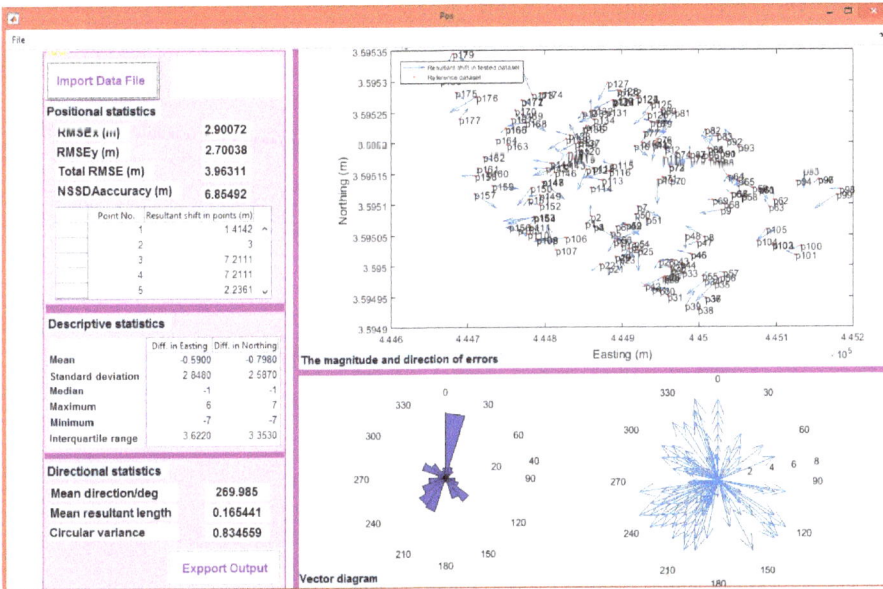

Figure 5. Calculation and visualization tool for RMSE (VGI points captured by GPS, compared to parcel coordinates from official mapping, Moharbeen urban area, Al-Hillah).

The calculated RMSE values were consistent for each data collection technology within a locality, but the summary statistics presented in Table 2 do indicate some variation. GPS accuracies suffered in urban areas due to limited satellite visibility and lack of stability for hand-held mobile devices. The iPad, with official base maps and QGIS functionality, appeared to be promising in terms of accuracy, although the rural communities could not be persuaded to use this technology: the majority preferred to annotate an analogue paper copy of the orthophoto, but misinterpretation (e.g., between shadows and land parcel edges) led to lower accuracies in some areas.

Table 2. Root mean square error (RMSE) for parcel corners for compared datasets aggregated by locality.

Study Area	No. of Points Tested	RMSE (Meters) cf. Official Data		
		Smartphone GPS	iPad Tablet	Analogue Paper Photo
Urban (4 sites)	778	4.364	1.357	2.615
Peri-urban (3 sites)	308	2.933	1.354	2.190
Rural (2 sites)	139	3.23	-	3.41

4.2. Completeness Results from VGI

Completeness was assessed by considering the total number of plots which are evident in the field, compared to those in the official records. A significant difference between the formal and volunteer totals has been found:

- Urban: 1235 plots on the official map; 2133 plots observed by the volunteers;
- Peri-Urban: 223 plots on the official map; 285 plots observed by the volunteers;
- Rural: 80 plots on the official map; 728 plots observed by the volunteers.

In the urban area there has been significant increase due to sub-division of plots. Occupying the same space, subdivided parcels are used for changes in land use, e.g., new shops; for developing new buildings within the parcel, e.g., additional houses; and for modified occupancy, e.g., new flats created from an original building. A typical example is of one land parcel, believed by the authorities to be a single plot, but identified by volunteers to have been sub-divided into three separate plots. Two of these were used for housing and the third consisted of several small shops, with a flat above. In another case, one owner of two individual large housing land plots had combined them to produce one plot, now containing five houses. Other locations highlighted a radical change of use, from purely residential to multiple use, including areas where all houses now have shops at the front. Peri-urban sites exhibit less change, because of lower land values (with less pressure to subdivide), and stable occupancy. In rural areas, some parts are very stable, but other locations have changed land use from agriculture to residential, and numerous housing plots have been created from one field. In one of the communities surveyed, a new irrigation scheme in an area where the official maps date from the 1950s, meant a completely different pattern and density of land parcels to that shown officially.

Table 3 shows the results of multiple crowdsourced VGI in testing the currency and validity of attribute data. Here, ownership details of sampled plots were to be determined: the official records of the name of a parcel's owner are incomplete and out-of-date, so each volunteer was asked for their opinion. In the vast majority of cases the 'crowd' of volunteers agreed on the name of the rightful owner: the percentage of disagreement in validating ownership data by consensus was low for each of the three different areas. It is concluded that ownership data obtained from groups of volunteers was correct, and could be used for validation and informing the official LAS organization.

Table 3. Verifying ownership data by crowdsource agreement.

Study Area	No. of Plots Tested	No. of Plots with Disagreement Recorded in Naming the Owner	Percentages of Disagreement
Urban (4 sites)	200	9	5%
Peri-urban (3 sites)	150	5	3%
Rural (2 sites)	80	2	2%

4.3. Experiences with Volunteers' Activity and Motivation

In addition to analysis of the data collected in this VGI project, further factors related to the citizens involved were assessed during the fieldwork, including their preferences, opinions, and motivations. This was derived from interviews with volunteers, observation of the work being done in the field, and analysis of the data captured. Feedback sessions were held with volunteers after their participation, recording their opinions and taking the opportunity to report back to them on their work.

It is clear that citizen volunteers have valuable information on their communities which, if added to the formal land administration system, can help to update it. For example, in one of the case study sites, volunteers reported that the official map is too old, because a river shown on the map has dried up and the area of land is occupied by some residents. Another example relates to permissions: the out-of-date official map shows a block of land parcels which appear to be available for development but, in reality, the community now considers the area of that block to be a heritage area.

After using the three different methods of data collection to identify their own or others' plots, volunteers were asked to specify which was the easiest and why. In effect, the usability of, and preference for, high-tech methods (smartphone/GPS for picking up coordinates; iPad for digital boundary demarcation), or low-tech methods (ordinary pen to delineate plot boundaries on paper-printed satellite/aerial image, topographic map, sketch map) was determined (Table 4).

The urban area volunteers preferred the iPad tablet configuration, which was also the most accurate for them; but in the peri-urban areas the less cumbersome and less obtrusive smartphone option was preferred. Those in rural areas had a distinct antipathy to the tablet, with a slight preference for the analogue method.

Table 4. Number of volunteer preferences for data collection methods in different communities.

Areas	GPS Enabled Smart Phone	iPad Tablet	Analogue Paper Photo
Urban	10	17	14
Peri-urban	17	7	13
Rural	13	0	14

At the end of the study, citizens, were asked about the reasons for their involvement. In Al-Hillah, as in many other areas of Iraq, and indeed other countries, the allocation, use, registration, occupancy, and transfer of land are extremely contentious issues and any formalised system of administration is open to all manner of professional incompetence, as well as fraud and profiteering. It was these issues which appeared to motivate the volunteers most strongly. Due to the localised ongoing unrest, long-term lack of resources, and basic inefficiency of the current system, several volunteers had direct experience, as land owners, of the land administration losing their documents inside their institute. Even when they are available, volunteers have been dissatisfied: a typical response was: *"I visited the land registration system to change the category of my parcel. However, I was shocked when I saw that the file was full of dust and in a very bad condition; a few of the papers were lost ... I was very depressed and angry at them for losing some documents in a place that we expected to be safe"*.

Volunteers were asked whether they were happy that they spent their time on the project and would they like to do more survey work (Table 5).

Table 5. Volunteer willingness to participate further.

Study Area	Yes	No	Not Sure
Urban (4 sites)	35	3	3
Peri-urban (3 sites)	31	2	4
Rural (2 sites)	26	0	1
Total	92	5	8

The majority answered that they would happy to be volunteers in the future for such projects. Reasons given for refusing, in contrast, were mainly related to the uncertain safety situation of the country and the threat of sectarian violence, along with a perception that data collection such as this was devious and an indirect means of government 'spying'. The recent years of conflict in civil society in Iraq, and its impact on societal structures, including the sense of community (which would promote volunteer activity such as this), does have an effect on projects such as the one described in this paper.

Further, some felt that their data collection work was challenging the official 'professionals', who had the sole responsibility and expertise to manage the LAS. One suggested that availability of VGI data *"may encourage some dishonest people to deceive our data for forgery purposes, which is currently common in Iraq"*.

This section has discussed the field exercises undertaken during this research project. The data collected, both positional and attribute, have been analysed in terms of its accuracy, and its utility in enhancing official records. Further, the effect of different types of environment (urban, peri-urban, rural) has been investigated, as has the impact of differing cohorts of volunteers. Varying technologies have been applied. The impact and utility of the VGI collected in stressed societal circumstances has been presented, along with an investigation of the social and community aspects of this VGI exercise. The final part of this paper concludes with comments on the significance of the study.

5. Discussion and Conclusions

This project relied overwhelmingly on the willingness of communities to engage with VGI collection and to consider the value of VGI when used within those communities. The identification of 'gatekeepers', representative and authoritative community leaders, was a major factor in successfully carrying out this project.

It was clear from initial interviews with the gatekeepers and with professional stakeholders, that the local municipality does not have the capacity to update or maintain the official land administration system. There had been a long-standing recognition of this situation, and an earlier attempt to sub-contract improvements to external consultants failed in 2006.

The potential for VGI within this system is, therefore, worthy of investigation. Analysis has shown the relative accuracy of different data collection methods in different contexts. It is argued here that, in some cases, it may be more important to collect some interim data, which the community can agree on and take ownership of, even if that means using a slightly less accurate method, than to focus simply on spatial accuracy. It can be concluded that in areas of conflict, or when official systems are under extreme stress, VGI may be the only realistic method of collecting usable data. In these cases, it may be more important to allow volunteers to choose a lower-accuracy method of data capture suited to their preferences and abilities.

Further conclusions relate to the advantages of incorporating VGI into the official land administration system.

1. Speed of data capture: the volunteer groups were able to gather more timely information within a few days than the hard-pressed official agency;
2. Lower costs: the use of basic technologies, including paper images and citizen-sourced annotation, has been shown to be sufficiently accurate for updating records in the official system. Even the more expensive methods, including GPS-enabled mobile phones and hand-held tablets, many already owned by volunteers, are cheaper than investing in agency-wide technologies relying on high-precision GPS or drone mapping programmes;
3. Updated registers of legally-acceptable standards: the speed and low cost of any VGI project of this type will result in significant amounts of valuable, contemporary information. This advantage is more contentious, as the definitive legal status of the VGI has not yet been formally confirmed (although the official agency has been encouraging), and it is also recognised that a more systematic approach to data collection will need to be developed, authorised, and monitored by the formal governmental body; and
4. Engagement of the community: encouraging the citizens and local stakeholders to 'take ownership' of the land registration process has significant societal benefits, and the community representatives (gatekeepers) were enthusiastic proponents of this research.

This paper has exemplified some of the issues involved in capturing VGI in circumstances where official law and order is limited and communities are not functioning in an ideal manner. Thus, in addition to the merits of using VGI, potential problems in enhancing or supplementing the land administration system with citizen-sourced VGI are recognized.

1. Embarking on a programme of data capture which relies on recruiting technically-aware and knowledgeable people, representative of a community, can be difficult: the gatekeepers were relied upon to find a willing cross-section of the local residents and business owners. Problems were encountered, for example, in encouraging female volunteers to use 'advanced' technologies. Technical skills in handling technology, training in filling out forms and recording the data required, and conflict resolution in small groups, all needed attention for successful VGI to be compiled;
2. Further, social problems were evident in contributors volunteering some of the information requested: questions such as 'who owns this land parcel?', 'is this land occupied illegally?', and 'how many people form the household in this property?' often proved uncomfortable for volunteers to ask;
3. There was also a perception, which was difficult to overcome, that this research was government-initiated, and the hostility of citizens to authority took much effort to overcome;

4. The final merging of captured VGI with the official data: in terms of required accuracy, this would not be problematic, but the legal standing of information captured by citizens, as opposed to official agencies, has not yet been tested.

This paper has concentrated on considering different methods of data collection that can suit different types of people, in varying geographical contexts. The research has given opportunity for a representative sample of citizens to volunteer and participate, with varying levels of education and experience. The promising levels of accuracy and completeness of the VGI data and their possible inclusion in a fit-for-purpose LAS, are of significant interest to the authorities of Al-Hillah. It has been shown that, despite challenging circumstances in engaging with citizens and acquiring good-quality data, there is potential for incorporating VGI into the land administration system of a poorly-documented, yet dynamic, area of a country which faces many problems.

Acknowledgments: The authors would like to thankfully acknowledge the contribution of the volunteers of Al-Hillah, the officials at the Land Administration Office in the municipality, Babylon University, and the Iraqi Cultural Attaché (London) for financial support.

Author Contributions: S.S.'s experience in urban environments in developing countries set the context for this study; D.F. directed the research programme and the writing of this paper; M.H. developed the methodology for the data collection exercise and conducted all the fieldwork in Iraq.

Conflicts of Interest: The authors declare no conflict of interest.

References

1. Goodchild, M. Citizens as sensors: The world of volunteered geography. *GeoJournal* **2007**, *69*, 211–221. [CrossRef]
2. Goodchild, M. Commentary: Whither VGI? *GeoJournal* **2008**, *72*, 239–244. [CrossRef]
3. US Agency for International Development (USAID), Iraqi Local Governance Program. *Land Registration and Property Rights in Iraq*; C. N: EDG-C-00-03-00010-00; RTI International: Baghdad, Iraq, 2005.
4. McLaren, R. Engaging the Land Sector Gatekeepers in Crowdsourced Land Administration. In Proceedings of the World Bank Land and Poverty Conference, Washington, DC, USA, 8–11 April 2013.
5. Enemark, S.; Bell, K.; Lemmen, C.; McLaren, R. Fit-for-Purpose Land Administration. In *FIG Publication No 60*; Enemark, S., Ed.; International Federation of Surveyors (FIG): Rome, Italy; World Bank: Copenhagen, Denmark, 2014; ISBN 978-87-9-285311-0.
6. Keenja, E.; De Vries, W.; Bennett, R.; Laarakker, P. Crowd sourcing for land administration: Perceptions within Netherlands Kadaster. In *FIG Working Week 2012: Knowing to Manage the Territory, Protect the Environment, Evaluate the Cultural Heritage, Rome, Italy, 6–10 May 2012*; International Federation of Surveyors (FIG): Rome, Italy, 2012.
7. Basiouka, S.; Potsiou, C.; Bakogiannis, E. OpenStreetMap for cadastral purposes: An application using VGI for official processes in urban areas. *Surv. Rev.* **2015**, *47*, 333–341. [CrossRef]
8. Grus, M.; Hogerwerf, J. VGI and Map Production in The Netherlands' Kadaster. In Proceedings of the VGI Workshop, Paris, France, 28 March 2014.
9. De Almeida, J.-P.; Haklay, M.; Ellul, C.; Carvalho, M. The role of Volunteered Geographic Information towards 3D Property Cadastral Systems. In Proceedings of the 4th International Workshop on 3D Cadastres, Dubai, UAE, 9–11 November 2014.
10. Niederer, S.; van Dijck, J. Wisdom of the Crowd or Technicity of Content? Wikipedia as a socio-technical system. *New Media Soc.* **2010**, *12*, 1368–1387. [CrossRef]
11. Bonney, R.; Cooper, C.; Dickinson, J.; Kelling, S.; Phillips, T.; Rosenberg, K.; Shirk, J. Citizen science: A developing tool for expanding science knowledge and scientific literacy. *BioScience* **2009**, *59*, 977–984. [CrossRef]
12. De Vries, W.; Bennett, R.; Zevenbergen, J. Neo-cadastres: Innovative solution for land users without state based land rights, or just reflections of institutional isomorphism? *Surv. Rev.* **2014**, *47*, 220–229. [CrossRef]
13. Seeger, C. The role of facilitated volunteered geographic information in the landscape planning and site design process. *GeoJournal* **2008**, *72*, 199–213. [CrossRef]

14. Coleman, D. Volunteered geographic information in spatial data infrastructure: An early look at opportunities and constraints. In Proceedings of the 12th GSDI Association World Conference, Singapore, 19–22 October 2010.

15. Brown, G.; Kelly, M.; Whitall, D. Which 'public'? Sampling effects in public participation GIS (PPGIS) and volunteered geographic information (VGI) systems for public lands management. *J. Environ. Plan. Manag.* **2014**, *57*, 190–214. [CrossRef]

16. Basiouka, S.; Potsiou, C. VGI in Cadastre: A Greek experiment to investigate the potential of crowd sourcing techniques in Cadastral Mapping. *Surv. Rev.* **2012**, *44*, 153–161. [CrossRef]

17. Cotfas, L.; Diosteanu, A. Evaluating Accessibility in Crowdsourcing GIS. *J. Appl. Collab. Syst.* **2010**, *2*, 45–49.

18. Tulloch, D. Many, many maps: Empowerment and online participatory mapping. *First Monday* **2007**, *12*, 2. [CrossRef]

19. Haklay, M.; Antoniou, V.; Basiouka, S.; Soden, R.; Mooney, P. *Crowdsourced Geographic Information Use in Government*; World Bank GFDRR: London, UK, 2014; 76p.

20. Berdou, E. *Mediating Voices and Communicating Realities. Using Information Crowd-Sourcing Tools, Open Data Initiatives and Digital Media to Support and Protect the Vulnerable and Marginalized*; Institute of Development Studies: Brighton, UK, 2011.

21. Bégin, D. Towards Integrating VGI and National Mapping Agency Operations: A Canadian Case Study. In Proceedings of the Workshop on the Role of Volunteered Geographic Information in Advancing Science: Quality and Credibility, Columbus, OH, USA, 18 September 2012.

22. Brandeis, M.; Nyerges, T. Assessing Resistance to Volunteered Geographic Information Reporting within Local Government. *Trans. GIS* **2016**, *20*, 203–220. [CrossRef]

23. Ciepłuch, B.; Jacob, R.; Mooney, P.; Winstanley, A. Comparison of the accuracy of OpenStreetMap for Ireland with Google Maps and Bing Maps. In Proceedings of the Ninth International Symposium on Spatial Accuracy Assessment in Natural Resuorces and Enviromental Sciences, International Spatial Accuracy Research Association, Leicester, UK, 20–23 July 2010.

24. Devillers, R.; Jeansoulin, R. Spatial data quality: Concepts. In *Fundamentals of Spatial Data Quality*; Rodolphe, D., Robert, J., Eds.; Wiley-ISTE: London, UK, 2010; pp. 31–42.

25. Shi, W.; Fisher, P.; Goodchild, M. *Spatial Data Quality*; Taylor & Francies: London, UK, 2003.

26. Veregin, H. Data quality parameters. In *Geographical Information Systems*, 2nd ed.; Longley, P., Goodchild, M., Maguire, D., Rhind, D., Eds.; John Wiley and Sons: New York, NY, USA, 1999; Chapter 12; Volume 1, pp. 177–189.

27. Hashemi, P.; Abbaspour, R. Assessment of Logical Consistency in OpenStreetMap Based on the Spatial Similarity Concept. In *OpenStreetMap in GIScience*; Lecture Notes in Geoinformation and Cartography; Arsanjani, J., Zipf, A., Mooney, P., Helbich, M., Eds.; Springer: Zug, Switzerland, 2015; pp. 19–36.

28. Fairbairn, D.; Al-Bakri, M. Using geometric properties to evaluate possible integration of authoritative and volunteered geographic information. *ISPRS Int. J. Geo-Inf.* **2013**, *2*, 349–370. [CrossRef]

29. Zielstra, D.; Zipf, A. A comparative study of proprietary geodata and volunteered geographic information for Germany. In Proceedings of the 13th AGILE International Conference on Geographic Information Science, Guimarães, Portugal, 11–14 May 2010.

30. Brassel, K.; Bucher, F.; Stephan, E.M.; Vckovski, A. Completeness. In *Elements of Spatial Data Quality*; Guptill, S., Morrison, J., Eds.; Elsevier: Oxford, UK, 1995; pp. 81–108.

31. Haklay, M. How good is volunteered geographical information? A comparative study of OpenStreetMap and Ordnance Survey datasets. *Environ. Plan. B* **2010**, *37*, 682–703. [CrossRef]

32. Jackson, S.; Mullen, W.; Agouris, P.; Crooks, A.; Croitoru, A.; Stefanidis, A. Assessing completeness and spatial error of features in volunteered geographic information. *ISPRS Int. J. Geo-Inf.* **2013**, *2*, 507–530. [CrossRef]

33. Adlington, G.; Tonchovska, R. Good Governance of Tenure. FAO and World Bank Support and Future Agendas. In Proceedings of the 4th Regional Conference for Cadastre and Spatial Data Infrastructure, Bled, Slovenia, 8–10 June 2010.

34. Lemmen, C. The Social Tenure Domain Model: A Pro-Poor Land Tool. In *FIG Publication No 52*; Uitermark, H., Lemmen, C., Eds.; International Federation of Surveyors (FIG): Copenhagen, Denmark, 2010; ISBN 978-87-9-090783-9.

35. Alemie, K.; Bennett, R.; Zevenbergen, J. Evolving urban cadastres in Ethiopia: The impacts on urban land governance. *Land Use Policy* **2015**, *42*, 695–705. [CrossRef]
36. Al-Bakri, M.; Fairbairn, D. Assessing similarity matching for possible integration of feature classifications of geospatial data from official and informal sources. *Int. J. Geogr. Inf. Sci.* **2012**, *26*, 1437–1456. [CrossRef]
37. Lanier, J. Digital Maoism: The hazards of the new online collectivism. *Edge* **2011**, *183*, 30.
38. Keen, A. *The Cult of the Amateur: How Blogs, MySpace, YouTube and the Rest of Today's User Generated Media Are Killing Our Culture*; Doubleday: New York, NY, USA, 2011.
39. Goodchild, M.; Li, L. Assuring the quality of volunteered geographic information. *Spat. Stat.* **2012**, *1*, 110–120. [CrossRef]
40. Navratil, G.; Frank, A. VGI for land administration—A quality perspective. *ISPRS-Int. Arch. Photogramm. Remote Sens. Spat. Inf. Sci.* **2013**, *1*, 159–163. [CrossRef]
41. Bishr, M.; Mantelas, L. A trust and reputation model for filtering and classifying knowledge about urban growth. *GeoJournal* **2008**, *72*, 229–237. [CrossRef]
42. Haklay, M.; Basiouka, S.; Antoniou, V.; Ather, A. How many volunteers does it take to map an area well? *Cartogr. J.* **2010**, *47*, 315–322. [CrossRef]
43. Maué, P. Reputation as tool to ensure validity of VGI. In Proceedings of the Workshop on Volunteered Geographic Information, Santa Barbara, CA, USA, 13–14 December 2007.
44. Forrester, J.; Cinderby, S. *Guide to Using Community Mapping and Participatory GIS*; NERC: Swindon, UK, 2014.

International Journal of
Geo-Information

MDPI

Article

A Citizen Science Approach for Collecting Toponyms

Aji Putra Perdana [1,2,*] and Frank O. Ostermann [1]

[1] Faculty of Geo-Information Science, University of Twente, 7500 AE Enschede, The Netherlands;
f.o.ostermann@utwente.nl

[2] Geospatial Information Agency (BIG), Jl. Raya Jakarta-Bogor Km. 46, Cibinong, Bogor 16911, Indonesia

* Correspondence: a.p.perdana@utwente.nl or aji.putra@big.go.id; Tel.: +31-610-27-6672

Received: 30 March 2018; Accepted: 13 June 2018; Published: 16 June 2018

Abstract: The emerging trends and technologies of surveying and mapping potentially enable local experts to contribute and share their local geographical knowledge of place names (toponyms). We can see the increasing numbers of toponyms in digital platforms, such as OpenStreetMap, Facebook Place Editor, Swarm Foursquare, and Google Local Guide. On the other hand, government agencies keep working to produce concise and complete gazetteers. Crowdsourced geographic information and citizen science approaches offer a new paradigm of toponym collection. This paper addresses issues in the advancing toponym practice. First, we systematically examined the current state of toponym collection and handling practice by multiple stakeholders, and we identified a recurring set of problems. Secondly, we developed a citizen science approach, based on a crowdsourcing level of participation, to collect toponyms. Thirdly, we examined the implementation in the context of an Indonesian case study. The results show that public participation in toponym collection is an approach with the potential to solve problems in toponym handling, such as limited human resources, accessibility, and completeness of toponym information. The lessons learnt include the knowledge that the success of this approach depends on the willingness of the government to advance their workflow, the degree of collaboration between stakeholders, and the presence of a communicative approach in introducing and sharing toponym guidelines with the community.

Keywords: citizen science; volunteered geographic information (VGI); toponym; crowdsourced data collection; data quality

1. Opportunities for New Approaches to Collect Place Names

Place names (known as toponyms) are an indispensable component of our communication about geographic features or regions, both natural and man-made [1,2]. They serve many purposes, including the obvious need for unambiguous identification for navigation, but also for current territorial claims and managing a society's past (e.g., to compare the renaming of streets or even entire cities following a regime change) [3–8]. Toponyms frequently have deeper meanings, often involving complicated semantics related to language and history [9–11], but many toponyms also describe the features they name. Some example toponyms from Indonesia are derived from folklore tales (Mount Tangkubanperahu, Banyuwangi), historical names (Jakarta from Jayakarta), or names of persons that have been adjusted to the local language (Malioboro from General Malborough, or Sampur from Zandvoort) [12–14]. Other (natural) features can cross multiple linguistic regions, for example, the river "Danube" has several names: "Donau" in Germany and Austria, "Dunaj" in Slovakia, "Duna" in Hungary, "Dunav" in Croatia and Serbia, "Dunav" and "Дунав" in Bulgaria, "Dunărea" in Romania and in Moldova, and "Dunaj" and Дунай" in the Ukraine [2]. Other toponyms originate from local geographical knowledge and history. Local citizens know places from their personal experiences and collectively agree and disagree in naming the places as part of their daily communication.

When surveying became a centralized and structured activity, the respective naming and mapping authorities (often part of the military forces) would collect, manage, and publish place names in the form of topographic maps, atlases, and gazetteers [15,16], sometimes taking control of local names. As part of this process, place names were standardized (at least within national boundaries) and, in case of ambiguities or multiple names, the authorities would officially approve names at the national level to be a part of a reference for worldwide communication.

In the last decade, the collection of toponyms has changed once again, potentially enabling the local population to have a more significant influence and contribution. The revolution of digital mapping and application allows citizens to contribute online through Web 2.0 technology and platforms, such as OpenStreetMap (OSM), Facebook Place Editor, Swarm Foursquare, and Google Local Guide. The absolute number of openly available toponyms increased due to the increase in crowdsourced and volunteered geographic information (VGI). Government agencies began to realize the potential use of citizens as scientists [15,17–20]. Researchers also explored crowdsourcing and gamification approaches in toponymic survey, place naming, and engaging the public in gazetteer creation [15,18,20–23].

Government agencies or toponymists (experts or researchers on the study of place names, or toponymy) are motivated to try such citizen science approaches for various reasons. One aim is to allow members of the general public to share indigenous or local geographical knowledge of place names. Another is to enable people to contribute to scientific investigation, ranging from data collection through analysis. More importantly, crowdsourced geographic information and citizen science approaches offer new opportunities for developing countries, particularly where existing gazetteers might be less complete, and where constraints on staff and resources are even more severe.

Nowadays, the national agency tasked with naming geographic features in Indonesia has been exploring potential approaches and technologies that can provide leverage for crowd involvement in toponym collection. The Geospatial Information Agency of Indonesia (Badan Informasi Geospasial (BIG)) conducted two pilot toponymic survey projects in 2015 and 2016. They then introduced a toponym data acquisition system in 2016 [24,25]. Usually, toponym collection is conducted in line with topographic mapping projects, and toponym standardization procedures are handled by naming authorities (national and regional committees for the standardization of toponyms) [24,26].

The pilot toponymic surveys were conducted in two distinct regions to examine the advantages of mobile, smartphone-based applications, when compared with GPS handhelds and maps, in recording toponyms. One survey in Yogyakarta (2015) collected toponyms of man-made features in urban areas. Another survey in Lombok (2016) gathered natural and man-made features in each district and region. The initial idea and motivation for the survey projects were to provide additional details or complete gazetteers. Group discussions with people in the field and members of toponymic survey projects revealed that local residents were eager to contribute to and learn about the use and impact of toponym collection.

This paper addresses issues in advancing toponym practice through three investigations. First, we systematically examined the current state of toponym collection and handling practice by multiple stakeholders, and we identified a recurring set of problems. Secondly, we developed a citizen science approach, based on participation, to collect toponyms. Thirdly, we examined their implementation in the context of an Indonesian case study. This research addresses identified problems in toponym collection, such as limited official staff in field surveys, the long procedure of the existing toponym practice, and issues of accessibility to all locations.

The following section addresses the first issue by examining the state of the art and deriving common problems. The subsequent two sections then describe a new framework that is capable of addressing the challenges, and show how the framework can be applied to a concrete, national case study (Indonesia). The last section discusses and summarizes our findings.

2. Current Challenges of Managing Toponyms—Citizens to the Rescue?

2.1. Systematic Evaluation of Challenges in Conventional Toponym Collection

UNGEGN (United Nations Group of Experts on Geographical Names) encourages nations to have national mapping agencies (NMA), or cadaster agencies, or to establish coordinating agencies for the standardization of toponyms in their countries [1,27]. So far, there has been no detailed investigation of the characteristics of UNGEGN countries regarding the coordination and regulation of the collection, or the maintenance and publication of place name databases. We explored the country reports and toponymic guidelines provided on the UNGEGN website to determine the current state of the art in toponym collection and maintenance. We selected documents from the 10th and 11th United Nations Conferences on The Standardization of Geographical Names (UNCSGN) in 2012 and 2017 [28,29]. We used UNCSGN 2012 as the baseline because of the discussion on VGI and crowdsourced geographic information proposed in this conference. From the perspective of data collection and maintenance, public authorities are responsible for collecting and standardizing place names, and publishing place name databases in a national gazetteer.

Our literature study revealed a range of problems encountered by current toponym collection practices. These ranged from high-level legislative framework and organizational issues, to concrete data-handling problems. Traditional toponym data handling typically featured lengthy and costly processing, with considerable delays between collection and publication, which further exacerbated the limits of human resources. Many national naming authorities have realized that crowdsourcing and citizen participation potentially can provide up-to-date and reliable geographic information based on local geographical knowledge. However, a naïve crowdsourcing approach would encounter challenges of credibility, legal issues (licensing, ownership, and copyright), and the sustainability of the system or project. In this paper, we suggest a taxonomy of problems in toponym collection identified from the literature, as can be seen in Table 1.

Table 1. Taxonomy of problems in toponym collection.

Category	Main Problems and Open Issues
Legal aspect	• Licensing, data ownership, and copyright • Data privacy and liability issues
Organizational issues	• The absence of a national naming authority • Coordination between public agencies • Collaboration with non-government sectors • Conflict resolution (potential for conflicts)
Funding	• No dedicated funding • Limited budgeting at local government
Procedures	• Inadequate regulatory procedures for the systematic approval and recording of place names • Insufficient training materials and guidelines on toponym collection • Long procedure, from collection until dissemination, of gazetteers
Personnel	• Limited human resources • Lack of trained staff • Language problems in interviews
Accessibility	• Insufficient transport infrastructure • Limited broadband and Internet services • Poor or bad weather conditions
Data Availability (Output)	• Incomplete place name database • Data uniformity issues (database structure and format file) • Duplicate places • Incorrect type of feature classes • Syntactic (data) integration (history of toponym records) • Semantic integration (meaning of places) • Spatial footprints (point-based location, bounding box (extent of features), and representation of vague places)

The data acquisition cycle can be identified as the main weakness of the processes in traditional toponym collection. For example, the toponym collection and verification cycle in Indonesia are generally conducted every 3 years to cover all 34 provinces for man-made features, except when there is an urgent case or a national priority. There are four main problems that cause this weakness: (1) extended procedures from data collection until dissemination, (2) limitations of human resources, (3) insufficient training materials, and (4) data uniformity issues and completeness. This assessment arises from a synthesis of the reports by governments on the situations in their countries, as presented in the 10th and 11th UNCSGN. If these problems can be tackled through collaborative approaches and using advanced technology, then government agencies can provide improved and complete gazetteers.

2.2. Bringing in the Power of Citizens

Collaboration among multiple stakeholders can be expected to help solve the above-mentioned problems. Several terms are being used interchangeably: crowdsourcing, VGI, or citizen science. A comprehensive review of these terms describes the role of citizens in crowdsourcing geographic information [30].

The term "crowdsourcing" is a combination of "crowd" and "outsourcing", coined by Howe [31]. Crowdsourcing is a process that involves outsourcing tasks to a distributed group of people. The GB1900 project is a successful example of a gamified crowdsourcing approach in toponym handling. Citizens participate online and share their knowledge of places (not only place names, but also place histories) through the transcription of toponyms and other features from maps on the GB1900 website [22]. The project and approach successfully tackled problems of limited human resources in field surveys.

VGI is defined as "the harnessing of tools to create, assemble, and disseminate geographic data provided voluntarily by individuals" [32]—in other words, geographic information produced by individuals and made available for the public. Public authorities and researchers also explored and tested mobile applications to collect vernacular place names, or urban names, which involved multiple stakeholders in several projects [20,21,33]. However, very little attention has been paid to the role and motivation of people's contribution as toponymists in digital place naming.

Public involvement and engagement in scientific projects is known as citizen science. Citizen science appeared in the mid-1990s, although the practice itself is older. Nowadays, many researchers have explored the definition, utilization, motivation, and typology of citizen science [34–38]. Citizen science projects have become increasingly attractive in natural and social science. People definitely can share their knowledge and receive feedback or obtain added value from it. Citizen science projects are based on volunteering and the contribution of information for the benefit of human knowledge and science [35]. The general public participates in scientific research activities and actively contributes to science. They provide experimental data and facilities for scientists. They raise new questions and help co-create a new scientific culture. They, themselves, become equipped with new learning and skills and receive a deeper understanding of scientific work in appealing ways [39].

The United States Geological Survey (USGS) conducted a successful story from crowdsourcing national topographic maps. The National Map Corps [19] brings future direction to improve and involve citizens collaboratively. National mapping and cadaster agencies in Europe have explored crowdsourcing and VGI approaches to update their topographical features [40]. For example, in Austria, people have contributed through a Web-GIS application and an additional survey conducted using paper-based maps with toponyms [18]. In the Netherlands, historical societies have been involved in the Dutch Kadaster project to improve toponym data as part of the new system of key registers for topography [15]. In Sweden, a crowdsourcing project among the Swedish NMA, Lantmäteriet, and the Swedish municipalities has developed a mobile application to collect toponyms (vernacular place names) and provide new toponym information in urban areas [20,33]. In Great Britain, the public have contributed and used the GB1900 Web application (provided by the National Library of Scotland, Edinburgh, UK) to help historians check and review place names and gather memories associated with places. This Web-based application was developed to collect toponyms and detailed information in old maps, such as base maps [22].

Many studies on VGI, crowdsourcing, citizen science, and geosocial media [18,19,41–43] have shown correlations between the power of where and public contributions. Several studies have shown that VGI and gamified crowdsourcing potentially are useful to collect and enrich (direct or indirectly) place name information. Investigations have studied the relationship between VGI, gamification, and geographic data collection [44]. Towns Conquer was one example of toponym collection using mobile apps. This mobile application was developed to collect vernacular names by updating or validating the existing place name database from the Spanish National Geographic Institute (IGN Spain) [21]. Collaboration between members of the public and toponymists, or a national naming authority, requires careful harmonizing, but this approach has the potential to complement the existing or traditional toponym practices. A legal framework on toponym collection for citizen participation could bring win-win solutions to toponym collection problems. Indonesia offers an example in law enforcement of geospatial information under their Indonesian Geospatial Information Act No. 4, 2011 and government regulations on the standardization of toponyms. The Indonesian government also has continued seeking and developing systems to involve communities or the general public in toponym collection [24].

3. An Approach to Integrate Citizen Science and Toponymy

3.1. Toponym Collection Framework

Following our assessment of the state of the art in toponym practice and crowdsourcing approach, we developed a framework to identify which problems could be addressed by citizen science approaches. In Figure 1, we depict the relationship among toponym challenges, opportunities, multiple stakeholders, and potential approaches. The center is the main goal—collecting toponyms—while the second layer consists of the existing approach and potentially collaborative approaches. The middle layer represents challenges and opportunities, while the outermost layer shows stakeholders. Generally speaking, the national naming authority has a legal mandate and is responsible for providing an accurate and complete gazetteer as authoritative data. The government should provide a legal framework that regulates data availability and organizational issues. Planning, implementing, and evaluating a collaborative approach can be a challenging project, especially for countries that have multi-dimensional problems, such as Indonesia, given its geographical, cultural, and language diversity.

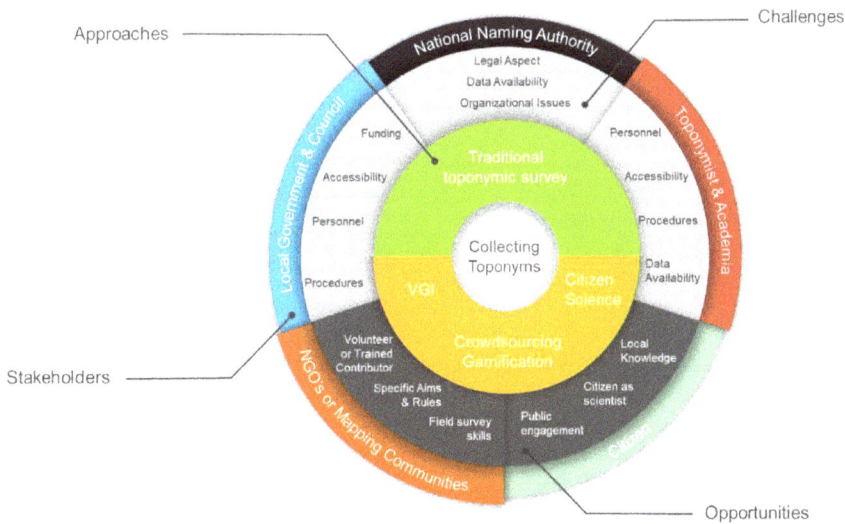

Figure 1. Challenges and opportunities to explore the potential use of a toponym collection approach with multiple stakeholders.

In the current toponym collection setup, a local government and council are responsible for their respective regions. At least four main challenges have to be handled by local authorities: funding, accessibility, personnel, and procedures. Often, there is no dedicated funding for toponym collection, which has to be linked to and integrated with other activities. Sometimes, the members of a regional council cannot approve the budget proposed by the local government, given other priorities in regional planning and development programs. In this case, it is the role of the national naming authority to establish a legal framework as a necessary foundation or reference in providing details, procedures, and budgeting for toponym collection. Inadequate regulatory procedures, especially at regional levels, are one of the challenges. Indonesia consists of 34 provinces and, currently, only one province (Special Region of Yogyakarta) provides a legal framework for place naming. The governor regulation on toponyms established by the Special Region of Yogyakarta provides details on the procedure of place naming to preserve local wisdom and history of place names. Eventually, this regulation may provide a solution to the problems of funding and limited personnel. Local governments can prepare detailed planning and implementation for fieldwork and training to improve data completeness and personnel capacity. On the other hand, local government can develop their general investment plans to tackle accessibility issues in their region, such as building infrastructure for fast Internet, roads, and bridges.

The quality of data and capacity-building activities might be maintained and improved by involving toponymists, researchers, or students to bridge the gap of information and knowledge of toponyms between local people and the government. They may be collaboratively involved in reviewing place name information from their scientific aspects, such as the writing, spelling, and meaning of place names, and the history of toponym records. However, the number of experts on geographical names or with academic discipline (for example: geography, history, and language backgrounds) interested in toponymy is limited. Sometimes, they have inadequate access to toponym data in rural areas or when trying to deal with problems of incomplete place names or integrate them with the meaning of names and history. Generally, toponymists and academia play an essential role in elaborating the problems associated with limited personnel, accessibility to data, lengthy procedures, and data availability.

VGI and crowdsourcing geographic information provided through digital platforms, such as OSM, Facebook Place Editor, Swarm Foursquare, and Google Local Guide, have indicated the potential resources from non-government organizations (NGOs), mapping communities, and citizens. Their presence also contributes to the documentation of toponyms. NGOs and mapping communities have their specific aims and rules, including procedures, and offer volunteers or trained contributors. OSM community members actively provide spatial databases (buildings, places, and point of interest) for disaster management. Many citizens have the geographical knowledge of places and willingness to share. There is an interesting opportunity for local governments to collaborate with OSM communities and citizens to produce complete toponym data. A citizen science approach for collecting toponyms will provide a more comprehensive place name database and elaborate the limitations on personnel. The emerging technologies offer some advantages, which enable people to contribute and reduce problems associated with the lack of staff. Mobile applications and Web-GIS for toponym data collection have been developed and explored by many researchers and governments [18,20,22,33].

The current state of toponym practice helps us to understand the potential position of advancing a toponymic survey project. We should consider how crowdsourced geographic information and citizen science approaches could tackle problems, such as: the long procedure, limited human resources, incomplete place name database, and integration of syntactic and semantic information. Some citizens are eager and able to enrich place name information. On the other hand, some NGOs and mapping communities in Indonesia are willing to follow the current standardization of toponyms, even though they have their aims and rules.

3.2. Existing Mobile and Web Applications for Toponym Collection

Fieldwork activity in toponym collection is a combination of collecting the geographic location of toponyms and providing textual information into a specific "name form" (questionnaire). Existing mobile and Web applications can help solve the problems of the lengthy procedure from data collection until dissemination, especially if the causes are a lack of trained contributors and limited availability of traditional toolkits (GPS handheld, voice recorder, and camera). The minimum requirements for mobile and Web applications for collecting toponyms consist of nine functionalities: (1) navigation, (2) marking GPS coordinates, (3) tracking, (4) displaying a map, (5) taking geotagged photos, (6) recording audio, (7) other geotagged notes or the ability for the generation of forms, (8) offline functionality, and (9) user-friendly and simple app.

GPS on mobile phones facilitates collecting toponyms because, previously, the availability of GPS handheld devices to be used in fieldwork was severely limited for local governments. Nowadays, there are many mature applications with different kinds of functionalities and navigating features. There are at least two promising GPS and navigation applications available to support toponym collection. First, there is GPS Essentials (http://www.gpsessentials.com/). This can enable local people to collect toponyms using its user-friendly, simple app and manual (also available in Bahasa Indonesia, developed by a local contributor and distributed through an online community). The second system is Maverick: GPS Navigation (https://wiki.openstreetmap.org/wiki/Maverick). This has offline functionality (use of offline maps and GPS) and a fully OSM-based offline navigation for Android.

Mobile phone applications for geographic data collection have emerged in many types and with many features. For instance, Humanitarian OSM Team (HOT) Indonesia developed Geo Data Collect (https://wiki.openstreetmap.org/wiki/Geo_Data_Collect) by integrating OSMTracker for Android (https://wiki.openstreetmap.org/wiki/OSMTracker_(Android)) and OpenDataKit (ODK) Collect (https://opendatakit.org/use/collect/). EpiCollect (http://www.epicollect.net/) is used by epidemiologists and ecologists, together with citizen scientists, for epidemiological data collection, collation, and visualization [45]. Meanwhile, the Towns Conquer game [21] was developed using Android SDK and ArcGIS SDK on the mobile client side, web services using PHP and SQL Server database on the server side. Another generic system architecture suitable for public participation using free open source software and mobile apps was studied [46,47]. In this system architecture, ODK Collect and the ODK Aggregate modules store data with a PostgreSQL database. EpiCollect and ODK provide functionality for creating forms for data collection. Survey123 for ArcGIS (https://survey123.arcgis.com/) and Fulcrum (https://www.fulcrumapp.com/) offer this functionality for fieldwork, with smart and simple questionnaires to collect data effectively.

There are three possible toponymic survey approaches using advanced technologies: (1) acquire toponym data using GPS Mobile apps, (2) build digital toponymic forms on mobile and Web applications for toponymic survey, and (3) develop new apps for toponyms data acquisition. The first could use a pilot study to focus on how mobile apps address the issues of limited human resources, time constraints, and data completeness. The second project would involve local governments and communities to participate in building the name form on apps. The main idea here would be to build interest and engage with them in the early stages of a toponymic survey project. The third project would require an evaluation of the urgency to develop new apps (based on evaluation of the two previous projects) and to evaluate the existing mobile and Web applications developed by the naming authority.

These three proposed projects might not solve some problems immediately, for example, the legal aspects, organizational issues, and funding. On the other hand, this kind of approach can increase general public participation and cover areas not exposed yet in national or regional programs. However, the use of a citizen science approach and the coordination among stakeholders are crucial to citizen motivation and contribution.

4. Indonesian Pilot Studies

4.1. Understanding Stakeholders in Toponym Collection

We argue that, in Indonesia, the organizational setting of toponym collection problems (see Table 1) is closely related to both the top-down and bottom-up approaches in decision-making and policy implementation. The national naming authority is focused on the learning process to manage these two approaches. They conduct annual meetings to get people at all levels actively involved by providing information, suggestions, and ideas to the policymaker. This organizational structure is shown in Figure 2. In a top-down approach, national naming authorities have the responsibility to initiate and set up the principles, policies, and procedures. Capacity building through training on toponyms is established by the national naming authority in coordination with local governments at the provincial level. The participants of toponymic training from the village, district, regency, or city level depend on the agenda of the training. In a bottom-up approach, local governments from villages and subdistricts up to district or city levels have the task of bringing all local actors to work together in order to promote and preserve local geographical knowledge of place names.

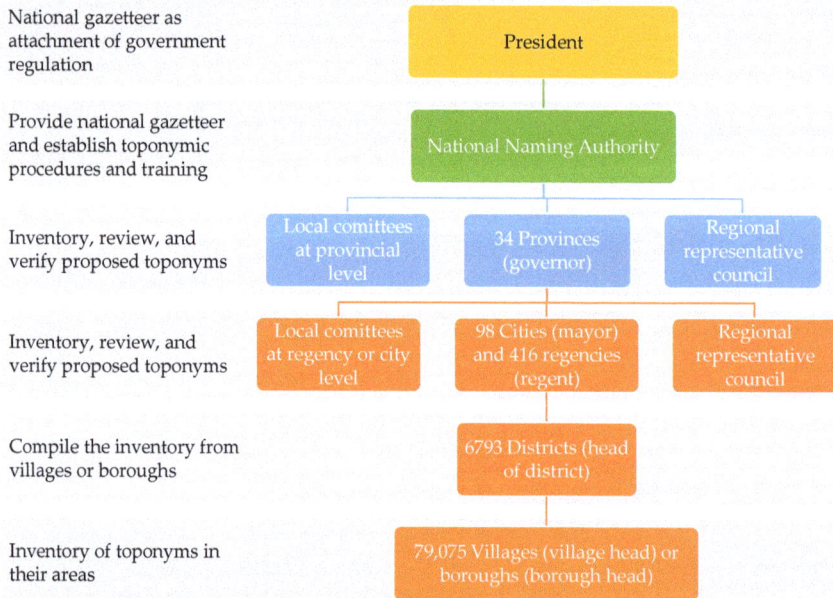

Figure 2. Organizational structure of public authorities for the standardization of toponyms in Indonesia.

Next, we examined toponym collection in Indonesia using stakeholder analysis. The main goal was to identify multiple stakeholders and learn their characteristics using data from interviews and observations during toponymic training and collection activities. We assigned scaled values and relative ranking in the measurement of interest and influence. Examples of questions and answers (Q and A) were:

- Q1: In few words, how would you describe the toponymy and toponym collection?

  A1: Toponymy is study of place names, while toponym collection is activity conducted by government or citizen to collect place names in their region and register the list of place names to naming authority.
- Q2: Do you know which institution is involved in local toponyms committee? Mention a few of the institutions if you know the information, regional and planning agency.

  A2: Yes, I know. Institutions in local toponym committee may consist of the governance bureau at regency or city level, head of district, and cadastral regional office.
- Q3: Are you ready to become a part of toponym committee or technical team to support the field survey?

  A3: Of course, I am ready because it is part of my task as official staff in the governance bureau.

  From interviews and observation results, we categorized the responses into ranks; an example being:

  3 = Has great interest and is ready to become involved and contribute in the workflow
  2 = Has the willingness to become involved, but does not know the procedure
  1 = Not interested

According to the existing organizational structure of toponyms practiced in Indonesia, the inventory of toponyms is conducted at the village and subdistrict level and coordinated by local committees on toponyms at district or city levels. After the inventory and review, the proposed toponyms are submitted to the higher level to be verified. In practice, this mechanism has not worked smoothly because of the lengthy bureaucratic procedure and limited budgeting at the local government level. The technical team or data collector and surveyors in topographic mapping activities provide toponym data to be used in the verification process. The stakeholder analysis matrix in Table 2 summarizes our investigation on the current constraints or findings, including their interest and potential influence in toponymic survey projects.

Table 2. Stakeholder analysis matrix.

Stakeholder	Motivations, Constraints, and Findings	Interest in Toponym Practice	Influence in Toponym Practice
Head of government (national to local level) [1]	Not interested in details, just results	Medium	Medium
National naming authority	Internal coordination (between public agencies)	High	High
Regional representative council	Lack of information on toponym practice	Low	Medium
Local committees	Budgeting and human resources	Medium	High
Surveyors [2]	Lack of skills and knowledge	Low	High
Traditional leaders	Frequent language barrier	Medium	High
Local residents	Expect to promote their neighborhood	High	High
Academia	Not entirely interested, it depends on the expertise	Low	Medium
Non-government organizations or mapping communities	Specific rules and platforms	Medium	Low

[1] President, governor, mayor/regent, head of district, village or borough head. [2] The technical team (data collector) at the local committee or surveyor in topographic mapping activities.

The Indonesian national naming authority remains committed to tackling the problems on toponym collection through seminars and toponymic training for local committees and relevant stakeholders. Nowadays, they also use media gatherings to promote issues and achievements of toponym collections in Indonesia to journalists. The next step is to optimize coordination among multiple stakeholders and crowd (citizen) participation.

4.2. Toponymic Survey Projects and Development of Toponymic Data Acquisition System

The pilot studies were conducted in two different regions and involved different participants. The first pilot study in Yogyakarta Special Province involved undergraduate students from Universitas Gadjah Mada (UGM)—Indonesia and provincial government. The second pilot study in Lombok Province involved provincial government and communities.

Table 3 presents the basic elements and steps of toponymic surveying. Planning was the first element, with the purpose to define the schedule, coverage of the study area, estimation of workload (volume, time, personnel), proposed methods, and work distribution. Then, a preliminary survey was conducted to establish communication and coordination with the local government, acquire permission and support letters down to the village level, and decide on the location for a base camp during the fieldwork. Data preparation consists of preparation of manuscript/printed maps and secondary data (such as points of interest and administrative boundaries from the local government). The participants in the Yogyakarta survey were 16 staff members from BIG and UGM. They were divided into eight teams of two surveyors each. Fieldwork was conducted in Kecamatan Gondomanan, Kota Yogyakarta. This location was selected because it has famous and historical buildings, such as the Fort Vredeburg Museum (official Indonesian name, Museum Benteng Vredeburg Yogyakarta), the Presidential Palace (Istana Yogyakarta or Gedung Agung), and Malioboro Street. The toponymic survey was conducted from 21 October 2015 to 26 October 2015, and was followed by data entry, editing, and compilation in the office. Every day, each team discussed and shared some suggestions to improve the quality of fieldwork. Based on their daily evaluation, the most challenging part was communication and data handling.

Table 3. The elements and steps of toponymic survey projects.

Elements	Steps
Preparation	Planning Preliminary Survey Data preparation
Fieldwork	Recording toponyms Interviews with local people
Office Treatment	Data entry and editing Data compilation
Verification	Review of place names Approval of place names
Data Publication	Create gazetteer Publish (printed and digital) gazetteer

In the region, most local residents spoke Javanese, even though several respondents could speak in Indonesian. In this case, UGM undergraduate students acted as translators during interviews. Each group was equipped with a GPS handheld (or mobile device with GPS navigation apps), camera, map, and name form for recording toponyms. It was optional for each group to use mobile devices, because GPS navigation apps were explored for the first time in this project. The geographical name form is shown in Figure 3. The national naming authority provided this (in a paper-based format) for recording detailed information, i.e., the place name used by the local government, alternative names, and more, including the meaning and the history of the name (if any). All data were recorded and compiled in GIS shapefile format.

The participants in the toponymic survey project in Yogyakarta were only able to collect 63 place names with information on their history, meaning, and alternative names from a total of 743 features (Figure 4a). It was difficult to interview or select a person who fully understood the meaning and history of each place. Support and coordination from the local government of the Special Region of Yogyakarta could probably help increase data completeness. Unfortunately, the local government was unable to support the survey adequately due to time constraints. However, in the preliminary survey, communication and coordination with the local government were done as part of the procedure.

Based on the preliminary survey, we improved the involvement of the local government and the community, as well as the equipment (tools and data management). The second survey was conducted in Kecamatan Pujut, Kabupaten Lombok Tengah, West Nusa Tenggara Province. We prepared 33 sheets of manuscript maps (with high-resolution satellite images at the scale of 1:5000). The surveyor team in this project consisted of eight persons from BIG and eight persons from the local government.

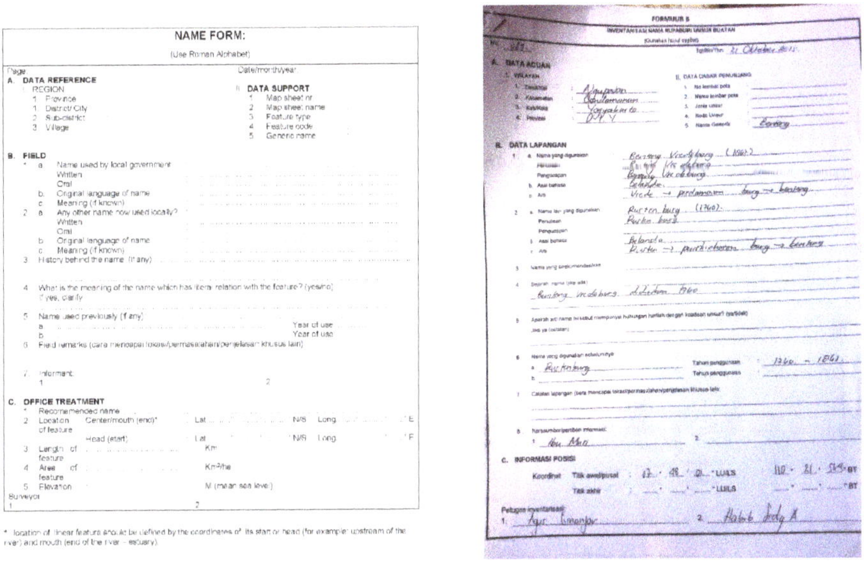

NAME FORM:

(Use Roman Alphabet)

Page Date/month/year

A. DATA REFERENCE

I REGION
1 Province
1 District/ City
2 Sub-district
3 Village

II DATA SUPPORT
1 Map sheet nr
2 Map sheet name
3 Feature type
4 Feature code
5 Generic name

B. FIELD

* a Name used by local government
Written
Oral
b. Original language of name
c Meaning (if known)
2 a Any other name now used locally?
Written
Oral
b Original language of name
c Meaning (if known)
3 History behind the name (if any)

4 What is the meaning of the name which has literal relation with the feature? (yes/no)
If yes, clarify

5 Name used previously (if any)
a Year of use
b Year of use
6 Field remarks (cara mencapai lokasi/permasalahan/penjelasan khusus lain)

7 Informant
1 2

C. OFFICE TREATMENT

* Recommended name
2 Location of feature Center/mouth (end)* Lat N/S Long E
Head (start) Lat N/S Long E
3 Length of feature Km
4 Area of feature Km²/ha
5 Elevation M (mean sea level)
Surveyor
1 2

* location of linear feature should be defined by the coordinates of its start or head (for example: upstream of the river) and mouth (end of the river – estuary).

(a) (b)

Figure 3. Names form used by the national naming authority (NNA) in Indonesia: (**a**) an example of the "Name Form" for collecting toponyms in the field; (**b**) complete name form from fieldwork in Yogyakarta. (Courtesy of Badan Informasi Geospasial).

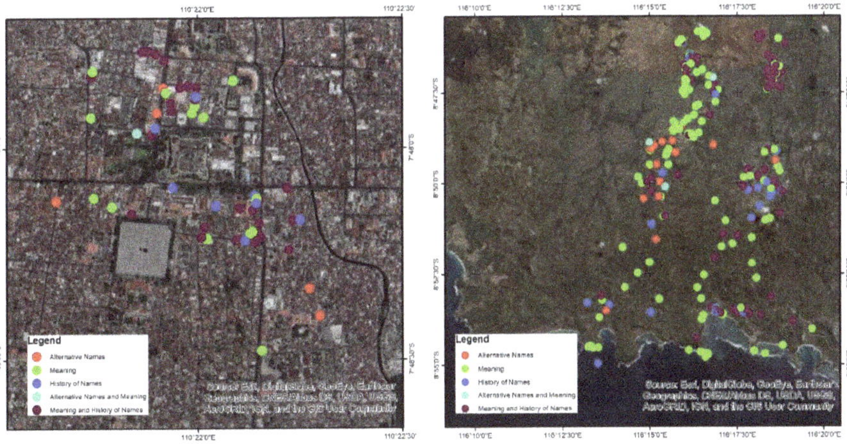

(a) (b)

Figure 4. Toponyms with alternative names, meaning, and history of names: (**a**) urban names in the case study of Yogyakarta provided 63 toponyms; (**b**) natural and man-made features in the case study of Lombok provided 367 toponyms.

Three main steps were conducted and improved in this survey: (1) collection and data entry, (2) verification with the local authority, and (3) data publication. A field survey was conducted for 11 days, from 24 September 2016 to 4 October 2016, by eight teams. Each team covered areas from three to six maps, depending on the characteristic of the region. Each team conducted data entry in the period from 30 September 2016 to 5 October 2016. In contrast to the previous project, the surveyors managed their data in Geodatabase file format and attached photos to this database. From a total of 1484 points collected in the Lombok project, only 367 place names had complete information on alternative names, meaning, and history behind the names (Figure 4b).

Data from the two pilot projects showed an increase in the number of data completeness due to the involvement of the local government and community in the second pilot project. In the first pilot project, we had ~8% of information about history/meaning/alternative names, versus ~25% in the second pilot project.

The verification process involved local people from the village (at least two local authorities or informants, usually the head of the village and traditional leader), subdistrict, and district level (Figure 5). To speed up the verification process, and based on the accessibility of villages, the team was divided into six groups. Each village representative checked place names in the compiled name forms and their geographic locations. The traditional leader and head of the village had the local geographical knowledge. They knew about the geography, history, and meaning, or possibly mythology, of places (if any).

(a)   (b)

Figure 5. Verification process in the toponymic survey in Lombok: (**a**) compilation of place names with approval from local authority; (**b**) respondents (local people) share their local geographical knowledge and put place names on the map. (Courtesy of Badan Informasi Geospasial).

These two pilot projects used existing GPS tools and navigation apps in the Android market, including GPS Essentials and Maverick GPS Navigation. The functionality of these GPS navigation apps was helpful if surveyors were navigating without a data connection and new to and unfamiliar with the study area. It also made the survey activity more effective and efficient. Data collected can be saved as notes in the apps with geographic coordinate location and geotagged photos. The result from GPS navigation apps can be exported to GIS format file and processed to the next step in gazetteer creation. Nowadays, local governments do not depend on the minimum availability of GPS handheld, voice recorder, and camera. They can use mobile phones supported with navigation apps to collect geographic location, record pronunciation, and take geotagged photos.

Competition between the teams in these two pilot projects was encouraged to maintain their motivation and improve the quality of data collected from fieldwork. Achievements calculated were based on working capacity (extent of the area of survey covered per day), data completeness, and difficulties in finding respondents and data management. Generally, the limiting factors were accessibility to location, weather, and density of geographic features in fieldwork areas. On the other hand, mobile applications could improve the surveyor's performance and increase public participation in the field surveys.

The national naming authority in Indonesia developed a toponym data acquisition application and introduced it in 2016 at an initial stage. According to the Indonesian report at the 11th UNCSGN (Eleventh United Nations Conference on the Standardization of Geographical Names) meeting in New York in 2017, the goal of developing the mobile app, called SAKTI (Sistem Akusisi Data Toponim Indonesia/Indonesian Toponymic Data Acquisition System), was to collect toponyms and send the data collected directly to the Badan Informasi Geospasial server [48]. Recently, in April 2018, BIG promoted and launched the new version of SAKTI mobile and Web-GIS applications (http://sakti.big.go.id/sakti/webgis/) to local governments in toponymic training for capacity building. The development of SAKTI sets out to standardize the procedure and database derived from field survey by local government (Figure 6). The benefit of the SAKTI mobile and Web-GIS applications are: (1) user-friendly and simple app, (2) displaying map (online base map provided by BIG), (3) effective (paperless and minimized error in writing coordinates), (4) safe (reduced risk of lost or damaged data in fieldwork), (5) standardized database (data structure based on standard toponymic database from BIG), and (6) time (expected to be faster than using paper-based survey). SAKTI mobile and Web-GIS applications do not provide sufficient offline functionality, but these applications have fulfilled the rest of minimum requirements for collecting toponyms. In the current version, the users should have an Internet connection to log in at the first attempt before beginning to collect toponyms.

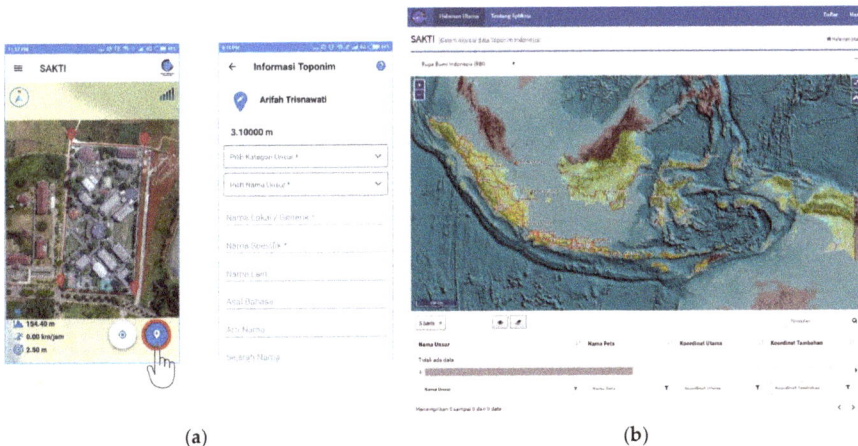

(a)   (b)

Figure 6. Selected screenshots of SAKTI (Sistem Akusisi Data Toponim Indonesia/Indonesian Toponymic Data Acquisition System): (**a**) "Name Form" for collecting toponyms in the field in SAKTI mobile application; (**b**) SAKTI Web-GIS (http://sakti.big.go.id/sakti/webgis/).

However, there is a limitation in this toponym workflow using mobile and Web-GIS applications developed by BIG regarding crowdsourced data. It is still limited to official staff (surveyors) who have to upload a letter of assignment from an authorized official or local committee. Users with guest accounts can use the apps, but they cannot submit their data to the server. To date, we cannot evaluate data collected from SAKTI because the introduction to local governments and the toponymic training have not been completed yet. It is expected that, by the end of 2018, SAKTI mobile and Web-GIS applications can provide toponym information and be used effectively by local governments to improve the toponym collection cycle in Indonesia.

5. Discussion and Recommendation

The first project in Yogyakarta faced problems in data management and post-fieldwork office treatment to produce gazetteers. Time management and realistic calculation of the capacity of human resources need careful consideration in the preparation step. Learning from Yogyakarta, the project in Lombok was equipped with guidelines and work plans for each team, including communication with stakeholders. Introduction and training on toponymic surveys were conducted before the actual fieldwork. In both projects, we could collect place name information on alternative names, meaning, and history of names for ~8% to ~25% of cases accompanied by agreed upon names between local residents and government.

Interviews and discussions with people involved in this project provided additional insights on problems. New ideas or strategies also suggested the need for more focused toponym collection and collaboration among stakeholders. From a technical perspective, mobile data collection and free open source software, such as QGIS, would be helpful to work with place-based geographic information and data management. The two main reasons why local government used free open source software were: (1) their cost-effective or cost-saving nature, as sometimes they have no dedicated funds to purchase commercially licensed software, and (2) the flexibility to use free open source software alongside any operating system and computer hardware.

We need to consider existing constraints, such as working time and staff members. Then, explore the willingness for contribution from citizens. The upcoming project will consider an evaluation of the previous toponym projects and establish more contributions from local people (the power of the crowd) as toponymists. The lessons learned from the Indonesian case studies for the development and fieldwork implementation include: (1) we have to start citizen science projects in other areas, and examine and prepare comparison analysis for improving the outcomes; (2) gamified citizen science can be a good means to maintain participant motivation and engage with different difficulty levels during data collection, analysis, and publication of toponyms, as the examples in Section 2.2 showed; and (3) we need to develop ways to ensure toponym collection can be more fun and, thereby, motivate the contributors, for example, by applying game theory elements.

Toponym collection and handling conducted by a national naming authority that has to involve local people as scientists is challenging. The new paradigm of a collaborative approach requires governments to adjust their usual business workflows. There are various types of problems and levels in toponym data handling and management. Several of the more developed countries already have focused on enriching their gazetteers, while developing countries are still dealing with trying to ensure that they have a sufficient base coverage of their entire territory. Conflicts of place names happen for various, and often particular, reasons, and in specific areas, for example, those associated with political or social issues or territorial ownership.

Recording alternative names, meaning, and histories of toponyms provides additional insights into place name information, as can be seen from our Indonesian case studies. A citizen science project on toponyms is open to a wide range of contributors and multiple stakeholders, and toponymists or national naming authorities are eager to establish a well-developed workflow and guidelines. Communication and technical skills to gather meaning and historical information of places and to manage spatial information needs to be improved. Local people would like to participate by marking and recording their places in Web or mobile application. In this sense, gamified toponym collection is a potential method of the toponymic survey to solve several problems at once, such as lack of human resources, tools, and data management.

In future work, we will investigate two types of toponymic survey projects. First, the collaborative project conducted by working together with government, scientists, and citizens. In this project, we will design data collection methods and develop digital forms of place name questionnaires. Second, we will evolve the co-created project as an independent toponymic survey conducted by mapping communities or local people. Some of the members of the public have actively handled a toponymic survey from the beginning until the end of the project (data publication). Overall, key aspects to a successful toponymic survey project are the willingness of the government to adapt their workflow, for collaboration between stakeholders to improve, and for a communicative approach to evolve in introducing and sharing toponym guidelines with the communities.

Author Contributions: A.P.P. and F.O.O. discussed the idea; A.P.P. undertook the fieldwork and analyses; A.P.P. and F.O.O. both contributed to the writing of this manuscript.

Acknowledgments: We would like to thank the Ministry of Finance for the Republic of Indonesia's Indonesian Endowment Fund for Education (LPDP) for supporting his PhD research. Badan Informasi Geospasial (BIG) supported this work as the National Naming Authority of Indonesia.

Conflicts of Interest: The authors declare no conflict of interest. The founding sponsors had no role in the design of the study; in the collection, analyses, or interpretation of data; in the writing of the manuscript, and in the decision to publish the results.

References

1. Kadmon, N. *Toponymy: The Lore, Laws, and Language of Geographical Names*; Vantage Press: New York, NY, USA, 2000; ISBN 0533135311.
2. Laurini, R. Gazetteers and Multilingualism. In *Geographic Knowledge Infrastructure*; Elsevier: New York, NY, USA, 2017; pp. 157–182. ISBN 9781785482434.
3. Madden, D.J. Pushed off the map: Toponymy and the politics of place in New York City. *Urban Stud.* **2017**. [CrossRef]
4. Alderman, D.H.; Inwood, J. Street naming and the politics of belonging: Spatial injustices in the toponymic commemoration of Martin Luther King Jr. *Soc. Cult. Geogr.* **2013**, *14*, 211–233. [CrossRef]
5. Plini, P.; Di Franco, S.; Salvatori, R. One name one place? Dealing with toponyms in WWI. *GeoJournal* **2016**, *83*, 1–13. [CrossRef]
6. Light, D. Tourism and toponymy: Commodifying and consuming place names. *Tour. Geogr.* **2014**, *16*, 141–156. [CrossRef]
7. Grossner, K.E.; Janowicz, K.; Keßler, C. Place, Period, and Setting for Linked Data Gazetteers. *Placing Names Enrich. Integr. Gaz.* **2014**, 1–17.
8. Choi, S.H.; Wong, C.U.I. Toponymy, place name conversion and wayfinding: South Korean independent tourists in Macau. *Tour. Manag. Perspect.* **2018**, *25*, 13–22. [CrossRef]
9. Ardanuy, M.C.; Sporleder, C. Toponym disambiguation in historical documents using semantic and geographic features. In Proceedings of the 2nd International Conference on Digital Access to Textual Cultural Heritage, DATeCH2017, Göttingen, Germany, 1–2 June 2017; pp. 175–180.
10. Ryan, C.; Grant, R.; Carragáin, E.Ó.; Collins, S.; Decker, S.; Lopes, N. Linked data authority records for Irish place names. *Int. J. Digit. Libr.* **2015**, *15*, 73–85. [CrossRef]
11. Capra, G.F.; Ganga, A.; Buondonno, A.; Grilli, E.; Gaviano, C.; Vacca, S. Ethnopedology in the study of toponyms connected to the indigenous knowledge on soil resource. *PLoS ONE* **2015**, *10*, 1–20. [CrossRef] [PubMed]
12. Cacciafoco, F.P. Mythical Place Names: Naming Process and Oral Tradition. Volume 23, pp. 25–35. Available online: https://geografie.uvt.ro/wp-content/uploads/2017/12/2_CACIAFOCCO_PERONO.pdf (accessed on 30 April 2018).
13. Carey, P.B.R.; Noorduyn, J.; Ricklefs, M.C. Asal Usul Nama Yogyakarta & Malioboro. Available online: https://www.goodreads.com/book/show/24632963-asal-usul-nama-yogyakarta-malioboro (accessed on 30 April 2018).
14. Ruchiat, R. Asal-Usul Nama Tempat di JAKARTA. Available online: https://www.goodreads.com/book/show/11143123-asal-usul-nama-tempat-di-jakarta (accessed on 30 April 2018).

15. Hogerwerf, J. Toponymic Data and Map Production in the Netherlands: From Field Work to Crowd Sourcing. Available online: https://unstats.un.org/unsd/geoinfo/UNGEGN/docs/11th-uncsgn-docs/E_Conf.105_87_CRP.87_9_Definitief_Toponymic%20data%20and%20map%20production%20in%20the%20Netherlands%20-%20from%20field%20work%20to%20crowdsourcing.pdf2017 (accessed on 30 April 2018).

16. Watt, B.; Konstanki, L.; Atkinson, R.; Box, P. The Four Faces of Toponymic Gazetteers. In Proceedings of the Tenth United Nations Conference on theStandardization of Geographical Names, New York, NY, USA, 31 July–9 August 2012; pp. 1–9.

17. Ahmouda, A.; Hochmair, H.H. Using Volunteered Geographic Information to measure name changes of artificial geographical features as a result of political changes: A Libya case study. *GeoJournal* **2017**, 1–19. [CrossRef]

18. Rampl, G. Crowdsourcing and GIS-based Methods in a Field Name survey in Tyrol; (Austria). 2014, pp. 1–7. Available online: https://unstats.un.org/unsd/geoinfo/UNGEGN/docs/28th-gegn-docs/WP/WP16_Field%20name%20survey%20in%20Tyrol.pdf (accessed on 30 April 2018).

19. McCartney, E.A.; Craun, K.J.; Korris, E.; Brostuen, D.A.; Moore, L.R. Crowdsourcing the National Map. *Cartogr. Geogr. Inf. Sci.* **2015**, *42*, S54–S57. [CrossRef]

20. Olteanu-Raimond, A.M.; Hart, G.; Foody, G.M.; Touya, G.; Kellenberger, T.; Demetriou, D. The Scale of VGI in Map Production: A Perspective on European National Mapping Agencies. *Trans. GIS* **2017**, *21*, 74–90. [CrossRef]

21. Castellote, J.; Huerta Guijarro, J.; Pescador, J.; Brown, M. Towns Conquer: A Gamified application to collect geographical names (vernacular names/toponyms). In Proceedings of the AGILE International Conference, Leuven, Belgium, 14–17 May 2013.

22. Southall, H.; Aucott, P.; Fleet, C.; Pert, T.; Stoner, M. GB1900: Engaging the Public in Very Large Scale Gazetteer Construction from the Ordnance Survey "County Series" 1:10,560 Mapping of Great Britain. *J. Map Geogr. Libr.* **2017**, *13*, 7–28. [CrossRef]

23. Cardoso, R.V.; Meijers, E.J. The metropolitan name game: The pathways to place naming shaping metropolitan regions. *Environ. Plan. A* **2017**, *49*, 703–721. [CrossRef]

24. Geospatial Information Agency of Indonesia Report of the Government of the Republic of Indonesia. In Proceedings of the Eleventh United Nations Conference on the Standardization of Geographical Names, New York, NY, USA, 8–17 August 2017.

25. Geospatial Information Agency of Indonesia Development of Toponym Data Acquisition System in Indonesia. In Proceedings of the Eleventh United Nations Conference on the Standardization of Geographical Names, New York, NY, USA, 8–17 August 2017.

26. Lauder, A.F. Ubiquitous place names Standardization and study in Indonesia. *Wacana* **2015**, *16*, 383–410. [CrossRef]

27. Kerfoot, H.; Närhi, E.M. *Manual for the National Standardization of Geographical Names*; Nova Iorque, United Nations Publication: New York, NY, USA, 2006; ISBN 92-1-161490-2.

28. UNGEGN 10th United Nations Conferences on The Standardization of Geographical Names (UNCSGN). Available online: https://unstats.un.org/UNSD/geoinfo/UNGEGN/ungegnConf10.html (accessed on 17 May 2018).

29. UNGEGN 11th United Nations Conferences on The Standardization of Geographical Names (UNCSGN). Available online: https://unstats.un.org/UNSD/geoinfo/UNGEGN/ungegnConf11.html (accessed on 17 May 2018).

30. See, L.; Mooney, P.; Foody, G.; Bastin, L.; Comber, A.; Estima, J.; Fritz, S.; Kerle, N.; Jiang, B.; Laakso, M.; et al. Crowdsourcing, Citizen Science or Volunteered Geographic Information? The Current State of Crowdsourced Geographic Information. *ISPRS Int. J. Geo-Inf.* **2016**, *5*, 55. [CrossRef]

31. Howe, J. The Rise of Crowdsourcing. *Wired Mag.* **2006**, *14*, 1–5. [CrossRef]

32. Goodchild, M.F. Citizens as Sensors: WEB 2.0 and the Volunteering of Geographic Information. *GeoFocus* **2007**, *7*, 8–10. [CrossRef]

33. United Nations Group of Experts on Geographical Names. *The Benefits of Geographical Names Standardization*; UNGEGN: New York, NY, USA, 2015; pp. 1–39.

34. Riesch, H.; Potter, C. Citizen science as seen by scientists: Methodological, epistemological and ethical dimensions. *Public Underst. Sci.* **2014**, *23*, 107–120. [CrossRef] [PubMed]

35. Haklay, M. Citizen science and volunteered geographic information: Overview and typology of participation. In *Crowdsourcing Geographic Knowledge: Volunteered Geographic Information (VGI) in Theory and Practice*; Springer: Dordrecht, The Nederland, 2013; Volume 9789400745, pp. 105–122. ISBN 9789400745872.
36. Vann-Sander, S.; Clifton, J.; Harvey, E. Can citizen science work? Perceptions of the role and utility of citizen science in a marine policy and management context. *Mar. Policy* **2016**, *72*, 82–93. [CrossRef]
37. Tinati, R.; Luczak-Roesch, M.; Simperl, E.; Hall, W. An investigation of player motivations in Eyewire, a gamified citizen science project. *Comput. Hum. Behav.* **2017**, *73*, 527–540. [CrossRef]
38. Schäfer, T.; Kieslinger, B. Supporting emerging forms of citizen science: A plea for diversity, creativity and social innovation. *J. Sci. Commun.* **2016**, *15*, 1–12.
39. Government of Canada. White Paper on Citizen Science for Europe. Available online: https://www.google.com/url?sa=t&rct=j&q=&esrc=s&source=web&cd=4&cad=rja&uact=8&ved=0ahUKEwjGsYvHxNLbAhWCF8AKHVTEAusQFgg5MAM&url=http%3A%2F%2Fsocientize.eu%2Fsites%2Fdefault%2Ffiles%2F04-socientize_final_conference.pdf&usg=AOvVaw3-pq0LIgN6mmOHYp8SRAXC (accessed on 30 April 2018).
40. Bol, D.; Grus, M.; Laakso, M.; Bol, D.; Registry, L.; Kadaster, M.A. Crowdsourcing and VGI in National Mapping Agency's Data Collection. In Proceedings of the 6th International Conference on Cartography and GIS, Albena, Bulgaria, 13–17 June 2016; pp. 13–17.
41. Kapenekakis, I.; Chorianopoulos, K. Citizen science for pedestrian cartography: Collection and moderation of walkable routes in cities through mobile gamification. *Hum.-Centric Comput. Inf. Sci.* **2017**, *7*, 10. [CrossRef]
42. Bordogna, G.; Carrara, P.; Criscuolo, L.; Pepe, M.; Rampini, A. A linguistic decision making approach to assess the quality of volunteer geographic information for citizen science. *Inf. Sci.* **2014**, *258*, 312–327. [CrossRef]
43. Granell, C.; Ostermann, F.O. Beyond data collection: Objectives and methods of research using VGI and geo-social media for disaster management. *Comput. Environ. Urban. Syst.* **2016**, *59*, 231–243. [CrossRef]
44. Odobasic, D.; Medak, D.; Miler, M. Gamification of Geographic Data Collection. In Proceedings of the Gi_Forum 2013 Creat Gisociety, Salzburg, Austria, 2–5 July 2013; pp. 328–337. [CrossRef]
45. Aanensen, D.M.; Huntley, D.M.; Feil, E.J.; Al-Own, F.; Spratt, B.G. EpiCollect: Linking smartphones to web applications for epidemiology, ecology and community data collection. *PLoS ONE* **2009**, *4*. [CrossRef] [PubMed]
46. Brovelli, M.A.; Minghini, M.; Zamboni, G. Public participation in GIS via mobile applications. *ISPRS J. Photogramm. Remote Sens.* **2016**, *114*, 306–315. [CrossRef]
47. Brovelli, M.A.; Minghini, M.; Zamboni, G. Public Participation GIS: A FOSS architecture enabling field-data collection. *Int. J. Digit. Earth* **2015**, *8*, 345–363. [CrossRef]
48. Tiangco, P.N. Report of the Asia South-East (ASE) Division. 2016, Volume 4. Available online: https://unstats.un.org/unsd/geoinfo/UNGEGN/docs/29th-gegn-docs/WP/WP32_4_29th%20UNGEGN%20Bangkok%20UNGEGN%20Asia%20South%20East%20Divisional%20Report.pdf (accessed on 30 April 2018).

International Journal of
Geo-Information

MDPI

Article

An Automatic User Grouping Model for a Group Recommender System in Location-Based Social Networks

Elahe Khazaei and Abbas Alimohammadi *

Department of Geospatial Information Systems, Faculty of Geodesy and Geomatics Engineering,
K. N. Toosi University of Technology, Tehran 19967 15433, Iran; Ekhazaei@mail.kntu.ac.ir
* Correspondence: alimoh_abb@kntu.ac.ir; Tel.: +98-21-8877-0218

Received: 29 December 2017; Accepted: 18 February 2018; Published: 21 February 2018

Abstract: Spatial group recommendation refers to suggesting places to a given set of users. In a group recommender system, members of a group should have similar preferences in order to increase the level of satisfaction. Location-based social networks (LBSNs) provide rich content, such as user interactions and location/event descriptions, which can be leveraged for group recommendations. In this paper, an automatic user grouping model is introduced that obtains information about users and their preferences through an LBSN. The preferences of the users, proximity of the places the users have visited in terms of spatial range, users' free days, and the social relationships among users are extracted automatically from location histories and users' profiles in the LBSN. These factors are combined to determine the similarities among users. The users are partitioned into groups based on these similarities. Group size is the key to coordinating group members and enhancing their satisfaction. Therefore, a modified k-medoids method is developed to cluster users into groups with specific sizes. To evaluate the efficiency of the proposed method, its mean intra-cluster distance and its distribution of cluster sizes are compared to those of general clustering algorithms. The results reveal that the proposed method compares favourably with general clustering approaches, such as k-medoids and spectral clustering, in separating users into groups of a specific size with a lower mean intra-cluster distance.

Keywords: location-based social networks (LBSNs); clustering; user preference; social relationship effect; spatial proximity

1. Introduction

The rapid development of the mobile Internet has enabled users to share their information on mobile phones. Recent advancements in location acquisition and wireless communication technologies have led to the development of location-based social networks (LBSNs). Location data bridge the gap between the physical and digital worlds and provide a deeper understanding of user preferences and behaviour. There are many real LBSN systems, such as Foursquare (www.foursquare.com), Gowalla, and GeoLife [1,2]. Moreover, recent studies on identifying user locations from traditional social networks, such as Twitter (www.twitter.com), have contributed to the development of various ways to obtain such information from real-world LBSNs [3].

In location-based social networks (LBSNs), users share information about their locations, the places they visit, and their movement alongside with other social information. Visits are reported explicitly (by user check-ins in known venues and locations) or implicitly by allowing for smartphone applications to report visited locations to the LBSN. This information is then shared with other users who are socially related (e.g., friends) [4].

With the development of social networks and online communities, an increasing number of activities are being performed in groups [5]. Web and information technologies should make our everyday life easier and more comfortable. In this regard, a recommender system contributes to reducing the information overload problem. Standard recommendation approaches, which have been used in various domains, mostly focus on a single user. However, there are many situations when the user interacts socially, with or without restraints. In some situations, we want to interact socially, (e.g., having dinner with friends), while in other situations we are forced to participate in groups, (e.g., mass transit). We are also a part of much larger social groups which form and adjust our behaviour and norms [6]. Nowadays, less attention is paid to social aspects of individuals and groups as units. Incorporating users' social links based on social networks and user personalities provides both the recommendation and grouping process with more realistic information modelling [7].

To support recommendation in social activities, group recommender systems were developed. LBSNs provide rich content (location, time-stamps) and social network information, which can help in modeling group dynamics for group recommendations [8]. There are cases where a group of people participates in a single activity. For instance, visiting a restaurant or a tourist attraction, watching a movie and selecting a holiday destination are examples of recommendations that are well suited for groups of people. Spatial group recommender systems provide suggestions about places when more than one person is involved in the recommendation process. Groups are composed of members with similar preferences that can have a similar recommendation. The more preferences that group members have in common, the more easily the group recommender system can suggest items that result in higher levels of satisfaction among the members. When groups do not already exist, another key aspect of group recommendation is related to groups identification [9]. Since the determination and coordination of group members is very time-consuming, in this paper an automatic selection process based on an unsupervised clustering approach is used to partition users into groups of a specific size with the most similar members.

The most popular approach for partitioning users into groups is the clustering algorithm. It is a fundamental research topic in data mining and is widely used for various applications in scientific fields such as artificial intelligence, statistics and social sciences. The objective of clustering is to partition the original data points into a number of groups so that data points within the same cluster are similar to each other, but are different from those in other clusters [10]. As the main objective of this study is to create groups of a specific size, there are several factors to be accounted for in similarity estimations. These are user preferences, social relationships, an individual's free days and spatial proximity, and these are also the key factors in creating a favourable space and maximizing user satisfaction. To achieve this aim, a modified *k*-medoids algorithm is developed and applied to user similarities, and consequently groups of a specific size are formed with similar members.

The main contributions of this study are as follows. (1) Taking into account the social relationships among users and their free days in group formation. These factors contribute to user satisfaction and raise the probability of recommendations being accepted by group members. Social relationships and free days, as well as user preferences are applied to characterize the similarities among users. (2) Considering the proximity of the visited locations as an index of the similarity of users. In reality, people tend to visit locations near their homes. In LBSNs, the spatial range of venues visited by the user is used to estimate his or her home location. For grouping users, the proximity of the locations visited by users, while considering their spatial range, is employed to compute similarities among users. Despite the significance of this factor, it has been either neglected or used ineffectively in previous group recommender systems for user grouping. In this study, however, this factor has been considered more effectively. (3) Automatic user grouping into groups of given sizes in LBSNs. Producing recommendations for a set of similar users allows the system to satisfy the individual users in a group and respect their constraints. In this context, an automatic group partitioning into groups of a given size in the form of unsupervised clustering is necessary.

The rest of the paper is organized as follows. Section 2 summarizes related work, followed by an overview of our system in Section 3. Section 4 present the two major parts of the proposed system: (1) similarity based on user preferences, social relationships, the user's free days, and spatial proximity, and (2) grouping users into groups of a given size. Further experimental results based on real data sets are provided in Section 5. Conclusions and key remarks are presented in Section 6.

2. Related Work

Group recommender systems usually consider predefined/a priori known groups, and only a few existing approaches are able to automatically identify groups [9]. With respect to the classification of existing systems, four different types of groups can be identified, which can be described as follows [11]:

- **Established group**: a number of individuals who explicitly choose to be part of a group, because of shared long-term interests. These groups have the property to be persistent and users actively join the group. Online communities that share preferences [12], people attending a party [13], and communities of like-minded users [14] are examples of this type of group.
- **Occasional group**: a group of people who occasionally do something together, for example, visiting a museum. Members have a common aim at a particular moment. They might not know each other, but they share interest for a common place. People who want to see a movie together [15], people traveling together [16], and people who want to dine together [17] are examples of the existing occasional groups.
- **Random group**: a group of people who share an environment at a particular moment without explicit interests that link them. Its nature is heterogeneous and its members might not share interests. People that browse the web together [18] and people in a public room [19] are some of the existing random groups.
- **Automatically identified group**: a group that is automatically detected considering the user preferences and/or the available resources. Such an approach is interesting for various reasons: (I) manual grouping can be very time consuming in large data sets, and (II) interests of people vary and usually change with time, so user grouping is a complex and continuous process requiring regular updates.

In automatic identification of groups, the goal is to find intrinsic communities of users. In 2004, an optimization function was introduced, known as the modularity [20], in which the generic partitioning of a set of nodes in the network is measured. In modularity, the number of internal edges in each partition is counted, with respect to the random case. The optimization of this function gives the natural community a network structure without a previous assessment of the number and the size of the partitions. Moreover, it is not necessary to embed the network in a metric space as in the case of the k-means algorithm. In addition, in this approach, the notion of distance or link weight can be introduced, but in a purely topological fashion [21]. Based on the optimization of the weighted modularity, a very efficient algorithm has been proposed to easily handle networks with millions of nodes. This algorithm generates a dendrogram, i.e., a community structure at various network resolutions [11,22].

The approach proposed in [23] aims to automatically discover communities of interest (CoIs) (i.e., a group of individuals who share and exchange ideas about a given interest), and produce recommendations for them. The CoI is identified through extraction of the preferences expressed by users in personal ontology-based profiles. Each profile measures the interest of a user via ontological concepts, and these expressed interests are used to cluster the concepts. User profiles are then split into subsets of interests, to link the preferences of each user with a specific cluster of concepts. Hence, it is possible to define relationships among users at different levels, obtaining a multilayered interest network that allows for multiple CoIs. Recommendations are built using a content-based CF approach.

In these approaches, detected communities have different sizes and there is no constraint on the community size. In this study, a method is developed according to which users are partitioned automatically into groups of a given size. This contributes to satisfying the preferences of each group by recommending preference-related places.

Li et al. (2014) proposed a group-coupon recommender system. For detecting similar group in this system, first the set of candidate customers is identified with a high willingness-to-purchase score, and then all the combinations of possible groups with specific size are listed. For each candidate group, its cohesion score is computed. Finally, the top-k groups with the highest cohesion score are selected as the recommended groups [24].

In 2014, Ganganath et al. introduced a modified *k*-means algorithm that obtains clusters with preferred sizes. Moreover, the modified algorithm makes use of prior knowledge about the given data set for selectively initializing the cluster centroids, which helps the algorithm to escape from local minima. In the assignment step, it assigns a new data point to the cluster whose centroid yields the least within-cluster sum of squares. Nevertheless, this is implemented only if the current cluster has not violated its size constraint. Otherwise, it passes to the next-best option until it reaches a cluster that has not yet exceeded its size constraint [25].

The exclusive lasso has been exploited to exert a balanced constraint and to introduce the ability to induce competition among different categories for the same data point. Chang et al. (2014) incorporated the exclusive lasso into *k*-means and min-cut clustering algorithms, and thus improved the ability of these two mainstream clustering algorithms to deal with balanced data points [10].

The approach proposed in [26] is a *k*-means-based clustering algorithm that optimizes the mean-square error for given cluster sizes. A straightforward application is balanced clustering, where the size of every cluster is the same. In the *k*-means assignment phase, the algorithm solves the assignment problem using a Hungarian algorithm. This is a novel approach, and results in an assignment-phase time complexity of $O(n^3)$, which is faster than the previous $O(k^{3.5}n^{3.5})$ achieved by linear programming in constrained *k*-means.

3. System Overview

This section first explains the data structures used in the paper, and then presents the application scenario and the architecture of the proposed method.

3.1. Preliminary

Figure 1 illustrates the relationships between five key data structures: user, venue, check-in, user location history, and category hierarchy. In an LBSN, a user records profile information, such as ID, name, age, gender, and home town. The user can also mark a visited venue, (e.g., a shop) and leave some comments, which is known in an LBSN as a check-in. A user can visit multiple locations and may generate a check-in for each visit (the solid arrows in Figure 1a). The location history of a user in the real world is obtained from all of the user's check-ins. A venue is a location that is associated with a pair of coordinates, indicating its geographical position and a set of categories denoting its functionalities. Venues are shown by squares on the map. The categories of venues have different granularities, usually represented by a category hierarchy as shown in the bottom part of Figure 1a [27]. For example, the "food" category includes "Chinese restaurant" and "Italian restaurant", and the "art and entertainment" category includes "art gallery" and "museum", etc. In the proposed system, a two-level category hierarchy obtained from Foursquare is used. In Figure 1b, the type of a category is shown, together with the number of sub-categories.

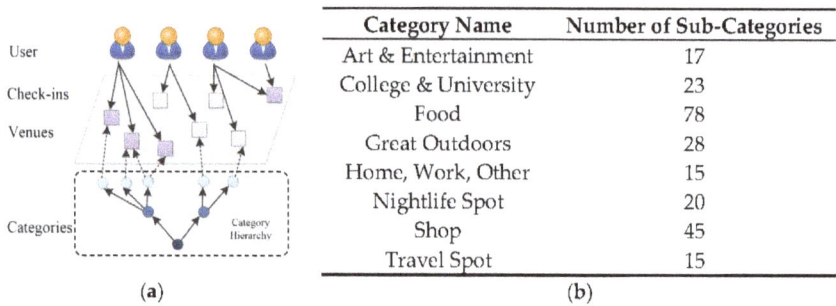

Category Name	Number of Sub-Categories
Art & Entertainment	17
College & University	23
Food	78
Great Outdoors	28
Home, Work, Other	15
Nightlife Spot	20
Shop	45
Travel Spot	15

(a) (b)

Figure 1. Data Structures in Location-Based Social Networks: (**a**) Overview of a location-based social network (adapted from [27]), (**b**) Detailed location category hierarchy in Foursquare.

3.2. Application Scenario

In a spatial group recommender system, a group is formed either by a predefined member or by the system itself, automatically. In automatic group detection, users are partitioned into the groups that have the most similar preferences. In addition to considering user preferences, social relationships among members of a group are also of significance for creating a pleasure space. Furthermore, the proximity of members' locations is essential for user convenience. Another significant factor for increasing the probability of accepting recommendations is the coordination of free days among group members. Thus, an individual's free days is a factor that has a key role in group member determination. The proposed system clusters users automatically with specific group sizes by considering common preferences, social relationships, similarity of users' free days, and spatial proximity. For instance, a possible application scenario in which spatial group recommendation can be applied, is when the user plans to spend free time. In this situation, coordinating and selecting members of a group is relatively difficult and time-consuming. In addition, individuals may like to become familiar with new people who share their interests, thus improving social relationships.

3.3. System Architecture

Our proposed system comprises six major components: (1) user preferences discovery, (2) social relationships effect, (3) spatial similarity, (4) similarity of users' free days, (5) user similarity, and (6) user clustering. The first component infers each user's expertise in each category according to the user's location histories. Given a predefined category hierarchy (Figure 1b), a user's location history in a city is sorted into groups of different location categories. Then, in each category, a group of location histories is modelled using a user location matrix, in which each entry denotes the user's number of visits to a physical location. Subsequently, each user's personal preferences are modelled by a weighted category hierarchy (WCH), taking advantage of the location category information of the user's location history, which helps to overcome the data sparsity problem. Specifically, a WCH is a subtree of the predefined category hierarchy, where the value of each node denotes the user's number of visits within a category. These values are further normalized on each layer of a WCH using the technique of term frequency-inverse document frequency (TF-IDF) [27]. TF-IDF is a numeric measure that is used to score the importance of a word in a document based on the frequency of appearance of that word in a given collection of documents. Finally, the similarity between two users is computed by applying a similarity function based on their WCHs.

The second component models the effect of social relationships among users. Social relations among users are considered as a graph in which the nodes and edges are users and social relations, respectively. The strength of the relationship between users is estimated based on the existing paths connecting the users. In addition, the system employs the users' common check-ins and social ties for measuring the relationship effect. The third component extracts the user's free days and computes the similarity of this parameter among users. The fourth component analyses the spatial proximity of users and computes similarity based on this factor. The fifth component combines the obtained similarities based on user preferences, social relationships, the user's free days, and spatial proximity to infer user likelihoods. The last component is the most significant part of the system. This component groups the users into groups with a specified number of members, so that each user is assigned to the group in which the user has the most similarity with other members. In the following sections, a further description of the system process is given. The system architecture is shown in Figure 2.

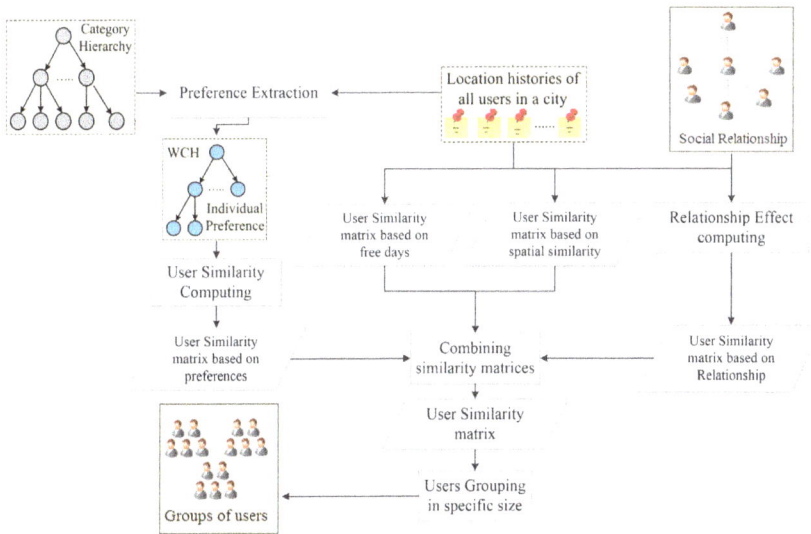

Figure 2. System architecture for automatic user grouping.

4. Materials and Methods

4.1. User Similarity

This section describes how user similarity is computed based on user preferences, the relationship effect, the user's free days, and spatial proximity.

4.1.1. Analysing User Preferences

User preferences are extracted according to the categories of his or her visited locations. First, a user location history is projected onto a predefined category hierarchy. As a result, each node receives a value representing the number of visits to a category. This is motivated by the fact that an individual's preferences are usually made up of multiple interests, such as shopping and visiting historical places, and these interests have different granularities, (e.g., "art and entertainment" → "museum"). Second, The TF-IDF value of each node in the hierarchy is calculated, where a user location history is regarded as a document and categories are considered as terms in the document. Intuitively, if a user likes a particular category, then he/she will visit more locations relating to that category. Furthermore, if a user visits locations within a category that other people use only rarely, it is more likely that this category is of greater interest to this user. For example, the number of visits to restaurants is generally higher than for other categories, such as art galleries in citizen location histories, but this does not imply that food should be ranked as the user's first interest. However, if a user is found to visit art galleries very frequently, the user may be truly interested in the arts.

Overall, a user's preference weight ($u.w_{c'}$) is calculated using Equation (1), where the first part of the equation is the TF value of category c in user u's location history and the second part denotes the IDF value of the category.

$$u.w_{c'} = \frac{|\{u.v_i : v_i.c = c'\}|}{|u.V|} \times lg\frac{|u|}{|\{u_j : c' \in u_j.C\}|} \tag{1}$$

In the above equation, $|\{u.v_i : v_i.c = c'\}|$ is user u's number of visits in category c', $u.V$ is the total number of the user's visits and $|\{u_j : c' \in u_j.C\}|$ counts the number of users who have visited category c' among all of the users U in the system. The WCH has several important advantages. It decreases concern about the different data scales of different users, it handles the data sparseness problem and it reduces the computational loads for computing further user similarities (from physical locations to categories). In addition, it enables similarity computation among users who do not have any common physical location histories; in other words, they live in different cities [3,27].

4.1.2. Similarity Based on User Preferences

Similarity computation is achieved via difference methods. In this paper, the cosine distance is used to estimate this value. For each user, a vector is created whose dimension is equal to the number of nodes on the first level of the WCH. The value of each item is the value of the corresponding node. The cosine distance is used to calculate the similarity of two users' vectors, according to Equation (2):

$$SimPreference\ (x_i, x_j) = \frac{x_i^T x_j}{\|x_i\| \|x_j\|}, \tag{2}$$

where x_i, x_j are the similarity vectors of two users.

4.1.3. Similarity Based on Relationship

Link prediction is an important research field in data mining with a wide range of scenarios. Many data mining tasks involve the relationships among objects. Link prediction can be used for recommendation systems, social networks, information retrieval, and many other fields [28].

Given that $G = \langle V, E \rangle$ is a graph of the social network, link prediction involves predicting the probability of the link between node V_i and node V_j. This can be considered as computing the "similarity" between nodes V_i and V_j, according to the network topology. In this paper, Katz's algorithm (1953) is used for measuring the social relationship effect. The idea of the method is that the existence of more paths between two nodes indicates a greater similarity between the two nodes. The Katz measure is defined as follows [28,29]:

$$Relationsim(u, v) = \sum_{l=1}^{l_{max}=\infty} \beta^l \cdot \left| path_{u,v}^l \right| \tag{3}$$

where $|path^l_{u,v}|$ is the number of paths between node u and node v, the length of the path is l, and β is a parameter taking values between zero and one. This parameter is used to control the contribution of a path to the similarity; the longer the path, the less contribution it makes to the similarity. To ensure that the Katz index converges, the value of β must be less than the inverse largest eigenvalue of the adjacency matrix ($\beta < 1/\lambda max$) [30]. The components of the adjacency matrix are defined as follows: if nodes i and j are connected in the network, then $a_{ij} = 1$; otherwise $a_{ij} = 0$.

One strategy for estimating the similarity between two friends is to calculate their common social circles [31]. For this purpose, the similarity between social friends is estimated using the following method:

$$Friendshipsim(u, v) = \begin{cases} \frac{|F(u) \cap F(v)|}{|F(u) \cup F(v)|} & , if\ u\ and\ v\ are\ friends \\ 0 & , otherwise \end{cases} \tag{4}$$

In Equation (4), $F(u)$ specifies a set of users who have a social relationship with user u.

Similar check-ins, i.e., check-ins at the same time and location, can also be considered as indicating the similarity of two users who have a social relationship, and can be used to calculate their similarity and social influence using the following method:

$$Checkinsim(u,v) = \begin{cases} \frac{|L(u) \cap L(v)|}{|L(u) \cup L(v)|} & , \textit{If } u \textit{ and } v \textit{ are friends} \\ 0 & , \textit{otherwise} \end{cases} \quad (5)$$

In Equation (5), $L(u)$ specifies a set of locations that were visited by user u. The final relationship similarity therefore is computed as:

$$SimRelation_f(u,v) = (1 - \alpha - \beta). \, Relationsim(u,v) + \alpha. \, Friendshipsim(u,v) + \\ \beta.Checkinsim(u,v) \quad (6)$$

4.1.4. Similarity Based on the User's Free Days

Free time provides citizens with time to spend outdoors and is associated with activities such as shopping, sightseeing, and socializing. These activities contribute to the expansion of relationships and new experiences. Free days vary from person to person, and accepting recommendations is more likely when members of the group share similar free time. This factor, meanwhile, is the key to coordinating group members in the grouping procedure. It can be extracted from location histories in LBSNs. For this purpose, a user-day matrix is computed from the user's visited locations on a specific day. To normalize the user-day matrix, the TF-IDF is calculated, where a user location history is regarded as a document and the day is considered as a term in the document. The cosine distance is used to calculate the similarity of two users' vectors.

4.1.5. Spatial Similarity

The geographical proximities of the locations influence the user's check-in behaviour. Usually, a user prefers to visit locations that are close to his or her residential address or office [4,32,33]. When the distance of the location from a user's home increases, the user's probability of visiting that location decreases. The home locations of users are usually not given in the check-in data set due to user privacy concerns. Nevertheless they can be estimated based on the assumption that check-ins are centred around the user's home location [34,35]. For this purpose, first, a minimum boundary box of the user's check-in locations is created. Then, this boundary is divided into small non-overlapping regions, and the check-ins are grouped based on those regions. The region with the maximum number of check-ins is considered to be the spatial range within which the user tends to visit venues. The average position of the check-ins inside the region is selected as the centre of the user's favourite spatial range and an approximation of the user's home location [35]. After the estimation of the approximate positions that the user has convenient access to, the distances between these positions are estimated. Finally, the users that are spatially closer to each other are considered to be more similar.

4.1.6. Combining Preferences, Relationships, Free Days, and Spatial Similarity

User preferences, social relationships, free days, and spatial proximity are criteria that are combined to compute the final similarity values between each pair of users, as follows:

$$Sim_f = \lambda. \, SimPreference(u,v) + \gamma. \, SimRelation_f(u,v) + \delta. \, SimSpatial(u,v) \\ + (1 - \gamma - \delta - \lambda). \, SimTemporal \quad (7)$$

where the parameters λ, γ and δ control the weights of user preference, relationship, and spatial similarity values, respectively.

4.2. User Grouping for a Given Group Size

For automatic selection of group members in a group recommender system, users are partitioned into groups with a specific group size, based on the similarity of their interests. In this regard, a modified *k*-medoids method is developed to cluster users into groups with specific sizes. In the proposed method, instead of using linear programming, a Hungarian algorithm is used in assignment phase of the *k*-medoids algorithm. In order to reduce the running time of the Hungarian algorithm for large data sets, multilevel *k*-way partitioning is used to divide the data set into the multiple parts. Then, with using parallel computing, the modified *k*-medoids method is applied for each part.

First, a brief description of the methods used in modified *k*-medoids algorithm is presented, and then details of the proposed method are described.

4.2.1. Multilevel *k*-Way Partitioning

The graph partitioning problem is the problem of partitioning the vertices of a graph into p roughly equal partitions so that the number of edges connecting vertices in different partitions is minimized. This approach has attracted great attention in areas such as parallel scientific computing, task scheduling, and VLSI design [36,37].

The *k*-way partitioning problem is generally solved by recursive bisection. That is, first, a two-way partitioning of V is obtained, and then a two-way partitioning of each resulting partition is determined recursively. After log (k) phases, graph G is partitioned into k partitions. Thus, the problem of performing a *k*-way partitioning is reduced to performing a sequence of bisections.

The multilevel recursive bisection (MLRB) algorithm has emerged as a highly effective method for computing the *k*-way partitioning of a graph. The basic structure of a multilevel bisection algorithm is very simple. The graph G is first reduced to a few hundred vertices, a bisection of this much smaller graph is computed, and then this partitioning is projected back to the original graph (with a higher number of vertices) by periodically refining the partitioning. Since the original graph has more degrees of freedom, these refinements decrease the edge cut. A detailed description of this algorithm can be found in [37].

4.2.2. Hungarian Algorithm

The assignment problem is one of the fundamental combinatorial optimization problems in the optimization or operations research branch of mathematics. It consists of finding a maximum weight matching (or minimum weight perfect matching) in a weighted bipartite graph. On the one hand, it is a special case of a more complex problem, such as the generalized assignment problem, the matching problem in graphs or the minimum-cost flow problem. On the other hand, real-world problems, such as the worker assignment problem, can be categorized as this type of problem. In its most general form, the problem can be stated as follows.

Consider a number of agents and tasks. Any agent can be assigned to perform any task, incurring some cost that may vary depending on the agent-task assignment. All of the tasks must be performed by assigning exactly one agent to each task and exactly one task to each agent, in such a way that the total cost of the assignment is minimized.

If the number of agents and tasks are equal and the total cost of the assignment for all of the tasks is equal to the sum of the costs for each agent (or the sum of the costs for each task, which is the same thing in this case), then this is called the linear assignment problem. The Hungarian algorithm is one of a group of algorithms that have been devised to solve the linear assignment problem within a certain time and bounded by a polynomial expression for the number of agents [38,39]. The Hungarian method of finding an optimal assignment is explained in more detail in [38].

4.2.3. *k*-Medoids Algorithm

In *k*-medoids methods, a cluster is represented by one of its points. This is an easy solution as it covers any attribute type, and the medoids have been proven to be resistant against outliers because of their insensitivity to peripheral cluster points. When medoids are selected, the clusters are defined as subsets of points close to their respective medoids, and the objective function is defined as the average distance, or another dissimilarity measure, between a point and its medoid [40,41]. The algorithm is as follows:

- Randomly select *k* data points as medoids.
- **Assignment step:** Assign each data point to the closest medoids.
- **Update step:** find new medoids of each cluster to minimize within cluster variance.
- Repeat assignment step and update step until the medoids do not change.

4.2.4. Modified *k*-Medoids for Grouping People into Groups of a Specific Size

The modified *k*-medoids method is the same as the standard *k*-medoids method, except that it guarantees specific cluster sizes. It is also a special case of the constrained *k*-means method, where cluster sizes are set to be equal or of a specific size. However, instead of using linear programming in the assignment phase, the partitioning is formulated as a pairing problem, which can be solved optimally by a Hungarian algorithm in time $O(n^3)$. For large data sets, executing a Hungarian algorithm takes a long time. In order to address this problem, first, multilevel *k*-way partitioning is used to divide a data set into multiple parts. Then, for each part, the modified *k*-medoids method is applied to cluster the users and parallel computing is used to reduce the running time.

For ease of expression of the proposed method, it is assumed that the number of users is n and all of the clusters have the same size (*k*). The process of the modified *k*-medoids method is similar to *k*-medoids, however, the assignment phase is different. In this method, instead of selecting the closest medoid, there are *n* pre-allocated slots (*n*/*k* slots per cluster), and data points can be assigned only to these slots (Figure 3). This will force all of the clusters to be of the same size assuming that $\lceil \frac{n}{k} \rceil = \lfloor \frac{n}{k} \rfloor = \frac{n}{k}$ ($\lceil x \rceil$ = ceiling (x), $\lfloor x \rfloor$ = floor (x)). Otherwise, there will be (*n* mod *k*) clusters of size $\lceil \frac{n}{k} \rceil$, and $k - (n \bmod k)$ clusters of size $\lfloor \frac{n}{k} \rfloor$. To find the assignment that minimizes the mean-square error (MSE), an assignment problem is solved via the Hungarian algorithm. Steps of implementation of the modified *k*-medoids method are as follows:

- Data set is divided in multiple parts with *k*-way partitioning. For each part, the following procedure is repeated.
- A bipartite graph is constructed consisting of *n* data points and *n* cluster slots (Figure 3).
- The cluster slots are partitioned into clusters with the largest possible even number of slots (it is assumed that all clusters have the same size, if different cluster size is given, cluster slots are divided based on different cluster size.)
- The initial medoids can be select randomly from all data points. (In this study, *k*-means++ is used to select the initial medoids from all data points.)
- **Assignment step:** The edge weight is the similarity between the point and the assigned cluster medoid. It is updated according to newly medoids. With using Hungarian algorithm, data points are assigned to cluster slots based on the edge weight.
- **Update step:** New medoid of each cluster are calculated based on similarity between the points and medoids. The update step is similar to that of the *k*-medoids method.
- The last two steps are repeated until the medoids do not change.

In contrast to the standard assignment problem with fixed weights, in this study weights are changed dynamically after each *k*-medoids iteration, according to the newly calculated medoids. Following this, the Hungarian algorithm is performed to obtain the minimal weight pairing. The similarities of the users are stored in an $n \times n$ matrix as the correct input format for the Hungarian algorithm. Algorithm 1 provides the pseudocode for the modified k-medoids method.

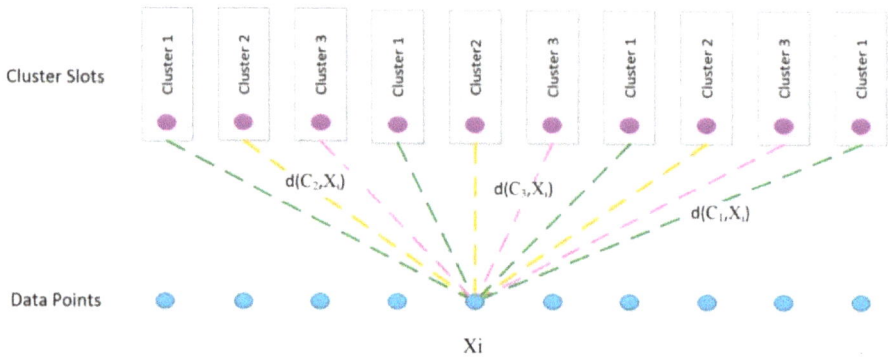

Figure 3. Minimum distance calculation with balanced clusters (adapted from [26]).

Algorithm 1. Modified k-medoids

Input: data set X, number of member in group
Output: partitioning of data set.
Partition data set to multi part with k-way partitioning
part $\leftarrow 0$
repeat
    Initialize medoid locations C^0 with k-means++
    $t \leftarrow 0$
    repeat
        Assignment step:
            Calculate edge weights. Solve an Assignment problem.
        Update step:
            Calculate new medoid locations C^{t+1}
        $t \leftarrow t + 1$
    Until medoid locations do not change.
Until all parts are clustering

4.3. Experimental Evaluation

In this section, first the settings of the experiments, including the data set, baseline approaches, and the evaluation method, are described. Results regarding both the effectiveness and the efficiency of the proposed system are presented and followed by a discussion.

Experimental Settings

Data sets. The two largest cities in the USA, New York City (NYC) and Los Angeles (LA), are considered in this study. Data sets from these cities, including the tips generated by users, are extracted from Foursquare [27]. Four data sets from the above-mentioned cities have been selected as follows. (1) Users whose home city is LA and who are visitors to places in LA, (2) users from New Jersey who visit LA, (3) users from New York visiting places within their city, and (4) users from New Jersey who are visiting New York. Statistics of experimental data sets are shown in Table 1. These data sets were collected during a period of 25 months from 1 February 2009 to 30 July 2011.

Foursquare has blocked the API for crawling a user's check-in data due to privacy concerns. However, the tips left by users are available for download. The proposed method could be more effective if check-in data was used, although it seems sensible to use tips as there are some associated advantages, such as the fact that they express a user's real interests. Sometimes, people check in at a venue without visiting the venue for any purpose. However, leaving a tip in connection with a venue usually means that the user has engaged in some essential activities, such as dining or shopping at the venue [27].

Table 1. Statistics of experimental data sets.

Home City	QUERY City	Total Users	Tips in City	Tips/User
LA	LA	977	11,700	11.9
NJ	LA	228	2553	11.20
NY	NY	3630	52,282	14.4
NJ	NY	2886	72,170	25.01

The following information is extracted: (1) user profile information, including the user ID, name, and home city, (2) the user's social relationships, including the user IDs of two-sided connections, (3) venue profile information, consisting of a venue's ID, name, address, GPS coordinates, and categories, and (4) user location histories, represented by all of the tips a user has left in the system. Each tip includes a venue ID, comments, and a timestamp. From the data set, the users who have over seven tips in a city are chosen as candidate query users.

Evaluation methods. For the evaluation of clustering solutions, validity indices are normally used. There are two types of validity indices: external indices and internal indices [42]. An external index is a measure of the agreement between two partitions where the first partition is the a priori known clustering structure, and the second results from the clustering procedure [43]. Internal indices are used to measure the quality of a clustering structure without external information. For internal indices, the results are evaluated using quantities and inherent features of the data set. In this paper, the ground truth labels are not known, therefore internal indices must be used. There are several internal indices for clustering evaluation.

As mentioned previously, users will be partitioned into groups of a specific size, in a process that differs from general clustering. It is mandatory that users are partitioned with similar preferences. Therefore, the intra-cluster distances are important for the evaluation of clustering, while the inter-cluster distances are not significant. The mean intra-cluster distance and the silhouette index are used for evaluating the proposed method. In addition, the three clustering methods of spectral clustering, k-medoids, and k-way partitioning are used for grouping users. The outcomes from these methods and from the proposed approach are compared and discussed.

Mean intra-cluster distance. In each cluster, the intra-cluster distances between points should be as small as possible. The mean intra-cluster distance for all of the clusters is an efficient index for evaluating the results of clustering.

Silhouette index. The silhouette refers to a method of interpretation and validation with respect to consistency within clusters of data. The silhouette value is a measure of how similar an object is to its own cluster (cohesion) compared to other clusters (separation). The silhouette is based on the mean score for every point in the data set (Equation (8)). Each point's individual score is based on the difference between the average distance of that point to other points in its cluster and the minimum average distance between that point and the other points of other clusters. This difference is then divided by a normalization term, which is the average with the larger value,

$$DB = 1/N \sum_{i=0}^{N} s_{x_i} \tag{8}$$

where, N is the number of points in the data set, and

$$s_{x_i} = (b_{q,i} - a_{p,i}) / max\{a_{p,i}, b_{p,i}\}$$

If x_i is a point in cluster p, then $b_{q,i} = mind_{q,i}$ where $d_{q,i}$ is the average distance between point x_i and every point of cluster q. On the other hand, $a_{p,i}$ is the average distance between point x_i and every other point of cluster p. The score range is between -1 and 1, indicating that as clustering improves, then the score will approach a value of 1 [44].

Parameter selection. The terms λ, γ and δ are parameters that stand for the weights of the user preference, relationship and spatial similarity values respectively. Subsequently, these parameters determine the weight of users' free day similarities. Due to the fact that the attractiveness of friendship with new individuals, visiting new places and the possibility to change free days may not be similar for all individuals, the λ, γ and δ parameters can vary from case to case. In this study, the values of these parameters are selected by a parameter space search with silhouette criteria, according to which λ, γ and δ parameter values are set to 0.3, 0.2, and 0.25, respectively.

5. Results and Discussion

For convenience, the group sizes are assumed to be equal with each group having six members. The proposed method is applied to the four selected data sets. For each data set, user preferences, social relationships, spatial proximity, users' free days, and final similarities, (where the latter is a combination of the first four factors with estimated weights), are considered separately for clustering. The mean intra-cluster distance and silhouette index values are calculated for the evaluation of clustering in each data set. Results of the evaluation for database #1 are shown in Table 2.

Table 2. Results of evaluation methods for automatic user grouping (database #1).

Parameters Value (λ, γ, δ)	$\lambda = 1$, $\gamma = 0$, $\delta = 0$	$\lambda = 0$, $\gamma = 1$, $\delta = 0$	$\lambda = 0$, $\gamma = 0$, $\delta = 1$	$\lambda = 0$, $\gamma = 0$, $\delta = 0$	$\lambda = 0.3$, $\gamma = 0.2$, $\delta = 0.25$
Silhouette Index	-0.066	-0.035	0.048	0.082	0.015
Mean intra-cluster distance	0.072	0.281	0.023	0.094	0.192

The silhouette index range is between -1 and 1, where a score that is closer to 1 indicates better clustering. As can be seen from Table 2, the silhouette score is near to zero because users are grouped in groups of specific size; this issue is different from the case of general clustering. Users with similar preferences are mandatorily partitioned where large clusters are forced to break up into clusters of a specific size. This causes similar individuals to be defined as different clusters, and consequently causes the distance to the nearest cluster to be reduced. In other words, the separations of the clusters are reduced, causing the silhouette score to be near zero. The positive silhouette score, however, indicates that the separation of the clusters is greater than the cohesion of the clusters over the majority of the points.

Similarity of clusters, as represented by low variances, is of greater importance than the distance between clusters, which decreases when the number of cluster members is reduced. For the mean intra-cluster distance, a lower value represents more cohesion within clusters. For example, in Table 2, the mean intra-cluster distance when only user preference similarity ($\lambda = 1$, $\gamma = 0$, $\delta = 0$) is considered for grouping the users, is estimated at 0.072. From Table 2, for the selected parameters ($\lambda = 0.3$, $\gamma = 0.2$, $\delta = 0.25$), the mean intra-cluster distance is 0.192. In the next phase, and in order to better interpret the values shown in Table 2, the mean intra-cluster distances of the other factors are estimated separately for previously defined clusters. These values are shown in Table 3 for database #1. In Table 3, column 1 indicates that only the user preference similarity is considered for partitioning users, and the mean intra-cluster distances of social relationships, spatial proximity, and free days (temporal distance) are measured for the determined groups. The other columns of Table 3 can be interpreted in a similar way.

The last column of Table 3 implies that by taking into account the user preferences, social relationships, users' free days, and spatial proximity, the mean intra-cluster distance is estimated at 0.191. After grouping the users, the mean intra-cluster distances of these factors are estimated at 0.170, 0.338, 0.242, and 0.112, respectively. According to Table 3, grouping users with one criterion decreases the mean intra-cluster distance value for that criterion, but this value then increases for the other factors. Table 3 shows that the value of the mean intra-cluster distance in social relationships is comparatively high, due to the lack of relationships among all of the users and a relatively small similarity value.

Table 3. The mean intra-cluster distances of user preference, spatial proximity, social relationships, and free days for database #1.

Mean Intra-Cluster Distance	Parameters Value (λ, γ, δ)				
	$\lambda = 1$, $\gamma = 0$, $\delta = 0$	$\lambda = 0$, $\gamma = 1$, $\delta = 0$	$\lambda = 0$, $\gamma = 0$, $\delta = 1$	$\lambda = 0$, $\gamma = 0$, $\delta = 0$	$\lambda = 0.3$, $\gamma = 0.2$, $\delta = 0.25$
User preferences distances	**0.072**	0.369	0.365	0.370	0.170
Social relationships distances	0.481	**0.281**	0.485	0.478	0.338
Spatial distances	0.265	0.259	**0.023**	0.273	0.112
Temporal distance	0.488	0.490	0.491	**0.094**	0.242
Final grouping					**0.191**

In order to evaluate the efficiency of the proposed method, the outcomes of this study are compared with other clustering algorithms. These algorithms, i.e., *k*-medoids and spectral methods, are two common clustering approaches that are applied to grouping people. In addition, multilevel *k*-way partitioning is used in this assessment because it creates a balanced partition. In these methods, the number of clusters must be specified. In this study, it is assumed that the group size is fixed, each group having six members. With this assumption, the number of desired clusters is achieved by dividing the number of users by the size of each cluster. The results show that the cluster sizes in the *k*-medoids and spectral clustering methods were not equal, so that either one point or a huge proportion of the data may be allocated to a single cluster, while the multilevel *k*-way partitioning creates balanced cluster sizes. In *k*-medoids, the number of clusters is less than the specified number of clusters; in some cases, some clusters do not even contain any points. In spectral clustering and multilevel *k*-way partitioning, the number of clusters is equal to the specified number of clusters.

In Table 4, the average of the cluster size distribution (per cent) in the proposed method, multilevel *k*-way partitioning, *k*-medoids, and spectral clustering methods are compared. As can be observed, in the *k*-medoids and spectral clustering methods, a high percentage of the clusters do not share the same specified size, while in the proposed method and in multilevel *k*-way partitioning a high percentage of the clusters are of equal size.

Table 4. The average of Cluster size distribution (per cent) resulting from the proposed and three existing clustering approaches.

Number of Group's Member	Proposed Method	Multilevel *k*-Way Partitioning	*k*-Medoids Clustering	Spectral Clustering
1			5.2	45.0
2			6.1	16.3
3			12.2	4.0
4			10.4	2.4
5			15.7	1.6
6	87.9	93.6	13.0	0.8
7	12.1	6.4	7.0	1.6
8			8.7	4.0
9			5.2	0.8
10+			16.5	23.4

The mean intra-cluster distances of the four different approaches for the four data sets are compared in Table 5. As the number of clusters with small sizes is outnumbered in *k*-medoids and spectral clustering, the mean intra-cluster distances of the methods are small. In order to compare the outcomes of the proposed method with those of *k*-medoids and spectral clustering, clusters with a size of less than four are removed in the mean intra-cluster distance calculation. The mean intra-cluster distance that is calculated by the proposed method is fairly small. According to Table 5, although multilevel *k*-way partitioning divided users into balanced cluster sizes, the mean intra-cluster distance in this method is higher than in the proposed method. Furthermore, in the multilevel *k*-way partitioning method, cluster sizes cannot change based on a predefined cluster size.

Table 5. The mean intra-cluster distance of the proposed method and three clustering approaches for the four data sets.

Method	Database #1	Database #2	Database #3	Database #4
Proposed method	0.191	0.187	0.198	0.173
Multilevel *k*-way partitioning	0.269	0.255	0.277	0.253
k-medoids clustering	0.171	0.165	0.183	0.162
k-medoids clustering without cluster size 1, 2, 3	0.292	0.268	0.305	0.281
Spectral clustering	0.098	0.096	0.112	0.094
Spectral clustering without cluster size 1, 2, 3	0.281	0.277	0.295	0.264

6. Conclusions

In a spatial group recommender system, the system recommends a place to a group of users. In this study, an automatic method for identifying groups of users with similar preferences, spatial proximity, free days, and social relationships has been proposed. Corresponding data sets for the parameters mentioned were obtained from the location histories and user profiles. Then, a modified *k*-medoids clustering algorithm was developed, which guarantees equal clusters or clusters of a specific size. The proposed method was evaluated using further experiments based on four data sets that were collected from Foursquare. The mean intra-cluster distance and the silhouette index were used for evaluating the proposed method. In addition, the three clustering methods of spectral clustering, *k*-medoids, and *k*-way partitioning were used for grouping users. The results of these methods and the results of the proposed approach were compared. The results showed that the proposed method can efficiently divide users into groups with a given group size. The mean intra-cluster distance for the proposed method is almost identical to that for the spectral clustering and *k*-medoids methods. However, the proposed method meets the objective of partitioning users into groups of a specific size. Although multilevel *k*-way partitioning created balanced cluster sizes, the proposed method has a comparatively lower mean intra-cluster distance. The proposed method is capable of partitioning users into clusters with specific predetermined sizes.

Foursquare is one of the most popular LBSNs worldwide, so data sets of this network have been used as an example and are representative of other LBSNs. So, results of the proposed approach for user grouping can be generalized to other LBSNs. Also, the proposed user grouping method can be used in other fields that needs user grouping, such as citizen science. In this study, location category is used for the determination of user preferences, and physical locations of users are ignored. Only the visited venue locations are used in order to calculate the spatial proximity of users. For future studies, inferring the spatial preferences of users by considering the physical locations and including the temporal influences and group sizes on the clustering results are aspects that are recommended for further investigation. Moreover, it is worth noting that in the context of the spatial group recommender system, a procedure for suggesting places according to preferences of the group members could be developed. Because of a lack of information about the uncertainty, reliability of the existing data sets has not been considered in this research, and it has been assumed that users' checks in at a place are correct and reflect their true preferences. Consideration of uncertainty and bias effects in crowdsourcing data is an important topic [45–47] and can be considered in future studies.

Author Contributions: Elahe Khazaei and Abbas Alimohammadi conceived and designed the experiments; Elahe Khazaei carried out model development, verification the models, and drafted the original version of the manuscript. Abbas Alimohammadi helped to revise the paper.

Conflicts of Interest: The authors declare no conflict of interest.

References

1. Bao, J.; Zheng, Y.; Wilkie, D.; Mokbel, M. Recommendations in location-based social networks: A survey. *Geoinformatica* **2015**, *19*, 525–565. [CrossRef]
2. Abbasi, O.; Alesheikh, A.; Sharif, M. Ranking the City: The Role of Location-Based Social Media Check-Ins in Collective Human Mobility Prediction. *ISPRS Int. J. Geo-Inf.* **2017**, *6*, 136. [CrossRef]
3. Wang, H.; Li, G.; Feng, J. Group-Based Personalized Location Recommendation on Social Networks. In *APWeb*; Chen, L., Jia, Y., Sellis, T., Liu, G., Eds.; Lecture Notes in Computer Science; Springer International Publishing: Cham, Switzerland, 2014; Volume 8709, pp. 68–80. ISBN 978-3-319-11115-5.
4. Wang, H.; Terrovitis, M.; Mamoulis, N. Location recommendation in location-based social networks using user check-in data. In *Proceedings of the 21st ACM SIGSPATIAL International Conference on Advances in Geographic Information Systems - SIGSPATIAL'13*; ACM Press: New York, New York, USA, 2013; pp. 364–373.
5. Guo, J.; Zhu, Y.; Li, A.; Wang, Q.; Han, W. A Social Influence Approach for Group User Modeling in Group Recommendation Systems. *IEEE Intell. Syst.* **2016**, *31*, 40–48. [CrossRef]
6. Butler, C.T.L.; Rothstein, A. *On Conflict and Consensus: A Handbook on Formal Consensus Decisionmaking*, 3rd ed.; Food Not Bombs: Santa Cruz, CA, USA, 2007.
7. Kompan, M.; Bielikova, M. Group Recommendations: Survey and Perspectives. *Comput. Inform.* **2014**, *33*, 446–476.
8. Purushotham, S.; Kuo, C.-C.J.; Shahabdeen, J.; Nachman, L. Collaborative Group-activity Recommendation in Location-based Social Networks. In *Proceedings of the 3rd ACM SIGSPATIAL International Workshop on Crowdsourced and Volunteered Geographic Information*; GeoCrowd'14; ACM: New York, NY, USA, 2014; pp. 8–15.
9. Ludovico, B.; Salvatore, C.; Satta, M. Groups identification and individual recommendations in group recommendation algorithms. In Proceedings of Workshop on the Practical Use of Recommender Systems, Algorithms and Technologies (PRSAT 2010), Barcelona, Spain, 30 September 2010.
10. Chang, X.; Nie, F.; Ma, Z.; Yang, Y. Balanced k-Means and Min-Cut Clustering. *arXiv preprint* **2014**, arXiv:1411.6235.
11. Boratto, L.; Carta, S. State-of-the-Art in Group Recommendation and New Approaches for Automatic Identification of Groups. In *Information Retrieval and Mining in Distributed Environments*; Soro, A., Vargiu, E., Armano, G., Paddeu, G., Eds.; Studies in Computational Intelligence; Springer Berlin Heidelberg: Berlin/Heidelberg, Germany, 2011; pp. 1–20.
12. Kim, J.K.; Kim, H.K.; Oh, H.Y.; Ryu, Y.U. A group recommendation system for online communities. *Int. J. Inf. Manag.* **2010**, *30*, 212–219. [CrossRef]
13. Pizzutilo, S.; De Carolis, B.; Cozzolongo, G.; Ambruoso, F. Group modeling in a public space: Methods, techniques, experiences. In Proceedings of the 5th WSEAS International Conference on Applied Informatics and Communications, Stevens Point, WI, USA, 15–17 September 2005; pp. 175–180.
14. Smyth, B.; Balfe, E. Anonymous personalization in collaborative web search. *Inf. Retr. Boston.* **2006**, *9*, 165–190. [CrossRef]
15. O'Connor, M.; Cosley, D.; Konstan, J.A.; Riedl, J. PolyLens: A recommender system for groups of users. In *ECSCW 2001: Proceedings of the Seventh European Conference on Computer Supported Cooperative Work 16–20 September 2001, Bonn, Germany*; Springer: Dordrecht, The Netherlands, 2001; pp. 199–218.
16. Ardissono, L.; Goy, A.; Petrone, G.; Segnan, M.; Torasso, P. Intrigue: Personalized recommendation of tourist attractions for desktop and hand held devices. *Appl. Artif. Intell.* **2003**, *17*, 687–714. [CrossRef]
17. McCarthy, J.F. Pocket Restaurant Finder: A situated recommender systems for groups. In Proceedings of the Workshop on Mobile Ad-Hoc Communication at the 2002 ACM Conference on Human Factors in Computer Systems, Minneapolis, MN, USA, 20–25 April 2002; pp. 1–10.

18. Lieberman, H.; Van Dyke, N.W.; Vivacqua, A.S. Let's Browse: A Collaborative Web Browsing Agent. In *Proceedings of the 4th International Conference on Intelligent User Interfaces*; IUI '99; ACM: New York, NY, USA, 1999; pp. 65–68.

19. Crossen, A.; Budzik, J.; Hammond, K. J. Flytrap. In *Proceedings of the 7th International Conference on Intelligent User Interfaces - IUI '02*; ACM Press: New York, NY, USA, 2002; p. 184.

20. Newman, M.E.J.; Girvan, M. Finding and evaluating community structure in networks. *Phys. Rev. E* **2004**, *69*, 26113. [CrossRef] [PubMed]

21. Newman, M.E.J. Analysis of weighted networks. *Phys. Rev.* **2004**, *70*. [CrossRef] [PubMed]

22. Blondel, V.D.; Guillaume, J.-L.; Lambiotte, R.; Lefebvre, E. Fast unfolding of communities in large networks. *J. Stat. Mech. Theory Exp.* **2008**, *2008*, P10008. [CrossRef]

23. Cantador, I.; Castells, P. Extracting multilayered Communities of Interest from semantic user profiles: Application to group modeling and hybrid recommendations. *Comput. Human Behav.* **2011**, *27*, 1321–1336. [CrossRef]

24. Li, Y.-M.; Chou, C.-L.; Lin, L.-F. A social recommender mechanism for location-based group commerce. *Inf. Sci.* **2014**, *274*, 125–142. [CrossRef]

25. Ganganath, N.; Cheng, C.-T.; Tse, C.K. Data Clustering with Cluster Size Constraints Using a Modified K-Means Algorithm. In Proceedings of the 2014 International Conference on Cyber-Enabled Distributed Computing and Knowledge Discovery, Shanghai, China, 13–15 October 2014; pp. 158–161.

26. Malinen, M. I.; Fränti, P. Balanced K-Means for Clustering. In *Structural, Syntactic, and Statistical Pattern Recognition*; Springer: Berlin/Heidelberg, Germany, 2014; pp. 32–41.

27. Bao, J.; Zheng, Y.; Mokbel, M. F. Location-based and preference-aware recommendation using sparse geo-social networking data. In *Proceedings of the 20th International Conference on Advances in Geographic Information Systems - SIGSPATIAL'12*; ACM Press: New York, NY, USA, 2012; p. 199.

28. Dong, L.; Li, Y.; Yin, H.; Le, H.; Rui, M.; Dong, L.; Li, Y.; Yin, H.; Le, H.; Rui, M. The Algorithm of Link Prediction on Social Network. *Math. Probl. Eng.* **2013**, *2013*, 1–7. [CrossRef]

29. Liben-Nowell, D.; Kleinberg, J. The Link Prediction Problem for Social Networks. *Proc. Twelfth Annu. ACM Int. Conf. Inf. Knowl. Manag.* **2003**, 556–559. [CrossRef]

30. Wu, J.; Hou, Y.; Jiao, Y.; Li, Y.; Li, X.; Jiao, L. Density shrinking algorithm for community detection with path based similarity. *Phys. A Stat. Mech. Appl.* **2015**, *433*, 218–228. [CrossRef]

31. Cheng, C.; Yang, H.; King, I.; Lyu, M.R. Fused matrix factorization with geographical and social influence in location-based social networks. In Proceedings of Twenty-Sixth AAAI Conference on Artificial Intelligence, Toronto, ON, Canada, 22–26 July 2012; pp. 17–23.

32. Ye, M.; Yin, P.; Lee, W.-C.; Lee, D.-L. Exploiting geographical influence for collaborative point-of-interest recommendation. In *Proceedings of the 34th International ACM SIGIR Conference on Research and Development in Information - SIGIR'11*; ACM Press: New York, NY, USA, 2011; p. 325.

33. Hu, L.; Sun, A.; Liu, Y. Your Neighbors Affect Your Ratings: On Geographical Neighborhood Influence to Rating Prediction. In *Proceedings of the 37th International ACM SIGIR Conference on Research & Development in Information Retrieval*; SIGIR '14; ACM: New York, NY, USA, 2014; pp. 345–354.

34. Rahimi, S.M.; Wang, X. Location Recommendation Based on Periodicity of Human Activities and Location Categories. In *Advances in Knowledge Discovery and Data Mining*; Springer: Berlin/ Heidelberg, Germany, 2013; pp. 377–389.

35. Zhou, D.; Rahimi, S.M.; Wang, X. Similarity-based probabilistic category-based location recommendation utilizing temporal and geographical influence. *Int. J. Data Sci. Anal.* **2016**, *1*, 111–121. [CrossRef]

36. Heith, M. T.; Raghavan, P. A Cartesian parallel nested dissection algorithm. *SIAM J. Matrix Anal. Appl.* **1992**, *19*, 235–253. [CrossRef]

37. Karypis, G.; Kumar, V. Multilevelk-way Partitioning Scheme for Irregular Graphs. *J. Parallel Distrib. Comput.* **1998**, *48*, 96–129. [CrossRef]

38. Kuhn, H.W. The Hungarian method for the assignment problem. *Nav. Res. Logist. Q.* **1955**, *2*, 83–97. [CrossRef]

39. Kuhn, H.W. Variants of the hungarian method for assignment problems. *Nav. Res. Logist. Q.* **1956**, *3*, 253–258. [CrossRef]

40. Berkhin, P. *Survey of Clustering Data Mining Techniques*; Technical Report; Accrue Software Inc.: San Jose, CA, USA, 2002.

41. Velmurugan, T.; Santhanam, T. Computational Complexity between K-Means and K-Medoids Clustering Algorithms for Normal and Uniform Distributions of Data Points. *J. Comput. Sci.* **2010**, *6*, 363–368. [CrossRef]

42. Wang, K.; Wang, B.; Peng, L. CVAP: Validation for Cluster Analyses. *Data Sci. J.* **2009**, *8*, 88–93. [CrossRef]

43. Dudoit, S.; Fridlyand, J. A prediction-based resampling method for estimating the number of clusters in a dataset. *Genome Biol.* **2002**, *3*, RESEARCH0036. [CrossRef]

44. Baarsch, J.; Celebi, M.E. Investigation of Internal validity measures for K-means clustering. In Proceedings of the International Multiconference of Engineers and Computer Scientists, Hong Kong, China, 14–16 March 2012; pp. 14–16.

45. Quattrone, G.; Capra, L.; De Meo, P. There's No Such Thing as the Perfect Map: Quantifying Bias in Spatial Crowd-sourcing Datasets. In Proceedings of the 18th ACM Conference on Computer Supported Cooperative Work & Social Computing, Vancouver, BC, Canada, 14–18 March 2015; pp. 1021–1032.

46. Zhang, J.; Sheng, V.S.; Li, Q.; Wu, J.; Wu, X. Consensus algorithms for biased labeling in crowdsourcing. *Inf. Sci.* **2017**, *382–383*, 254–273. [CrossRef]

47. Chakraborty, A.; Messias, J.; Benevenuto, F.; Ghosh, S.; Ganguly, N.; Gummadi, K.P. Who Makes Trends? Understanding Demographic Biases in Crowdsourced Recommendations. In Proceedings of the 11th AAAI International Conference on Web and Social Media (ICWSM 2017), Montreal, CA, USA, 15–18 May 2017.

International Journal of
Geo-Information

isprs

MDPI

Article

hackAIR: Towards Raising Awareness about Air Quality in Europe by Developing a Collective Online Platform

Evangelos Kosmidis [1,*], Panagiota Syropoulou [1], Stavros Tekes [1], Philipp Schneider [2],
Eleftherios Spyromitros-Xioufis [3], Marina Riga [3], Polychronis Charitidis [3],
Anastasia Moumtzidou [3], Symeon Papadopoulos [3], Stefanos Vrochidis [3], Ioannis Kompatsiaris [3],
Ilias Stavrakas [4], George Hloupis [4], Andronikos Loukidis [4], Konstantinos Kourtidis [5],
Aristeidis K. Georgoulias [5] and Georgia Alexandri [5]

[1] DRAXIS Environmental S.A., Mitropoleos 62, 54623 Thessaloniki, Greece; syropoulou.p@draxis.gr (P.S.);
    stavros@draxis.gr (S.T.)
[2] NILU-Norwegian Institute for Air Research, Instituttveien 18, 2007 Kjeller, Norway;
    Philipp.Schneider@nilu.no
[3] Information Technologies Institute—Centre for Research & Technology Hellas, 6th Km Charilaou-Thermi
    Road, 57001 Thermi, Greece; espyromi@iti.gr (E.S.-X.); mriga@iti.gr (M.R.); charitidis@iti.gr (P.C.);
    moumtzid@iti.gr (A.M.); papadop@iti.gr (S.P.); stefanos@iti.gr (S.V.); ikom@iti.gr (I.K.)
[4] Technological Educational Institute of Athens, Agiou Spiridonos 28, 12243 Egaleo, Greece;
    ilias@ee.teiath.gr (I.S.); hloupis@teiath.gr (G.H.); andronikos.loukidis@gmail.com (A.L.)
[5] Department of Environmental Engineering, School of Engineering, Democritus University of Thrace,
    Vasilissis Sofias 12, 67100 Xanthi, Greece; kourtidi@env.duth.gr (K.K.); argeor@env.duth.gr (A.K.G.);
    alexang@auth.gr (G.A.)
*   Correspondence: kosmidis@draxis.gr; Tel.: +30-23-1027-4566

Received: 28 March 2018; Accepted: 7 May 2018; Published: 12 May 2018

Abstract: Although air pollution is one of the most significant environmental factors posing a threat to human health worldwide, air quality data are scarce or not easily accessible in most European countries. The current work aims to develop a centralized air quality data hub that enables citizens to contribute to air quality monitoring. In this work, data from official air quality monitoring stations are combined with air pollution estimates from sky-depicting photos and from low-cost sensing devices that citizens build on their own so that citizens receive improved information about the quality of the air they breathe. Additionally, a data fusion algorithm merges air quality information from various sources to provide information in areas where no air quality measurements exist.

Keywords: air quality estimation; air pollution; citizen science; sky images; social media; data fusion

1. Introduction

At present, air pollution is one of the most significant factors posing a threat to health worldwide. According to the World Health Organization [1], ambient air pollution was responsible for the premature deaths of 3.7 million people under the age of 60 in 2016. Europe's most problematic air pollutant in terms of human health is particulate matter [2].

Particulate matter (PM) can have significant effects on human health including asthma, lung cancer, and cardiovascular issues. Particulate matter up to 10 micrometers in diameter (PM_{10}) is able to penetrate the bronchi, while particulate matter with diameter up to 2.5 micrometers ($PM_{2.5}$) can penetrate the lungs and enter the circulatory system [3]. The International Agency for Research on Cancer (IARC) concluded in 2013 that particulate matter is carcinogenic to humans [4]. PM is also harmful to the environment, and its effects include increased acidity of lakes and streams, nutrient balance changes in coastal waters and river basins, reduced levels of nutrients in soil, damage to forests and crops, reduced diversity in ecosystems, damage to stone and other materials, and reduced visibility (haze) [2].

Although there is a general consensus that air pollution is affecting human life and well-being worldwide, there is little awareness of the role that each one of us can play to mitigate this problem. Awareness primarily requires access to information, which should be widely available and easily understandable, as required by the Aarhus Convention (which provides for the right of everyone to receive environmental information that is held by public authorities), and by the Air Quality Directives (2004/107/EC and 2008/50/EC). However, although such information is available, it is generally not easily accessible by citizens. Common problems reported are [5]

- inadequate air quality monitoring networks in some areas, consisting of insufficient numbers and/or inappropriately located, old, and unreliable monitoring stations [5];
- the difficulty for citizens to interpret data published long after breaches of limit values have occurred in highly technical formats.

Even when people are aware of the issue, they usually do not associate their own individual behavior with these outcomes. However, the air is a public good [6]. Particularly for air pollution, it is collective, rather than individual action which is necessary to mitigate the problem, as the air we breathe is a resource common to everybody.

The hackAIR project (www.hackair.eu) aims to develop an air quality data hub by developing an open platform that enables communities of citizens to easily set up low-cost air quality monitoring networks and engage their members in measuring and publishing outdoor air pollution levels. By combining official data with air quality estimates from sky-depicting images and from sensing devices that citizens can build on their own, hackAIR provides citizens with improved and easily accessible information about localized air pollution levels. In order to also provide information in areas where no air quality monitoring stations exist, a data fusion algorithm for merging air quality information from various sources has been developed and air-quality-aware personalized services (e.g., outdoor activity recommendations) are provided to the public.

The general issue of raising awareness of the air pollution problem and its impacts on health has received considerable attention by several initiatives in the past. The official database of the European Environment Agency (www.eea.europa.eu/themes/air/air-quality-index/index), Air Pollution in World (http://aqicn.org/map/world), AirNow (www.airnow.gov) in the US and Canada, and the London Air Quality Network (www.londonair.org.uk) are some of the existing websites providing information on air pollution levels. In addition to independent initiatives, several projects funded by the European Commission, such as ObsAIRve (www.obsairveyourbusiness.eu), CITEAIR (www.citeair.eu) and PASODOBLE (http://cordis.europa.eu/project/rcn/94372_en.html), have dealt with the issue of air pollution. The majority of these solutions are available only in regions with existing monitoring stations because they require official measurements as input. To overcome this issue, hackAIR acquires air pollution data from various sources and offers tools to citizens to contribute their own measurements. Thus, it offers much richer information on air pollution levels—whether official monitoring stations are available or not—accompanied by activity recommendations adjusted to each user's personal profile. These recommendations enable citizens to identify areas with better air quality within the city and safeguard their health.

Citizen observatories for air pollution monitoring with sensors are increasingly viewed as an essential tool for better observing and understanding our environment [7]. CITI-SENSE (www.citi-sense.eu) and EveryAware (www.everyaware.eu) are two of the existing EU funded projects aiming to empower citizens to participate in air pollution monitoring, enhancing their awareness of the problem and promoting behavioral change. Under the same concept, many commercial solutions have also been developed. The Envi4All mobile application (http://envi4all.com/) provides real-time and forecast air quality data and enables users to report how they perceive the quality of the outdoor air at that specific moment. PlumeLabs (https://plumelabs.com/) also offers this information and gives the opportunity to citizens to track and report air pollution data with a PlumeLabs sensor. The Air Quality Egg (http://airqualityegg.com) promotes the creation of a sensing device that could be used by hobbyists and can monitor NO_2, CO, CO_2, SO_2, $PM_{2.5}$, and VOCs in the atmosphere. Several do-it-yourself initiatives have been also organized as part of recent efforts to democratize air quality monitoring. Indicatively, step-by-step instructions can be found for a wide variety of projects via ExploreInstructables.com, SparkFun.com, and PublicLab.org. The main limitation of such approaches is their failure to attract the interest of a critical mass of users that would provide them with a sufficient amount of air quality information. In order to overcome this limitation, hackAIR addresses the issue of data collection by including additional sources of data which do not necessarily require action by volunteers. These include the estimation of approximate air quality from sky-depicting images from social media and open data on air quality.

There have been various research efforts aiming at developing alternative, easy-to-produce, and inexpensive instruments for air pollution monitoring. Such an approach was introduced by Wang et al. [8] aiming to quantify the loading of black carbon on quartz fiber filters with digital photographic methods. Field results demonstrated that this approach provides measurements as precise and accurate as the ones acquired with expensive instruments, while it exhibits short analysis time and easy operation. Ramanathan et al. [9] introduced a similar approach, integrating an aerosol filter in a cellphone to collect filter images. The images are analyzed in real time to determine the current concentrations of black carbon.

From the different components of hackAIR, air pollution estimation from sky-depicting photos has been scarcely used in commercial offerings. This approach has been explored in previous research works. For example, Babari et al. [10] developed model-driven approaches to monitor and estimate atmospheric visibility distance and air pollution levels through the use of ordinary cameras. Another approach is iSPEX (http://ispex-eu.org/), a low-cost mass-producible optical add-on for smartphones with a corresponding app, which turns a smartphone camera into a spectropolarimeter [11]. However, the aforementioned approaches remained in the state of a prototype and did not achieve commercial realization nor acquisition of a critical mass of data.

2. The hackAIR Methodology

The hackAIR solution has the primary goal of enabling communities of citizens to easily set up air quality monitoring networks and to engage their members in measuring and publishing outdoor air pollution levels, leveraging the power of online social networks, mobile and open hardware technologies, and engagement strategies. hackAIR allows for the collection of data from publicly available sources (measurements from ground-based stations, open data, and sky-depicting images uploaded to social media) and contributions from the hackAIR community (sky-depicting photos taken with the hackAIR app and measurements from low-cost sensing devices). Besides the platform, a social media monitoring tool has been developed to enable the discovery of social media accounts that belong to users or organizations interested in air quality issues and thus help with user engagement in the hackAIR activities.

In the following, we provide an overview of the various methodological elements which, in combination, represent the hackAIR solution. In summary, hackAIR collects readily available air quality measurements from the web and publicly available images of the sky from social media and webcams. For these images, machine learning algorithms automatically detect and extract clear sky regions that can be used for air quality estimation. Then, an algorithm that estimates the levels of air pollution based on the color of the extracted sky regions is applied. In parallel, the hackAIR solution offers guidelines to users on how they can build their own low-cost sensing devices to monitor their local air pollution and contribute actual measurements. All the above data are combined with official air pollution measurements from ground-based stations and satellite observations and feed a data fusion system which offers estimates of air quality even in areas with no measurements. Finally, based on each user's preferences and sensitivities, hackAIR provides recommendations on how they can protect themselves from air pollution. All these services are offered through an integrated platform available as a web and mobile application that contributes to the creation of an improved knowledge base for air quality data in Europe and a change towards more proactive and environmentally friendly behavior.

2.1. Environmental Data Discovery and Indexing

Data discovery and collection in hackAIR focuses on two data types: (a) readily available air quality measurements and (b) publicly available images of the sky that can be used for image-based air quality estimation using the techniques described in Sections 2.2 and 2.3. As mentioned above, with respect to readily available air quality measurements, the collection includes both measurements from official air quality stations (e.g., established by governmental organizations) and measurements from low-cost sensor networks. With respect to publicly available images of the sky, two sources are explored: a) social media images and b) images from public webcams.

The collection of official air quality measurements in hackAIR relies on the OpenAQ (https://openaq.org) open data platform. OpenAQ aggregates and shares (via an open Application Programming Interface (API)) high-quality data about air quality from multiple official sources around the world (e.g., the European Environmental Agency) including more than 30 countries in Europe, which is currently the focus of the hackAIR framework. Importantly, the OpenAQ system checks each data source for updated information every 10 min, which guarantees that the data will be available from the platform almost immediately after they are published by the original data providers. In Europe, OpenAQ currently provides PM_{10} and $PM_{2.5}$ data from about 1800 and 800 locations, respectively. The hackAIR data collection framework retrieves up-to-date information from each location by regularly performing appropriate queries to the Representation State Transfer (REST) API provided by OpenAQ.

Besides measurements from official air quality stations, the hackAIR solution involves the collection of measurements from personal low-cost air quality stations established by citizens and promoted by a number of air quality initiatives, similar to hackAIR. One such initiative is luftdaten.info which currently comprises more than 4000 sensors in Europe. The sensors of the luftdaten.info network constitute an ideal data source for hackAIR as they measure both PM_{10} and $PM_{2.5}$ concentrations and their latest data are always available through an open API. The hackAIR data collection framework uses relevant APIs to retrieve up-to-date information from each sensor on an hourly basis.

As far as the collection of publicly available sky images is concerned, hackAIR explores the possibility of using publicly shared geotagged images from social media platforms. According to the KPCB Internet Trends Reports 2016 (www.kpcb.com/blog/2016-internet-trends-report), which provides an overview of the trends related to image sharing for 2005–2015, the most popular image sharing platforms are Snapchat, Facebook Messenger, Instagram, WhatsApp, and Facebook. Unfortunately, all these platforms either do not distribute user-contributed images through a free API, or pose very strict limitations that prohibit their practical usage. Therefore, hackAIR uses Flickr, the next most popular image sharing platform that also provides a free open API.

To collect images from Flickr, the hackAIR data collection framework implements a collector that periodically calls the appropriate Flickr API endpoints to retrieve the URLs, timestamps, and geolocations of all images captured within the last 24 h in Europe. This leads to the collection of about 5000 geotagged images on a daily basis.

In addition to Flickr images, the hackAIR data collection component incorporates images from public webcams. Webcams offer the advantage of a continuous image stream from fixed known locations. To this end, two large webcam repositories are considered—AMOS [12,13] and webcams.travel [14]. Combined, the two repositories provide access to about 3500 webcams in Europe: about 2500 come from AMOS and 1000 from webcams.travel (the latter source actually provides access to a significantly larger number of webcams (>20,000) but limitations of their free API prohibit the regular collection of images from more than 1000 webcams per day). In the case of AMOS, a customized web data extraction framework was developed, while in the case of webcams.travel, data is retrieved through a client application for the provided API. In both cases, data collection is performed 4 times per day during daytime.

Efficient indexing and storage of the information retrieved from all the previously described sources relies on a MongoDB instance. This offers efficient mechanisms for handling geographical data and performing geospatial queries.

2.2. Image Processing for Sky Detection

As soon as the Flickr and webcam images are retrieved, a number of automated image processing operations are launched in order to (a) determine whether sky is depicted in the image and (b) localize the clear (of non-sky elements, clouds, or humidity) sky area and compute the color statistics that are used as input to the air quality estimation method described in Section 2.3.

To automatically detect the presence of sky in an image, we build a visual concept detection model using Deep Convolutional Neural Networks (DCNNs). In particular, we finetune a state-of-the-art pretrained DCNN model (Inception-v3 [15]) by replacing the last layer of the network with a new layer trained on a manually annotated (for the sky concept) set of 2500 Flickr and webcam images. This model achieves an accuracy of 96.2% when evaluated on an independent test set.

In the next processing step, all images recognized as sky-depicting are further processed in order to determine the exact location of sky. To this end, two alternative sky localization approaches are considered: one that applies deep learning for image segmentation and one that is based on simple image processing heuristics.

The first approach that we apply is the Fully Convolutional Network (FCN) [16]. Building upon the recent advances in deep and transfer learning, FCN applies fully convolutional finetuning using whole-image inputs and per-pixel ground truth labels to adapt deep networks pretrained for image classification to image segmentation tasks. In hackAIR, we use a publicly available FCN model (https://github.com/shelhamer/fcn.berkeleyvision.org/tree/master/siftflow-fcn16s) (FCN-16) pretrained on the SIFT Flow dataset [17] (which includes annotations for the sky class) that was found to achieve an average pixel precision of 94.3%.

The second approach for sky localization consists of a set of heuristic rules that aim to identify pixels that meet certain criteria with respect to their color values and the color values of neighboring pixels. In rough terms (a more detailed description of this algorithm can be found in [18]), if R, G, and B denote the Red, Green, and Blue values of each pixel, sky pixels must satisfy the following three conditions:

$$\frac{R}{G} \in [0.5, 1], \frac{G}{B} \in [0.5, 1], and \frac{B}{R} > 1.25$$

and, moreover, occupy large contiguous areas in the upper part of the image.

To compare the two approaches, we first evaluated their performance on the SUN benchmark database [19] and found that the FCN approach performs significantly better in recognizing the sky regions, obtaining a pixel precision of 91.77% versus 82.45% for the heuristic approach. However, we noticed that the ground truth annotations of the SUN database (as well as those of the SIFT Flow database on which the FCN model is trained) are noisy, in the sense that the regions annotated as sky often contain non-sky areas (e.g., clouds, the sun, or small objects). The presence of such areas is problematic as it may result in noisy pixel color statistics and consequently compromise the validity of the air quality estimations that rely on these statistics.

Therefore, we conducted an additional evaluation of the two methods, tailored to assessing their performance on the more specific task of identifying sky regions that are suitable for air quality estimation (i.e., do not contain non-sky elements). In particular, we performed sky localization with each approach on 200 randomly selected Flickr and webcam images and manually assessed through inspection the validity of the sky regions detected by each approach. Note that in this type of evaluation, it is expected that both approaches will achieve significantly lower performance scores as even a small non-sky area within the detected sky region deems the whole region usable. Indeed, we found that the sky regions extracted by FCN are suitable in only 24.8% of the cases versus 47.9% for the heuristic approach.

A careful examination of the extracted regions suggests that the heuristic approach owes its superior performance to the extraction of more fine-grained regions and, therefore, more cases where the detected region consists only of sky pixels. More importantly, the visual examination also revealed that the two approaches can work in a complementary way. We found that the FCN approach is better at avoiding significant errors (e.g., sea or building windows recognized as sky), while the heuristic approach can successfully filter out small non-sky elements such as small objects and text overlays that are particularly common in webcam images. Motivated by this complementarity, we tested a hybrid sky localization approach that first extracts the sky region using the FCN approach and then refines it by applying the heuristic approach considering only the pixels recognized as sky by the FCN approach. This hybrid approach manages to extract suitable sky regions in 80.3% of the images and is the method of choice for the hackAIR solution.

The final processing step consists of computing the mean red-to-green (R/G) and green-to-blue (G/B) ratios from the parts of the images identified as sky and providing them as input to the air quality estimation method described in the following section. Since images commonly undergo various types of transformations by users (e.g., artistic filters, color enhancements) before being uploaded to social media platforms such as Flickr, we studied how such transformations affect the results of the image analysis (i.e., the calculated R/G and G/B ratios). To this end, we selected a set of 87 sky-depicting (untransformed) Flickr images and applied 24 popular image transformations (resizing, color level effects, artistic filters, etc.) on each image using the image manipulation API of Cloudinary (http://cloudinary.com). Original images and their transformed versions were then processed independently for sky detection and localization, and R/G and G/B ratios were extracted from all images with a usable sky region. By measuring the Pearson correlation between the R/G and G/B ratios of original and transformed images, we found that both ratios are very robust against most transformations (Pearson correlation > 0.9 for 18 out of 24 transformations), with only 2 very intense transformations (brightness increase $\geq$ 50%, red-rock artistic filter) causing a significant distortion on the calculated ratios and 2 transformations rendering all images unusable for air quality estimation. These results (the interested reader is referred to [20] for more details on the experimental setup and the detailed results of this analysis) suggest that image transformations are expected to have a negligible impact on image-based air quality estimations and this pertains only to a small fraction of the images used by the hackAIR framework since webcam images and images captured with the hackAIR app are typically untransformed.

2.3. Estimation of Air Quality from User-Generated Sky Photos

Atmospheric visibility is a very useful indicator of the so-called aerosol optical depth (AOD: a measure of the extinction of radiation due to scattering and absorption by aerosols) and, subsequently, of air pollution resulting from suspended particulates, especially in drier climates [21,22]. Passive remote sensing instruments (e.g., sunphotometers, spectrophotometers) retrieve aerosol optical properties, such as AOD, by measuring the incident radiation on the ground at specific wavelengths. To assign the light intensities measured by the instruments to AOD values, a Look-Up Table (LUT) approach is usually followed. LUTs are produced with the use of a radiative transfer model (RTM). RTMs allow for the calculation of the intensity of the light transferred in the atmosphere under different user-specified scenarios. These scenarios may include information about the position of sun relative to Earth (solar zenith angle) and atmospheric parameters which are related to clouds, aerosols, water vapor, ozone, surface albedo, etc. This way, one knows what light intensity should be expected under specific atmospheric conditions. Comparison of the measured spectral light intensities with those from a LUT allows for retrieval of the AOD.

The color (Red–Green–Blue) of the sky is partly determined by atmospheric aerosols (amount and type). Several scientific efforts around the world to retrieve atmospheric aerosol-related properties from images taken from digital cameras and paintings [21] have returned promising results so far. Previous employments of similar methods, i.e., using digital cameras, have shown that the comparison of AOD derived from sky images with those retrieved with sunphotometers operated side by side showed differences similar to the nominal error claimed in the AOD sunphotometer networks [23]. A relevant field experiment showed that AOD from sky images is highly correlated with AOD from sunphotometers, with a correlation coefficient of 0.95 and an average retrieval error of around 7% [24]. Furthermore, sky radiances obtained by digital cameras were compared with CIMEL sunphotometer radiances, finding mean absolute differences between 2% and 15% except for pixels near the sun and high scattering angles [25]. The method used in hackAIR is based on the comparison of the R/G and G/B ratios from images and precalculated LUTs to retrieve AOD [18].

2.4. Design of Guidelines for Low-Cost Air Quality Sensing Devices

In order to increase air pollution awareness and attract open software and hardware communities to provide added value to the hackAIR project, low-cost sensing devices were developed. The sensing devices consist of three main elements: a sensor, a processing unit, and a communication module. Sensors selected for the current project are based on the optical determination of particles by means of a light scattering method. A typical sensor includes a light source (IR LED or diode laser), a photo-sensing device (photodiode or phototransistor), and a focusing lens. The particles pass through the light beam, and scatter and absorb the light. The detected light intensity is directly correlated with the concentration of particles. Air flow is ensured by forced flow using mini blowers or heated elements. Two classes of sensors were tested and selected: LED based (Shinyei PPD42NS) and Laser based (DFRobot SEN0177, Inonafit SDS011, and Plantower PMS5003).

The communication capabilities of the proposed devices dictate the selection of appropriate processing units that usually fall into two categories: For distributed networking solutions (Ethernet, WiFi), we adopt the Arduino ecosystem in typical implementations (Arduino UNO with Ethernet shield or with a dedicated hackAIR WiFi shield, Wemos, and compatible NodeMCU implementations). For personal networking, we adopt the Bluetooth low-energy (BLE) protocol, which is implemented by means of Cypress® Programmable System-on-Chip (PSoc)/Programmed Random occurrence (PRoc) modules. Arduino and Wemos solutions are designed to be placed in closed cases at predefined places while their locations are provided during the registration at the hackAIR portal. The PSoC/PRoC sensing device is designed to be portable and implemented using the BLE protocol in beacon format packets while additional information like the geolocation and user credentials are attached in the submitted network packet from the mobile device that is responsible for sending the data over the internet to the hackAIR portal. The portability of the proposed devices is further extended since the power supply can be provided by portable power banks.

Recently, temperature and humidity measurement capability was included on the designed hardware and software. Specifically, DHT11 and DHT22 sensor boards were used in order to increase the measurement reliability, and software code was added in the provided libraries accordingly. Detailed hardware descriptions (i.e., schematics and printed circuit boards (PCBs)) of all the above sensors and the corresponding software (open access codes and libraries) are available on the project's GitHub site (https://hackair-project.github.io/hackAir-Arduino/general/, https://github.com/hackair-project/hackAIR-PSoC/wiki).

All designs incorporate power saving algorithms and the sensors are shut down during the time intervals between measurements in order to reduce power consumption and increase the expected lifetime of the sensors' laser system.

Furthermore, an air quality characterization sensor made of Commercial Off-the-Shelf (COTS) materials based on well-established image processing and computer vision techniques aiming to provide qualitative PM concentrations was developed. The sensor comprises a test surface where the

PMs are collected. For this purpose, a 5 cm square piece of the aluminium side of a Tetra Pak food packaging carton is used. Along the diagonal axis of the user's choice, two small dots are made using a 0.7 mm mechanical pencil, near the center of the test surface.

Subsequently, a thin layer of petroleum jelly is applied in order to trap the PMs. The hackAIR cardboard sensor is then exposed outdoors for 24 h. Afterwards, the cardboard sensor is retrieved and a set of five images of the area of interest is captured using a cell phone camera and a macro lens of at least ×12 magnification. For the characterization of the air quality, the hackAIR cardboard sensor implements an algorithm that is described in the flowchart depicted in Figure 1. To estimate the number of blobs which correspond to PM_{10} or greater air particles trapped on the petroleum jelly layer of the test surface, well-established image processing and computer vision techniques are run. Such algorithms involve Otsu's method for thresholding [26], Contrast Limited Adaptive Histogram Equalization—CLAHE [27], and Moore's tracing algorithm for extracting.

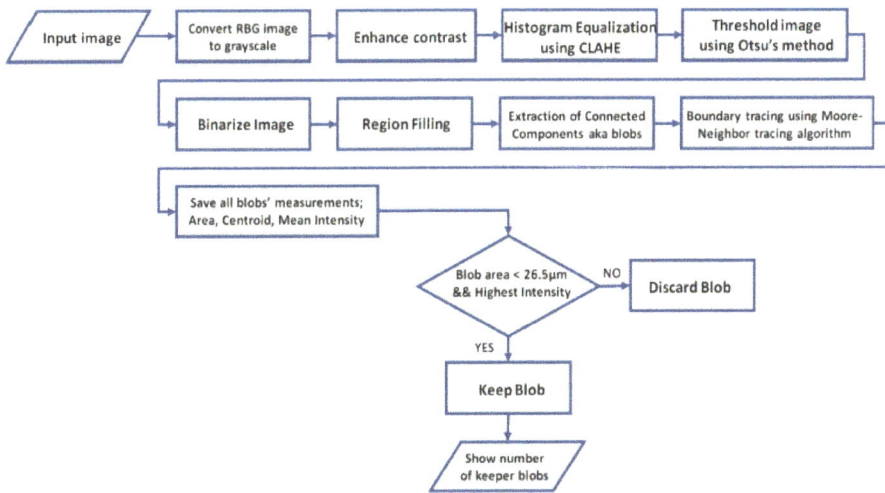

Figure 1. Particulate matter (PM) concentration estimation algorithm flowchart.

2.5. Data Fusion Tools

In order to add value to the hackAIR observations and to provide the platform users with estimates of the air quality at any given location, even if no measurements were made there, a data fusion system was developed. This system has the primary objective of interpolating the point-based observations in space such that air quality estimates are available at any point within the domain. Data fusion is a subset of data assimilation [28] and when used with observations and a model, it allows for spatially interpolating point observations in a mathematically objective way while at the same time constraining the model. Data fusion has been successfully applied in such a fashion for real-time urban-scale air quality mapping by combining observations from a low-cost sensor network [29] with information from a high-resolution local-scale dispersion model [30,31]. We use data fusion here primarily with the goal of spatially interpolating the observations, which are often subject to significant uncertainty, in a meaningful way. As such, the model is used as auxiliary information to guide the interpolation in areas where no observations are available.

The input data to the data fusion system are hereby twofold: the observational data is primarily composed of the measurements made by the participants, e.g., aerosol optical depth estimates from sky-depicting images and particulate matter observations using optical particle counters mounted in open platforms. As model information, we use the operationally available data from the Copernicus

Atmosphere Monitoring System (CAMS). More specifically, we use the daily ensemble forecast of the CAMS regional modelling system, representing the median of the ensemble of seven participating models [32].

The data fusion system used here is conceptually similar to the one described in [30] and is based on geostatistics [33–35]. More specifically, the technique uses universal kriging for interpolating the observations in space using model information as a spatial proxy to guide the interpolation. The interpolation is carried out at a country level with a spatial resolution of 5 km by 5 km. Currently, the focus lies on the two study sites—Germany and Norway—but the method can be readily extended to larger regions. The resulting fused maps are updated once every hour using observations and model information averaged over a period which can be adjusted based on the number of available observations. One of the major advantages of using a geostatistical framework as a data fusion technique is that the underlying kriging methods by default provide pixel-level uncertainty estimates. This can be important for propagating sensor-level uncertainties all the way to the end users. Given the availability of urban-scale modelling output with spatial resolutions of 100 m × 100 m and less, the methodology can also be applied for street-level maps of air quality [30,31] and experiments along these lines with the observations collected as part of hackAIR are planned for the second part of the project.

2.6. Personalized Recommendations Based on Environmental Data

Several initiatives, including projects (PESCaDO (http://pescado-project.upf.edu/) [36]) and applications (AirForU (http://newsroom.ucla.edu/releases/new-app-lets-you-check-air-quality-as-easily-as-checking-the-weather), Clean Air Nation (https://play.google.com/store/apps/details?id=io.gonative.android.robzl&hl=en), Air Visual (https://airvisual.com/app), Breezometer (https://breezometer.com), etc.), demonstrate the added value of up-to-date, spatiotemporally defined air-quality-related information and recommendation provision [37]. However, the above applications produce recommendations that generally apply to sensitive people, without any specialization to specific individuals' needs. Our aim is to implement a user-driven decision support (DS) framework that takes into account variations in people's air-quality-related vulnerabilities, as well as the environmental dynamics and relations, and produces recommendations, either in the form of general tips on how to reduce their ecological footprint or of personalized advice on how individuals may respond to existing atmospheric conditions, upon request for decision support.

The preferences that each user reported in the hackAIR system are classified according to age, cardiovascular or respiratory diseases (e.g., asthma), the performance of outdoor activities that cause breathing extension (e.g., jogging), and some other states (e.g., pregnancy, working outdoors). For each of the above user types and air pollution levels, specific messages with fixed content were defined by the hackAIR environmental experts based on a review of the literature on individuals' susceptibility. Therefore, the problem of providing personalized inferences is transformed into a classification task where different recommendations are given to the users according to a specific set of classes in the ontology that they belong to.

For the common representation and orchestration of heterogeneous information (environmental-, health-, user-profile-related data) and their existing relations, as those were defined by environmental and health experts of the project, we make use of ontologies—the state-of-the-art Semantic Web technology for structuring and semantically integrating data [38]. Ontologies have been extensively adopted in separate parts of the decision-making process [39–41], mostly for the formal representation of the domain of interest. On the contrary, our developed framework demonstrates the extensive use of ontologies and of relevant reasoning technologies for handling both the static (representation) and dynamic (realization, inference) processes of a DS system.

More specifically, we implemented a multilayered, ontology-based DS framework that comprises the following components [42] (Figure 2):

- The knowledge base (KB)—A set of interconnected ontologies which define the abstract schema (TBox layer) and actual assertions (ABox layer). It serves as a formulation and storage module of involved data, represented by the following main concepts: Person, Health_Sensitivity, Activity, Location, Environmental_Data, Request, and Recommendation, and their related subclasses and relationships.
- The recommendation module (rules layer)—a rule-based inference mechanism, implemented by fully exploiting the SPIN (SPARQL Inferencing Notation) -rules framework [43] (http://spinrdf.org/). It runs on top of the above ontological definitions, so as to interpret the existing data, produce new knowledge, and thus infer appropriate recommendations for the users, with respect to the existing air quality conditions and specific profile characteristics (age, health status, preferred outdoor activities) which implicitly define the different levels of individuals' vulnerability under severe atmospheric conditions, according to the project's experts.

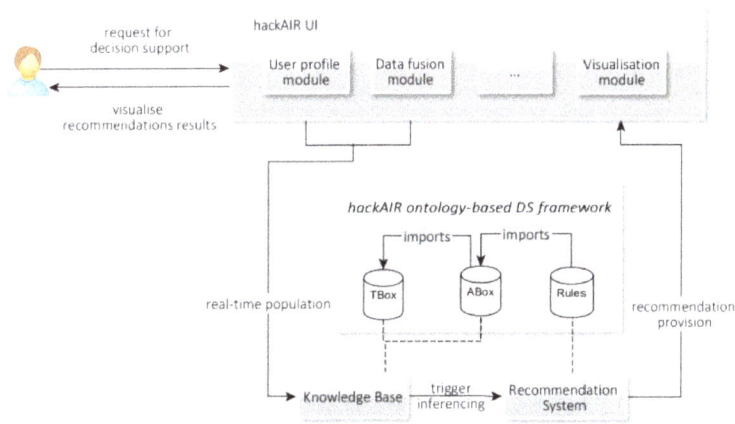

Figure 2. User-triggered processing pipeline for decision support (DS).

Representative scenarios defined within the hackAIR project, with the support of the project partners and environmental experts, are analyzed in depth in project deliverable D4.2 [44]. Here, for demonstration purposes, we consider the following hackAIR users: (1) a pregnant woman with respiratory diseases who likes to take long walks; and (2) a 65-year-old person who daily performs some outdoor work. Both live in an area supported by the hackAIR platform and request personalized recommendations through the hackAIR app. The KB receives via the implemented web service each request and stores the involved data (user profile details and fused air quality data) in the form of triples, conforming to the developed ontology schema. Appropriate ontology rules are triggered automatically, classifying (i) users in relevant basic or complex user profile classes; e.g., *User1 as Pregnant_and_SensitiveHealth_Person and Pregnant_and_SportsWalking_Person, and User2 as Elderly_and_Outdoor Job_Person*; and (ii) air quality measurements in relevant Air Quality Index (AQI) values. The above inferences feed the second level of ontology rules responsible for the proper matching between user profiles and defined recommendations. Thus, each user receives one or more recommendations according to his/her classification results and also to existing air quality conditions; e.g., User1 receives the suggestion: "You should go for a walk in an area with cleaner air.", while User2 receives: "You should consider with your supervisor (if any) the possibility of changing your working environment for today" when atmospheric pollution is extremely high.

The proposed ontology-based implementation facilitates the extensibility (enriching the KB with new concepts, rules, and recommendation messages per case), modularity (customizing existing schema and rules), and adoptability of the system (well-established, formal representation of data, seamless communication with third-party modules).

2.7. Development of the Integrated HackAIR Platform

The hackAIR platform follows a Service-Oriented Architecture (SOA) for the integration of all platform components. The principles of service orientation are independent of any vendor, product, or technology. Services are unassociated, loosely coupled units of functionality that are self-contained and each service implements at least one action. This makes it possible to reuse the code in different ways throughout the application by changing only the way an individual service interoperates with other services that make up the application, versus making code changes to the service itself. Furthermore, the hackAIR Platform implements a REST Web Services design approach to sustain integration among subcomponents and other integrations like mobile devices.

The hackAIR platform's architecture is composed of the following three layers: Data, Business Logic, and Application. The communication between these layers is bi-directional and always passes through the business logic layer. The communication between the Application and the business logic layers is handled through a RESTful API presented in Figure 3, where AQ represents air quality and COTS the hackAIR Commercial Off-the-Shelf solution.

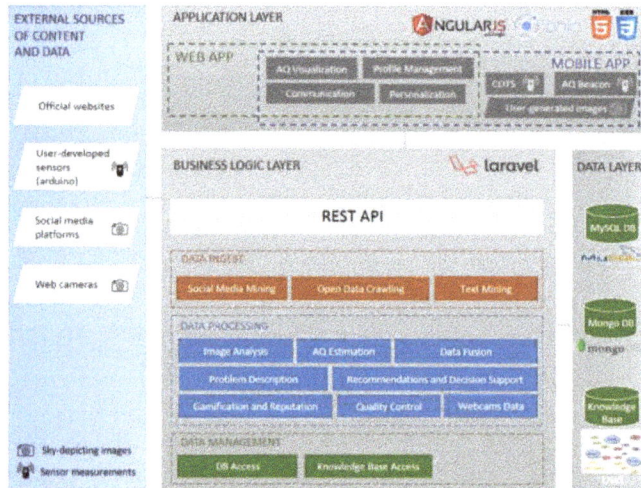

Figure 3. The hackAIR platform's architecture.

Data layer

The data layer includes the data persistence mechanisms that are responsible for storing and retrieving the data used by the hackAIR platform. It comprises a database for storing the basic hackAIR data entities and a Knowledge Base for storing semantically enriched information.

Business Logic layer

The business logic layer handles the requests coming from the application layer by applying the business logic rules and replying securely with proper content to the client applications (web and mobile applications). In addition to that, it receives measurements taken by the Arduino devices through the API and routes them to the appropriate module. Finally, the business logic layer communicates with external components including official websites, social media platforms, and webcams to retrieve air quality data.

The business logic layer includes various modules of the hackAIR platform, which all fall under the following three categories:

- Data ingest (responsible for acquiring data from various sources);
- Data processing (responsible for information transformation, processing, and implementation of business rules),
- Data access (responsible for communicating with the hackAIR platform's database systems).

Application layer

The application layer hosts the user's interaction interface available for access via the web and mobile and also via a community portal. The applications provided through this layer were developed using the Angular JS framework, HTML5, and CSS3. The mobile application was built and wrapped using the Ionic framework in order to provide all features required for the Android and iOS platforms, offering full end-user access to all hackAIR services. Users not only consume data, but can also provide data like personal information and air quality estimations (through photos and air quality measurements coming from hardware devices).

The interface of the platform was designed by User Interface/User Experience (UI/UX) experts to ensure the best experience of users (Figures 4 and 5).

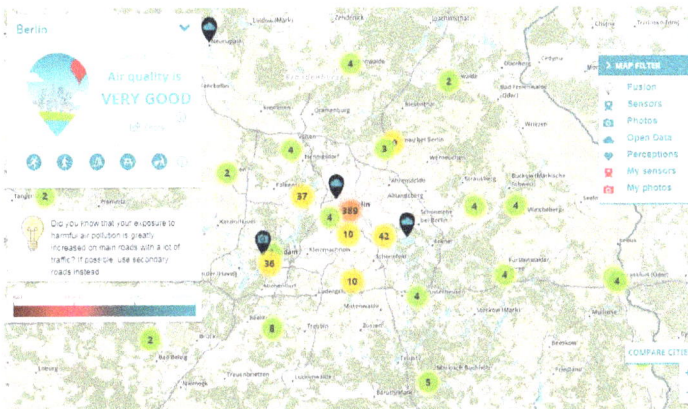

Figure 4. The map interface of the hackAIR web platform.

Figure 5. The map interface of the hackAIR mobile platform.

Besides the integrated hackAIR platform which implements all the core hackAIR technologies, a standalone social media monitoring tool has also been developed within hackAIR with the aim of increasing the outreach and adoption of the platform. This tool aims to support the discovery of social media accounts that belong to users or organizations that are interested in air quality issues and could therefore be approached and engaged.

The tool builds upon and extends an open-source software framework (https://github.com/ MKLab-ITI/mmdemo-dockerized) [45] for monitoring, analysis, and search over multiple social media platforms. The framework offers powerful tools for (a) tracking social media discussions around keywords and user accounts of interest; (b) obtaining real-time analytics over the tracked keywords/accounts; (c) identifying new accounts; and (d) analyzing the structure of communities. Within hackAIR, two important extensions of the framework are implemented to cover the needs of the project with respect to (a) account discovery and (b) audience analysis.

The first extension is the development of a sophisticated methodology for discovering hackAIR-relevant accounts on different social media channels. The methodology is based on the collection of a large initial set of candidate user accounts from different social media platforms, and then on the application of automatic classifiers to separate between relevant and irrelevant accounts. The collection of candidate accounts is based on querying the respective platform APIs with a list of air-quality-related keywords, while additional candidate accounts are discovered with the help of an iterative exploration process based on Twitter lists and by monitoring air-quality-related streams of tweets. The classification of accounts into relevant and irrelevant is based on classification models trained on text features extracted from the accounts' profile metadata (description) and their posts.

The second extension concerns the development of a solid methodology for analyzing the social media audience of hackAIR. Moving beyond the existing analytics tools of social media platforms, this methodology leverages social network analysis methods to create a more structured view of the hackAIR social media audience. Specifically, the network of hackAIR followers and their followers is analyzed with the help of a community detection algorithm to extract groups of Twitter accounts that are more densely connected to each other, and therefore correspond to topically and geographically focused communities. This result combined with the analysis of network connectivity of Twitter accounts leads to the discovery of influencers, i.e., accounts that are followed by many other accounts and that have an important role in their community.

Approaching and engaging with such accounts is expected to contribute significantly in increasing the outreach and adoption of hackAIR.

3. Conclusions

The project describes an attempt to address the need of communities to engage their members in outdoor air quality monitoring and tackle the challenge of using citizen science data. The potential of using user-generated air quality data for enhancing the available datasets and acquiring information in areas where no official air quality monitoring stations exist is a trend that receives increasing acceptance.

The described work and the various systems that have been developed and tested provide the basis for a new approach in air quality monitoring and health impact assessment, while at the same time the developed platform gives a great opportunity for similar approaches to be "applied" on the same principles and reasoning. The following results show at a glance that air quality can be monitored and presented with a usable manner with alternative means.

- Low-cost laser sensors are a reliable alternative that can provide the air pollution trends in an area. Even tests with the Commercial Off-the-Shelf solution show promising findings.
- Sky-depicting images can also successfully represent the current air quality scale.

Different users can depict different pieces of information at no additional effort while groups of interest can intuitively be motivated to offer their support and time to the development of new sensors, thus increasing their engagement and their overall awareness. Finally, hackAIR paves the way for stakeholders from different scientific domains and business sectors to work together for better air quality monitoring.

Author Contributions: E.K. and P.S. coordinated the current work and the writing of this paper, and they were also responsible for the background research and the personalization component. S.T. was the technical leader of the hackAIR platform development. P.S. designed and developed the hackAIR data fusion model. E.S.-X., P.C., S.P., S.V. and I.K. developed the environmental data discovery and indexing and the image processing for sky detection methodologies, while M.R. and A.M. designed the ontological framework and developed the decision support module for the provision of personalized recommendations. I.S., G.H. and A.L. designed and validated the hackAIR low-cost sensing devices. K.K., A.K.G., and G.A. developed and tested the algorithm for air quality estimation from user-generated sky photos.

Acknowledgments: The work was carried out under the framework of the hackAIR project (GA 688363) supported through the EU programme on "Collective Awareness Platforms for Sustainability and Social Innovation" and funded through the Horizon 2020 Research and Innovation Programme until December 2018.

Conflicts of Interest: The authors declare no conflict of interest.

References

1. World Health Organization. *Review of Evidence on Health Aspects of Air Pollution—REVIHAAP Project*; Technical Report; World Health Organization, Regional Office for Europe: Copenhagen, Denmark, 2013.

2. EEA. *Air Quality in Europe—2017 Report*; EEA Report No 13/2017; European Environment Agency: Copenhagen, Denmark, 2018.

3. Pérez, L.; Medina-Ramón, M.; Künzli, N.; Alastuey, A.; Pey, J.; Pérez, N.; García, R.; Tobías, A.; Querol, X.; Sunyer, J. Size fractionate particulate matter, vehicle traffic, and case-specific daily mortality in Barcelona (Spain). *Environ. Sci. Technol.* **2009**, *43*, 4707–4714. [CrossRef] [PubMed]

4. International Agency for Research on Cancer. *Outdoor Air Pollution a Leading Environmental Cause of Cancer Deaths*; Press Release No 221; WHO: Copenhagen, Denmark, 2013.

5. Andrews, A. *The Clean Air Handbook: A Practical Guide to EU Air Quality Law*; Clean Air Project (Co-Financed by LIFE+); ClientEarth: London, UK, 2015.

6. Biscaye, P.; Clarke, J.; Fowle, M.; Anderson, C.L.; Reynolds, T. *Global Public Goods*; Evans School Policy Analysis and Research, Brief #325; University of Washington: Seattle, WA, USA, 2016.

7. Liu, H.; Kobernus, M.; Broday, D.; Bartonova, A. A conceptual approach to a citizens' observatory—Supporting community-based environmental governance. *Environ. Health* **2014**, *13*, 107. [CrossRef] [PubMed]

8. Du, K.; Wang, Y.; Chen, B.; Wang, K.; Chen, J.; Zhang, F. Digital photographic method to quantify black carbon in ambient aerosols. *Atmos. Environ.* **2011**, *45*, 7113–7120. [CrossRef]

9. Ramanathan, N.; Lukac, M.; Ahmed, T.; Kar, A.; Siva, P.; Honles, T.; Leong, I.; Rehman, I.H.; Schauer, J.; Ramanathan, V. A cellphone based system for large scale monitoring of black carbon. *Atmos. Environ.* **2011**, *45*, 4481–4487. [CrossRef]

10. Babari, R.; Hautiere, N.; Dumont, E.; Paparoditis, N.; Misener, J. Visibility monitoring using conventional roadside cameras—Emerging applications. *Transp. Res. Part C Emerg. Technol.* **2012**, *22*, 17–28. [CrossRef]

11. Snik, F.; Rietjens, J.H.H.; Apituley, A.; Volten, H.; Mijling, B.; Di Noia, A.; Heikamp, S.; Heinsbroek, R.C.; Hasekamp, O.P.; Smit, J.M.; et al. Mapping atmospheric aerosols with a citizen science network of smartphone spectropolarimeters. *GRL* **2010**, *41*, 7351–7358. [CrossRef]

12. AMOS. Available online: http://amos.cse.wustl.edu (accessed on 1 May 2018).

13. Jacobs, N.; Roman, N.; Pless, R. Consistent Temporal Variations in Many Outdoor Scenes. In Proceedings of the IEEE Conference on Computer Vision and Pattern Recognition (CVPR), Minneapolis, MN, USA, 17–22 June 2007; pp. 1–6. [CrossRef]

14. webcams.travel. Available online: https://www.webcams.travel (accessed on 1 May 2018).

15. Szegedy, C.; Vanhoucke, V.; Ioffe, S.; Shlens, J.; Wojna, Z. Rethinking the inception architecture for computer vision. In Proceedings of the IEEE Conference on Computer Vision and Pattern Recognition, Las Vegas, NV, USA, 27–30 June 2016; Curran Associates Inc.: New York, NY, USA, 2016.

16. Long, J.; Shelhamer, E.; Darrell, T. Fully convolutional networks for semantic segmentation. In Proceedings of the IEEE Conference on Computer Vision and Pattern Recognition, Boston, MA, USA, 7–12 June 2015; Curran Associates Inc.: New York, NY, USA, 2015.

17. Liu, C.; Yuen, J.; Torralba, A. Sift flow: Dense correspondence across scenes and its applications. *IEEE Trans. Pattern Anal. Mach. Intell.* **2011**, *33*, 978–994. [CrossRef] [PubMed]

18. Spyromitros-Xioufis, E.; Moumtzidou, A.; Papadopoulos, S.; Vrochidis, S.; Kompatsiaris, Y.; Georgoulias, A.K.; Alexandri, G.; Kourtidis, K. Towards improved air quality monitoring using publicly available sky images. *Multimedia Technol. Environ. Biodivers. Inform.* **2018**. accepted for publication.

19. Jianxiong, X.; Hays, J.; Ehinger, K.A.; Oliva, A.; Torralba, A. Sun database: Large-scale scene recognition from abbey to zoo. In Proceedings of the 20th Conference on Computer Vision and Pattern Recognition (CVPR), Istanbul, Turkey, 23–26 August 2010; Curran Associates Inc.: New York, NY, USA, 2010.

20. Spyromitros-Xioufis, E.; Papadopolos, S.; Moumtzidou, A.; Vrochidis, S.; Kompatsiaris, Y. hackAIR Deliverable D3: 2nd Environmental Node Discovery Indexing and Data Acquisition. 2017. Available online: https://www.researchgate.net/publication/324594192_hackAIR_deliverable_D32_2nd_environmental_node_discovery_indexing_and_data_acquisition (accessed on 1 May 2018).

21. Zerefos, C.S.; Tetsis, P.; Kazantzidis, A.; Amiridis, V.; Zerefos, S.C.; Luterbacher, J.; Eleftheratos, K.; Gerasopoulos, E.; Kazadzis, S.; Papayannis, A. Further evidence of important environmental information content in red-to-green ratios as depicted in paintings by great masters. *Atmos. Chem. Phys.* **2014**, *14*, 2987–3015. [CrossRef]

22. Riffler, M.; Schneider, C.; Popp, C.; Wunderle, S. Deriving atmospheric visibility from satellite retrieved aerosol optical depth. In Proceedings of the EGU General Assembly 2009, Vienna, Austria, 19–24 April 2009. Available online: http://meetings.copernicus.org/egu2009 (accessed on 1 May 2018).

23. Olmo, F.J.; Cazorla, A.; Alados-Arboledas, L.; López-Álvarez, M.A.; Hernández-Andrés, J.; Romero, J. Retrieval of the optical depth using an all-sky CCD camera. *Appl. Opt.* **2008**, *47*, H182–H189. [CrossRef] [PubMed]

24. Huo, J.; Lü, D. Preliminary retrieval of aerosol optical depth from all-sky images. *Adv. Atmos. Sci.* **2010**, *27*, 421–426. [CrossRef]

25. Roman, R.; Anton, M.; Cazorla, A.; de Miguel, A.; Olmo, F.J.; Bilbao, J.; Alados-Arboledas, L. Calibration of an all-sky camera for obtaining sky radiance at three wavelengths. *Atmos. Meas. Tech.* **2012**, *5*, 2013–2024. [CrossRef]

26. Senthilkumaran, N.; Vaithegi, S. Image segmentation by using thresholding techniques for medical images. *Comput. Sci. Eng.* **2016**, *6*, 1–13. [CrossRef]

27. Pisano, E.D.; Zong, S.; Hemminger, B.M.; DeLuca, M.; Johnston, R.E.; Muller, K.; Braeuning, M.P.; Pizer, S.M. Contrast limited adaptive histogram equalization image processing to improve the detection of simulated speculations in dense mammograms. *J. Digit. Imaging* **1988**, *11*, 193–200. [CrossRef]

28. Lahoz, W.A.; Schneider, P. Data assimilation: Making sense of Earth Observation. *Front. Environ. Sci.* **2014**, *2*, 1–28. [CrossRef]

29. Castell, N.; Schneider, P.; Grossberndt, S.; Fredriksen, M.F.; Sousa-Santos, G.; Vogt, M.; Bartonova, A. Localized real-time information on outdoor air quality at kindergartens in Oslo, Norway using low-cost sensor nodes. *Environ. Res.* **2017**. [CrossRef] [PubMed]

30. Schneider, P.; Castell, N.; Vogt, M.; Dauge, F.R.; Lahoz, W.A.; Bartonova, A. Mapping urban air quality in near real-time using observations from low-cost sensors and model information. *Environ. Int.* **2017**, *106*, 234–247. [CrossRef] [PubMed]

31. Schneider, P.; Castell, N.; Dauge, F.R.; Vogt, M.; Lahoz, W.A.; Bartonova, A. A Network of Low-Cost Air Quality Sensors and Its Use for Mapping Urban Air Quality. In *Mobile Information Systems Leveraging Volunteered Geographic Information for Earth Observation*; Bordogna, G., Carrara, P., Eds.; Springer International Publishing: Basel, Switzerland, 2018; pp. 93–110. [CrossRef]

32. Marécal, V.; Peuch, V.H.; Andersson, C.; Andersson, S.; Arteta, J.; Beekmann, M.; Ung, A. A regional air quality forecasting system over Europe: The MACC-II daily ensemble production. *Geosci. Model Dev.* **2015**, *8*, 2777–2813. [CrossRef]

33. Chilès, J.-P.; Delfiner, P. *Geostatistics: Modeling Spatial Uncertainty*, 2nd ed.; John Wiley & Sons: Hoboken, NJ, USA, 2012; ISBN-10 0470183152.

34. Goovaerts, P. *Geostatistics for Natural Resources Evaluation*, 1st ed.; Oxford University Press: New York, NY, USA, 1997; ISBN-10 0195115384.

35. Isaaks, E.H.; Srivastava, R.M. *An Introduction to Applied Geostatistics*, 1st ed.; Oxford University Press: New York, NY, USA, 1989; ISBN-10 0195050134.

36.   Wanner, L.; Rospocher, M.; Vrochidis, S.; Johansson, L.; Bouayad-Agha, N.; Casamayor, G.; Karppinen, A.; Kompatsiaris, I.; Mille, S.; Moumtzidou, A.; et al. Ontology-centered environmental information delivery for personalised decision support. *Experts Syst. Appl.* **2015**, *42*, 5032–5046. [CrossRef]

37.   Moumtzidou, A.; Papadopoulos, S.; Vrochidis, S.; Kompatsiaris, I.; Kourtidis, K.; Hloupis, G.; Stavrakas, I.; Papachristopoulou, K.; Keratidis, C. Towards Air Quality Estimation Using Collected Multimodal Environmental Data. In Proceedings of the IFIN 2016 and First International Workshop on Internet and Social Media for Environmental Monitoring (ISEM 2016), Florence, Italy, 12 September 2016; Volume 10078, pp. 147–156. [CrossRef]

38.   Gruber, T.R. A translation approach to portable ontology specifications. *Knowl. Acquis. J.* **1993**, *5*, 199–220. [CrossRef]

39.   Rospocher, M.; Serafini, L. An Ontological Framework for Decision Support. *JIST* **2012**, *7774*, 239–254. [CrossRef]

40.   Kontopoulos, E.; Martinopoulos, G.; Lazarou, D.; Bassiliades, N. An ontology-based decision support tool for optimising domestic solar hot water system selection. *J. Clean. Prod.* **2016**, *112*, 4636–4646. [CrossRef]

41.   Wetz, P.; Trinh, T.D.; Do, B.L.; Anjomshoaa, A.; Kiesling, E.; Tjoa, A.M. Towards an Environmental Information System for Semantic Stream Data. In Proceedings of the 28th EnviroInfo Conference, Oldenburg, Germany, 10–12 September 2014; Gomez, M.J., Sonnenschein, M., Vogel, U., Winter, A., Rapp, B., Giesen, N., Eds.; BIS-Verlag: Oldenburg, Germany, 2014; pp. 637–644.

42.   Riga, M.; Kontopoulos, E.; Karatzas, K.; Vrochidis, S.; Kompatsiaris, I. An Ontology-based Decision Support Framework for Personalized Quality of Life Recommendations. In Proceedings of the 4th International Conference on Decision Support System Technology, (ICDSST 2018), LNBIP 313, Heraklion, Greece, 22–25 May 2018; pp. 1–14, (to appear). [CrossRef]

43.   Knublauch, H.; Hendler, J.A.; Idehen, K. SPIN—Overview and Motivation. W3C Member Submission. 2011. Available online: https://www.w3.org/Submission/spin-overview/ (accessed on 9 May 2018).

44.   hackAIR Consortium. Deliverable 4.2: Semantic Integration and Reasoning of Environmental Data. 2017. Available online: http://www.hackair.eu/wp-content/uploads/2016/03/d4.2-semantic_integration_and_reasoning_of_environmental_data.pdf (accessed on 9 May 2018).

45.   Schinas, M.; Papadopoulos, S.; Apostolidis, L.; Kompatsiaris, Y.; Mitkas, P.A. Open-Source Monitoring, Search and Analytics Over Social Media. In Proceedings of the International Conference on Internet Science, Thessaloniki, Greece, 22–24 November 2017; Springer: Cham, Switzerland, 2017; pp. 361–369.

International Journal of
Geo-Information

MDPI

Article

Coupling Traditional Monitoring and Citizen Science to Disentangle the Invasion of *Halyomorpha halys*

Robert Malek [1,2,*], **Clara Tattoni** [1], **Marco Ciolli** [1], **Stefano Corradini** [3], **Daniele Andreis** [3], **Aya Ibrahim** [2,4], **Valerio Mazzoni** [2], **Anna Eriksson** [2] **and Gianfranco Anfora** [2,5]

[1] Department of Civil, Environmental and Mechanical Engineering, University of Trento, 38123 Trento, Italy; clara.tattoni@unitn.it (C.T.); marco.ciolli@unitn.it (M.C.)
[2] Research and Innovation Center, Edmund Mach Foundation, 38010 San Michele all'Adige, Italy; ibrahim.aya@spes.uniud.it (A.I.); valerio.mazzoni@fmach.it (V.M.); anna.eriksson@fmach.it (A.E.); gianfranco.anfora@fmach.it (G.A.)
[3] Technology Transfer Center, Edmund Mach Foundation, 38010 San Michele all'Adige, Italy; stefano.corradini@fmach.it (S.C.); daniele.andreis@fmach.it (D.A.)
[4] Department of Agricultural, Food, Environmental and Animal Sciences, University of Udine, 33100 Udine, Italy
[5] Center for Agriculture, Food and Environment, University of Trento, 38010 San Michele all'Adige, Italy
* Correspondence: robertnehme.malek@unitn.it; Tel.: +39-0461-615-509

Received: 28 March 2018; Accepted: 30 April 2018; Published: 4 May 2018

Abstract: The brown marmorated stink bug, *Halyomorpha halys* Stål (Hemiptera: Pentatomidae), is an invasive pest that has expanded its range outside of its original confinements in Eastern Asia, spreading through the United States, Canada and most of the European and Eurasian countries. The invasiveness of this agricultural and public nuisance pest is facilitated by the availability of an array of suitable hosts, an *r*-selected life history and the release from natural enemies in the invaded zones. Traditional monitoring methods are usually impeded by the lack of time and resources to sufficiently cover large geographical ranges. Therefore, the citizen science initiative "BugMap" was conceived to complement and assist researchers in breaking down the behavior of this invasive pest via a user-friendly, freely available mobile application. The collected data were employed to forecast its predicted distribution and to identify the areas at risk in Trentino, Northern Italy. Moreover, they permitted the uncovering of the seasonal invasion dynamics of this insect, besides providing insight into its phenological patterns, life cycle and potential management methods. Hence, the outcomes of this work emphasize the need to further integrate citizens in scientific endeavors to resolve ecological complications and reduce the gap between the public and science.

Keywords: Pentatomidae; Environmental niche modeling; citizen science; crowdsourcing; MaxEnt; QGIS; brown marmorated stink bug

---

1. Introduction

As defined by the Oxford Dictionary [1], citizen science (CS) is 'the collection and analysis of data relating to the natural world by members of the general public, typically as part of a collaborative project with professional scientists'. Oftentimes in ecological studies, there is a large amount of data to process or an extensive geographic range to cover. This poses a problem for a single researcher or even a small team of researchers [2]. Citizen scientists could help fill this role if provided with the capabilities to effectively assemble and share data.

Having citizens participate in gathering scientific data has several benefits, including improved science and technology literacy among participants and reduced costs [3]. Studies also suggest that engaging citizen volunteers makes it more likely that programs collect data relevant to local conservation and management issues [3,4]. Such data may improve professional predictions on species' future distributions, allowing the timely dissemination of these results to an educated public [5]. Volunteering citizens may also have access to lands that may not be accessible to professional scientists, allowing them to discover invasive species not yet detected elsewhere [6].

The field of ornithology has the longest history of CS [7], with thousands of amateur and professional ornithologists worldwide. One would assume that arthropods might not be as alluring for the ordinary citizen as much as birds are. Nevertheless, some of the more colorful insects have indeed caught the public's eye. The North American Butterfly Association (NABA) has initiated a program to monitor butterflies, in order to better quantify their range and abundance. Moreover, crowdsourced records on the periodical cicada, *Magicicada* spp., through the website www.magicicada.org, have been used to build mapped distributions of this insect to detect its range changes [8]. Mosquito Alert is another CS project developed in recent years to assist in the monitoring and management of disease carrying mosquitoes, *Aedes albopictus* Skuse (1894) and *Aedes aegypti* Linnaeus (1762). Citizens are invited to report sightings of the insects or of potential breeding sites; this information is communicated to public health managers to monitor and control the spread and damage caused by these "urban *Aedes*" [9].

One of the most documented expressions of global anthropogenic forcing is the human-induced movement of non-native species [10]. This phenomenon usually refers to the voluntary or accidental introduction of taxa or genotypes far from their historical distributional areas as a result of trade, tourism, agriculture or biological control programs [11,12].

The brown marmorated stink bug (BMSB), *Halyomorpha halys* Stål (Hemiptera: Pentatomidae) is an invasive pest that was introduced into the United States from Asia in the mid-1990s [13]. It has spread throughout most of the United States, as well as into Canada [14]. In Europe, BMSB was first detected in 2007 in Zurich, Switzerland [15]; its range has now expanded to include most of the European and Eurasian countries [16]. It was first detected in Italy in 2012 in the province of Modena [17]. BMSB feeding on pome fruits results in deformed, symptomatic produce with indents on the surface and corky spots in the flesh, debilitating their marketability [18].

Over $21 billion worth of crops in the United States alone have been estimated to be threatened by *H. halys* feeding damage [19]. Additional irritation by this pest lies in its overwintering behavior where it tends to aggregate in man-made structures [20], rendering it a pervasive residential nuisance.

Some of the most severe agricultural and annoyance problems have been recorded in Italy [17]. In the fall of 2017, the Friuli-Venezia Giulia region in North Eastern Italy, witnessed one of the gravest anthropogenic aggregations of the bug in recent years (http://www.udinetoday.it/cronaca/invasione-cimici-marmorata-asiatica-talmassons-medio-basso-friuli.html). In Trentino Alto Adige region in Northern Italy, BMSB was first detected in the spring of 2016 [21]. Its presence in this region poses an imminent threat to vineyards and especially to the apple industry, which accounts for 65% of the Italian apple production [22].

The recording, mapping and monitoring of invasive species are prerequisites for successful biological invasion risk management [23]. Thus BugMap, a mobile application, was designed with this purpose in mind. It is a CS approach that aims at collecting crowdsourced data on the occurrence of the alien BMSB in a newly invaded range. Obtained reports allow species distribution modelling (SDM), which aims to predict the areas where environmental conditions are suitable for the survival and establishment of the pest [24]. For invasive species management, habitat suitability maps identify areas where invasive species (1) may actually be present but are not yet detected and (2) may disperse to in the future, thus providing assistance for planning and prioritizing areas for surveillance. Such information can also assist in determining the extent, cost and likelihood of the success of a control program [25].

However, invasive species distribution models (iSDMs) face special challenges because (1) they typically violate SDM assumption that the organism is in equilibrium with its environment; and (2) species absence data are often unavailable or believed to be too difficult to interpret [26]. In general, these modelling methods combine species locality data (geo-referenced coordinates of latitude and longitude from confirmed presence) with environmental variables to create a model of species requirements for the examined variables [25].

Geographic information system (GIS) technologies are enhancing our ability to study and understand the large-scale spatial structure and dynamics of insect populations, as influenced by heterogeneous environments. In the past 20 years, advancements in mapping technology and access to tools that allow us to geo-reference our location have allowed for increased acquisition and accuracy of data [27]. The ubiquity of the internet, cell phones and wireless technology has led to increasing importance of mobile GIS as a mode of data acquisition, which promoted increased interest in CS and crowdsourcing data [28]. These technologies offer great potential in entomological research and contribute to the refining of monitoring and management methods of invasive alien pests [29,30].

The scope of this study was to evaluate whether the contribution of volunteers would improve the existing monitoring strategies of an alien stink bug, freshly invading their territory and menacing their agricultural production. The effect of user training on the accuracy of citizen reports was evaluated and the amount of crowdsourced data was quantified and compared to reports obtained through traditional monitoring methods. Moreover, we used the collective data registered by both parties through BugMap to disentangle the invasion dynamics and phenology of BMSB in Trentino, Northern Italy. In addition, we mapped the potential distribution of this invasive pest based on the integration of both citizen monitoring and traditional methods. We expect this work to provide insight into the importance of such projects and the utility of combining crowdsourced and traditional survey data, for the improvement of ecological monitoring, species distribution mapping and invasive species management programs.

2. Materials and Methods

2.1. Data Acquisition

2.1.1. Study Area

The study area is located in Trentino, North Eastern Italy, covering 6214 km^2 of land south of the Alps. It is a mountainous region influenced by a continental climate, with most of the territory lying 1000 m above sea level and around 55% covered by coniferous and deciduous forests. Trentino includes developed touristic, agricultural, industrial and commercial areas that are connected by main roads and railway transport infrastructures, with a population of 537,000 inhabitants concentrated in the plain areas and in the valley floors [31]. Despite its mountainous nature, agriculture remains one of the most important contributors to this region's economy with over €800 million of agricultural produce sold in 2013 [32]. A significant proportion of this agricultural production is at risk from the establishment and spread of BMSB.

2.1.2. BugMap, a Mobile-Based Application for Crowdsourced BMSB Reports

BugMap is a free mobile application that was designed by Edmund Mach Foundation bio-informaticians, initiated in autumn 2016 and compatible with both iPhone (Apple Inc., Cupertino, CA, USA) and Android (Google LLC., San Francisco, CA, USA) operating systems. It is a user-friendly platform that allows citizens to report the presence of this pest whenever encountered. A guidance section was added to familiarize the users with the morphological features of the different life stages of the bug. Notes on its invasive history, potential hosts, overwintering behavior and induced symptoms on host plants caused by BMSB feeding are also included. BugMap allows the gathering of information regarding how (trap, beating or visual), when (date) and where (location) the insect was observed.

Reporting users start by (1) either indicating their location on the map in the application or by allowing their geographical coordinates to be automatically registered by BugMap. Next, a simple form must be filled out with respect to (2) the date of the sighting. Mandatory segments also include (3) the number of specimens, (4) the phenological stage (adult, nymph, both or unknown), (5) type of sighting (visual, trap or beating methods) and (6) location (inside or outside buildings, garden-hedges, green urban areas, means of transport, bushes, wild areas or agricultural settings). Most importantly, the form needs to be accompanied by (7) a photograph of the suspected insect.

Five experts swiftly assess the reports once submitted, as valid, invalid or unsure (in the case of unclear photographs). To reduce the evaluation bias, each validation is double-checked and amended in case of any doubt by the other experts. Additionally, a feedback section allows experts to send back a message through the application to the users, thanking them for their contribution, explaining the differences between the reported species and BMSB in the case of invalidity and in some instances requesting a clearer photograph when diagnosis cannot be made on the basis of the current one.

2.1.3. BugMap Campaign

An advertisement campaign was initiated for BugMap shortly after the final design of the platform. Talks were delivered at the University of Trento to Bachelor and Masters' students of applied ecology. In the school of Edmund Mach Foundation, technical days were planned to involve students and technicians in the identification and reporting process of BMSB. An exposition in the Museum of Science of Trento (MUSE) was organized during the "notte dei ricercatori", where the application was introduced to scientists from various fields and to the general public. Presentations and abstracts in conferences (IPM 3.0, First Italian Citizen Science conference) allowed the international community of citizen science and integrated pest management to familiarize themselves with BugMap and understand its significance from an ecological and a social perspective. The appearances of co-authors on Italian television channels helped the dissemination of BugMap to a larger audience outside of the study region.

Various social media platforms such as Instagram (#bugmap) and Facebook (https://www.facebook.com/Bugmap-1926843807640177/) were also employed in order to further the spread of the initiative, with pages created and managed to facilitate the learning of citizens about this ecological monitoring method and the menace posed by the invasive bug in question. Moreover, flyers (Figure S1) depicting the identification and reporting process of BMSB were designed and spread in all of the above-mentioned locations.

2.1.4. Pheromone Traps

When the insect was detected in Trentino alto-Adige in the spring of 2016, phytosanitary services of the Trento province, along with Edmund Mach Foundation placed pheromone traps (RESCUE!®, Sterling International, Inc., Spokane, WA, USA) in different areas of the province. The traps ($n = 18$) were positioned in various green urban areas, apple and pear orchards and large parking spaces, in an attempt to elucidate the insect's hitchhiking behavior (Figure 1). They were used for capturing the stink bug starting from May–June 2016 until September–October 2017, with bi-weekly control. Traps are amended with BMSB aggregation pheromone components, along with a synergist that improves the attraction properties of the mixture, and functions in a 30 m radius. These traps have a non-toxic mode of action, capturing males, females and all stages of BMSB life cycle from nymphs to adults.

2.2. Modeling Current and Potential Distribution of BMSB in Trentino

2.2.1. Environmental Predictors

A distinct set of environmental parameters with potential effects on BMSB distribution was selected, as described by Capinha and Anastácio [33]. Digital Elevation Model, land-use, hydrography, road network and forest tracks were employed (Table 1) and are all freely available from the PAT cartographic portal of the Autonomous Province of Trento [34].

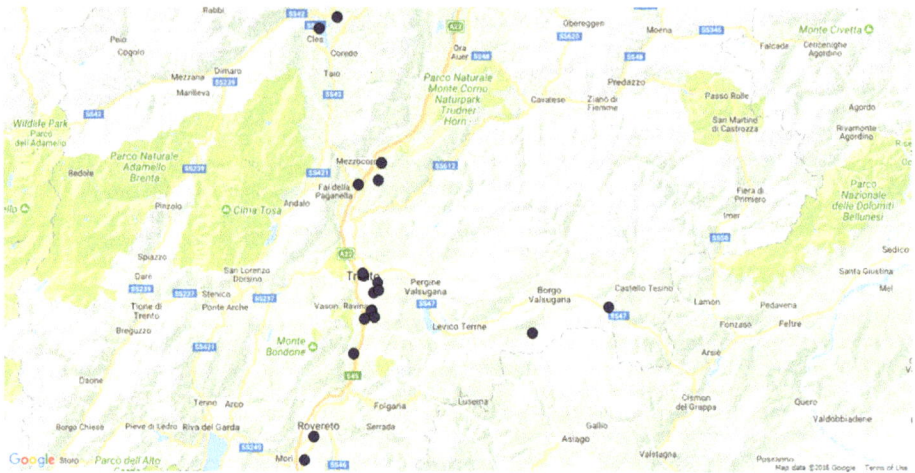

Figure 1. An Open Street Map of the area monitored by pheromone traps, with dark-blue dots representing the coordinates of the traps placed in orchards, field crops, public parks, outdoor parking spaces, bushes and anthropogenic settings.

Table 1. List of the environmental GIS layers included in the analysis.

Index	Spatial Resolution	Parameter
Digital elevation model (DEM) [a]	10 m	Slope
		Aspect
		Average hours of sun per season
Land use [b]	10 m	Continuous urban fabric
		Green urban areas
		Fruit trees and berry plantations
Hydrography	10 m	Distance from rivers
		Distance from lakes

[a] From the digital elevation model the slope, aspect and hours of sun were calculated and derived using GRASS GIS version 7. These three variables are a good proxy for temperature in a mountainous environment. [b] Land use was classified by GRASS GIS into 30 classes (Table S1) to avoid co-linearity and account for the species' ecology.

2.2.2. BMSB MaxEnt Distribution Modeling

For modeling the species' distribution, the software MaxEnt was used (MaxEnt version 3.3.3; http://www.cs.princeton.edu/wschapire/maxent/), which is a machine learning algorithm that applies the principle of maximum entropy to predict the potential distribution of species from presence only (PO) data and environmental variables [26,35]. PO data collected from BugMap and pheromone traps were input into MaxEnt, as well as the set of environmental predictors across the Trentino landscape. The program attempts to estimate a probability distribution of species occurrence that is closest to uniform while still subject to environmental constraints [36]. All data were resampled at 100 m resolution to increase the speed of calculation in MaxEnt using the jackknife test for determining variables that reduce the model reliability when omitted. Previous models assessing the potential distribution of BMSB worldwide also utilized MaxEnt, using a resolution of 4.5 Km at the equator and strictly employing bioclimatic variables [37].

2.2.3. Accounting for BugMap Sampling Bias

According to Fourcade et al. [38], the best methods to increase overall model performance with a travel-time biased data set were (1) Systematic Sampling and (2) providing a bias file to MaxEnt as recommended by the manual [36]. The bias file was calculated in GRASS (Geographic Resources Analysis Support System) GIS version 7 [39], using the module for computing a Gaussian kernel from data points with a radius of 500 m and then rescaled to a range of 1 to 20 [40]. This raster file represents the sampling effort, and it is an input for MaxEnt utilized to weigh the random background data [36].

The subsampling grids of 500 m and 1000 m were generated by the QGIS (version 2.18.16) vector toolbox, and then a single point for each square was randomly sampled and used as input Four different MaxEnt models were run with default parameters settings, using 30% of data for training as follows: (1) full dataset with no bias file, (2) full dataset with bias file, (3) systematic sampling over a grid of 500 m, (4) systematic sampling over a grid of 1000 m. Receiver operating characteristics analysis (ROC) was performed in R [41] to compare the Area Under the Curve (AUC) of all the models in order to identify the best bias treatment solution for our case study [42].

2.2.4. Setting the Threshold of BMSB Distribution Model

The output from the MaxEnt models is a map of logistic values ranging from 0 to 1; however, the interpretation of this continuous output is not straightforward, so it is a common practice to reclassify the map in a binary format according to a cut-off value. The use of the default 0.5 cut-off value is not recommended [36], especially when presence and background data are unbalanced as in our case [42]. The ROC plot-based approach was adopted, as suggested by [42,43]. The analysis of the ROC curve allows us to identify the point that maximizes the sensitivity against 1- specificity and that is considered the best classifier for the data [43].

3. Results and Discussion

3.1. Citizens' Impact

3.1.1. Citizen Science *VS.* Traditional Monitoring

A total of 306 valid BMSB registrations were split between those done visually by citizens and those obtained by traditional monitoring activities such as installment of pheromone traps and tree-beating methods. Volunteers contributed to the monitoring with 250 reports, compared to 56 by technicians (Figure 2). The majority of citizen-sightings (214) reported <10 BMSB, compared to 31 by technicians. Large aggregations of the insect (>50) were reported by both citizens and technicians.

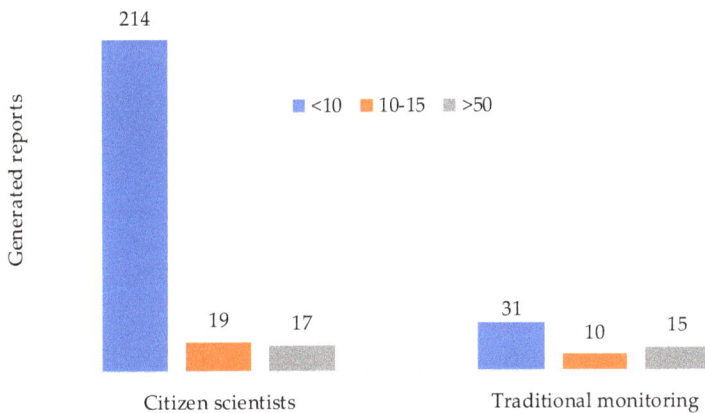

Figure 2. A representation of the different BugMap reports registered by citizens and those registered by technicians, adopting traditional monitoring methods (Pheromone traps, tree-beating). Blue bars: <10 specimens; orange bars: 10–15 specimens; grey bars: >50 specimens.

Valid BMSB reports were also classified according to the location of the sighting and to the reporting entity: in our case, citizen scientists and technicians (traditional monitoring). All the reports from means of transport and almost all reports from buildings (99%) were performed by citizens (Figure 3). Similarly, citizen reports far exceeded those registered by technicians in gardens, agricultural and wild areas. On the other hand, traditional monitoring methods reported 63% of the total sightings from green urban areas.

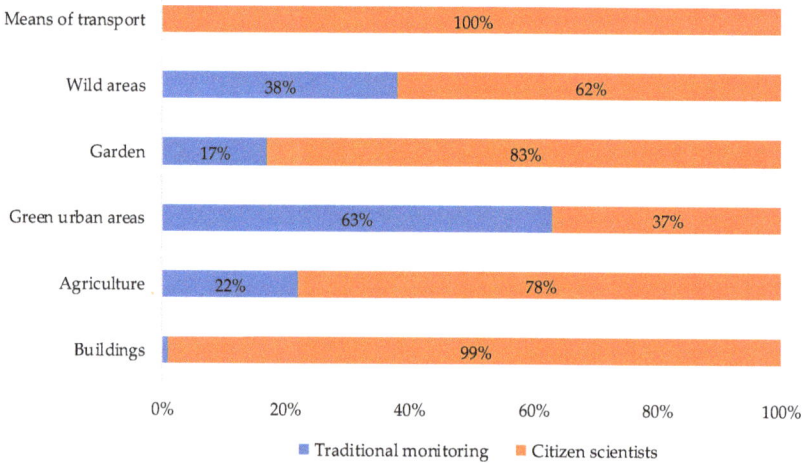

Figure 3. A comparison between traditional monitoring and citizen contribution, taking into account the percentage of reports generated by the two parties from various sighting locations.

These results highlight the differences between the two contributing parties, with citizens generating a larger number of reports from more diversified geographical areas. Reports from the general public allowed the identification of a heavily infested nucleus in a public park "Parco Gocciadoro", outside the city of Trento. This knowledge puts forward the possibility to either treat this zone with selective insecticides, or increase trap density to maximize catches, thus reducing the risk of diffusion to nearby agricultural settings. The real-time tracking of the spread and distribution of insects is usually impeded by technical hitches (lack of time, traps and technicians). Acquainted citizens provided insight on the presence of the insect by performing a more intensive sampling of premises that are difficult to reach by a team of scientists. Thus, the complications aforementioned have been alleviated thanks to the involvement of citizens in the monitoring activity (Figures S2 and S3). These outcomes are in alignment with the proposition of Dickinson et al. [2], who stated that properly trained volunteers could help fill the role of professional scientists regarding the prediction of species distribution.

The highest number of registrations originated from buildings (157), which might be due to the possibility that citizens are far more likely to encounter the bug in their lodgings during the colder months of the year, and to the association between urban development and the initial phase of BMSB establishment and dispersal [44]. A total of 64 reports were registered in agricultural areas, followed by green urban areas (52), gardens (23), wild areas (8) and means of transport (2). Of the 52 registrations from green urban areas, 20 reported large aggregations of insects (>50), along with 11 sightings of 10–15 insects. This can be explained by the availability of an array of desired host plants in public parks that serve as feeding and breeding grounds for BMSB, i.e., *Robinia pseudoacacia*, *Fraxinus* sp., *Acer* sp., *Cornus* sp., and *Corylus avellana*, as well as by the proximity of these hosts to overwintering shelters [45].

A possible means for reducing BMSB populations is to have citizen volunteers deploy small, pyramid-style pheromone traps to maximize the insect catches in nearby urban settings [46]. These preventive control measures should be diapause-aware, meaning that they could be executed in spring and late-fall, to gradually reduce the pest population and minimize the damage caused to sensitive crops throughout the season [47]. BMSB is an adept hitchhiker, often detected in vehicles and freight shipment [48,49]. Although BugMap reports from means of transportation were low (1%), the first detection of BMSB in Trentino can be traced back to a family entering via an infested rental car from the neighboring, pest-ridden Veneto region [50]. This illustrates how the stowaway behavior of BMSB can generate a cascade of social, economic and ecological losses.

3.1.2. The Effect of Training on User Performance

BugMap users can register through Facebook (Fcb), the Edmund Mach Foundation (FEM) or remain anonymous. A total of 125 users were registered to the application, 73 of them were active and participated in reporting the insect, while the other 52 were inactive. In order to compare the effect of training on the accuracy of reporting the target insect, users were split between those registered through FEM and those through Fcb. FEM users produced 174 reports, 79% of which were accurate, compared to 71 reports by Fcb users, of which 64% correctly identified BMSB. The higher accuracy and performance of FEM users could be due to the hands-on training they received and to their familiarity in dealing with arthropods, which stems from working in a scientific and agricultural environment. Our results are in accordance with the study of Crall et al. [51], who noticed that in-person training improved the data collected on invasive species by volunteers. Facebook users performed fairly well, with a reporting accuracy of 64%, meaning that they are generally aware of the alien invasion and that BugMap-based educational tools are helpful, but may need further refining.

BMSB adults were often confused with native pentatomids i.e., *Raphigaster nebulosa* Poda, *Dolycoris baccarum* Linnaeus, both phytophagous species that may be competing with the invasive pest for the occupation of similar ecological niches. In addition, *Troilus luridus* Fabricius, *Arma custos* Fabricius and *Pentatoma rufipes* Linnaeus, also caused confusion; however, these species are predators of eggs and juvenile plant pests and may contribute to the biological control of BMSB. All the latter organisms look alike to the untrained eye; therefore, it comes as no surprise that both FEM and Fcb users generated false reports, corresponding to (21%) and (36%), respectively. This outcome is similar to what was observed by Maistrello et al. [45], through their citizen science initiative to track BMSB via the website of the University of Modena and Reggio Emilia.

3.2. BMSB Invasion

3.2.1. Invasion Dynamics in Trentino

The valid BugMap reports were classified seasonally (Figure 4), based on the locality of the sighting. A total of 306 valid reports were received during the two-year period since the initiation of BugMap until mid-February 2018. Few reports ($n = 17$) were registered in 2016, the first year of the application release and local invasion, compared to 289 registrations in 2017. This might be due to the rising familiarity of the public with both the insects' invasion and BugMap, or an increasing population of the pest. BMSB possesses biological characteristics that are common among successful colonists across taxa, including an r-selected life history and association with human-modified ecosystems [52]. Its population growth can also be due to successful establishment and spread in Trentino by benefiting from host availability and being released from natural enemy pressure, leading to increased population density and fitness [53]. Similar observations in New Jersey, USA, concluded that BMSB underwent a population increase of 75% each year, during the period spanning from 2004 to 2011 [54]. An alternative mechanism in invasion ecology that could be linked to BMSB success is the Evolution of Increased Competitive Ability (EICA). This hypothesis states that a reduction in natural enemies could result in the selection for invasive populations that invest less in defense mechanisms and shift resources towards improving growth and fecundity, thereby achieving a competitive advantage over native species [55,56].

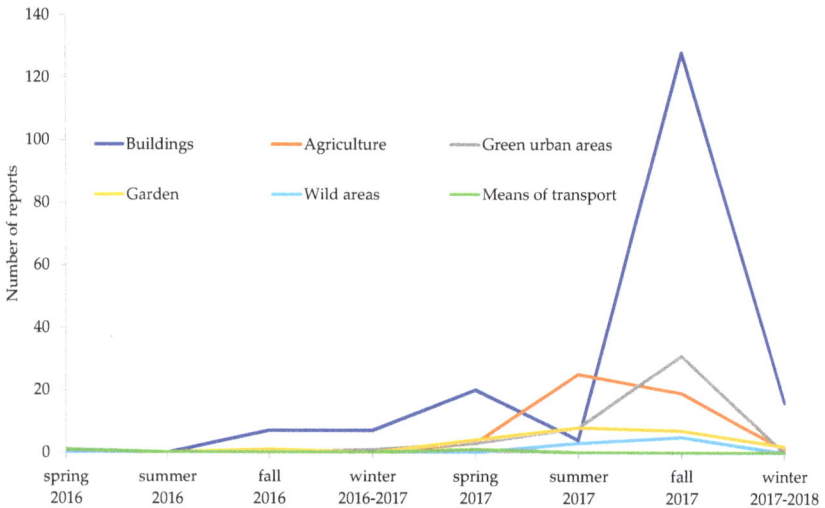

Figure 4. A representation of the varying number of valid BMSB reports in different localities since the initiation of BugMap in spring 2016 up until February 2018.

The first sightings in 2017 were registered in February-March, with individuals and small aggregations of adults being reported in buildings. Such registrations kept increasing (n = 20) until spring 2017. Afterwards, a decline in reports from anthropogenic structures coincided with an increase of sightings in agricultural areas (n = 25) in the summer of 2017, indicating a possible exit from overwintering sites and dispersal onto host plants for feeding and mating. During late-summer and early-fall, there was an upsurge of reports from buildings (n = 128) and green urban areas (n = 31). This trend can be explained by the possibility that woody ornamental plants in public parks and in the vicinity of man-made shelters provide early-and late-season resources for adults emerging from and returning to overwintering sites [57].

These findings are analogous with the knowledge on the dynamics of BMSB, being a landscape-level pest that moves across habitats throughout the season [58]. It aggregates, enters in diapause and spends the winter in dead or standing trees and prevalently in anthropogenic structures, a behavior which may result in reduced overwintering mortality [20]. Rising temperatures in spring are believed to be responsible for breaking diapause, whereby insects start moving out of man-made structures into adjacent fields in search for nutritive hosts [59]. During fall, declining temperatures and shorter days trigger the shelter-seeking behavior of the insects [60], which could explain the peak of reports from buildings and green urban areas.

3.2.2. Seasonal Phenology of BMSB

BMSB adults were consistently reported year-long; however, their density dramatically increased during fall 2017, with 102 generated reports (Figure 5). The first appearance of nymphs happened in June 2017, with 13 reports during summer, increasing in fall to 38 and disappearing in winter. Reports of both life stages occurred in summer (16) and their incidence increased in fall (34). The number and density of adults from trap catches during fall (13 reports; 5 were >50 insects) were far greater than those generated in spring (3 reports of individual insects) and summer (5 reports; none were >50 insects). It has been previously noted that all BMSB life stages are attracted to the pheromone season-long [61]. Our observations of increasing numbers and catch density during fall can probably be explained either by a rising population, or that BMSB sensitivity might be higher during this season, given that their ability to aggregate in suitable shelters during that period directly affects their winter survival.

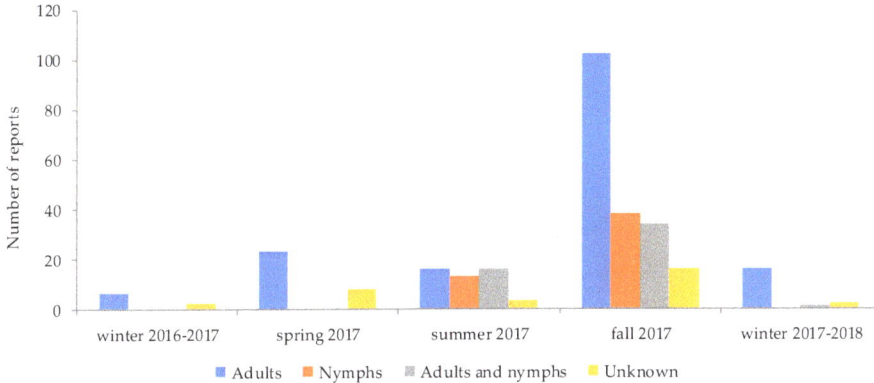

Figure 5. The number of sightings of different BMSB life stages is showcased with respect to the date of the report. Life stages are classified according to the original reports; "unknowns" are specimens that users were unable to classify.

Our results are in accordance with other observations on the phenology of BMSB, in the climatically similar mid-Atlantic region in the United States, where spring-adults also emerge from overwintering sites in late-spring [62]. Females are believed to be reproductively immature in early-spring, resulting in a delay in reproduction [18], hence the first appearance of nymphs in Trentino followed in summer. The decrease of adult observations during summer is probably due to their dispersion onto host plants within the forest edge for early-season feeding and perhaps oviposition [13]. Sightings of fall-adults peaked in the period of September to November; this can be related to a seasonal population increase [44], with spring-and summer-adults mating and laying eggs that in turn develop into mature stages. The occurrence and overlapping of adults and nymphal stages during summer and fall hint at a bi-voltine life cycle in Trentino, which is consistent with biological studies in other Italian regions that characterized two BMSB generations per year [45]. Based on this data, BugMap can accurately estimate the emergence of the 1st and 2nd generation adults. The rapid decline of nymphal populations in late fall is probably due to fifth instars molting to the adult stage or mortality due to frost [18].

3.2.3. Menace in Agricultural Areas

In 2016, there were no reports of the bug from cultivated zones; BMSB was first sighted in agricultural areas in the spring of 2017, representing the first Trentino case of an open field crop infestation. A total of 64 reports were registered from cultivated areas. Apple orchards constitute the zones with the highest number of BMSB sightings (Figure 6). Other crops include cherry, peaches and small fruits (17%), vegetable crops (5%) and vineyards (4%). All life stages of the stink bug (adults and nymphs) were found in orchards, vegetable crops and vineyards.

Most reports (79%) from agricultural areas were registered during late-summer and early-fall (September and October). This can be related to the fact that this period, with temperatures ranging between 17 and 19 °C, represents the most preferable feeding time of the year for BMSB [63]. Studies on the severity and damage inflicted on key crops found that late-season apple is more susceptible to economic injury caused by BMSB feeding [64,65]. The overlap between these two notions is particularly alarming, given the ubiquity of late-season apple orchards throughout the Trentino territory, raising the need to control this pest before harvest time to minimize losses. A possible means of control would be the exploitation of the behavioral ecology of this insect, through the adoption of a border-based attract-and-kill technique [61]. The latter enhances the strong 'edge-effect' exhibited by BMSB (the tendency to inhabit trees at the orchard perimeter), by baiting select border row trees with pheromone traps, and subsequently treating them with effective insecticides. This method was found efficient in arresting BMSB in a 2.5 m radius around baited trees, and damage to fruit was significantly reduced in the remainder of the apple orchard.

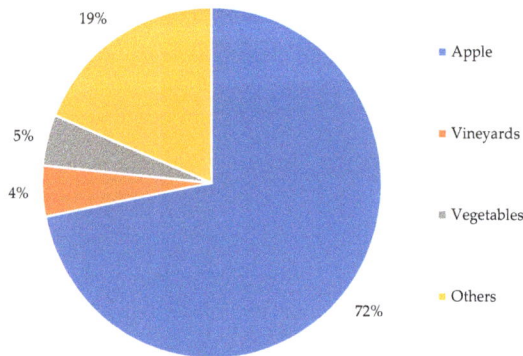

Figure 6. A display of the proportion of reports from different agricultural cultivations, with n = 64 reports from agricultural cultivations.

3.3. BMSB Projected Distribution in Trentino

3.3.1. MaxEnt-Generated Suitability Map

The occurrence data collected on BugMap platform exhibit a well-known geographical bias pattern termed travel-time bias (Table S2) [38,66]. Reports are especially concentrated in the urban area of Trento and in the neighboring villages, while records from open field are rare. After trials with different bias treatment methods, the best performing MaxEnt model corresponded to systematic sampling over a grid of 500 m, utilizing 144 valid reports that were used to train (70%) and to test (30%) the model. The aforementioned model had the highest AUC value for training data with 0.982, AUC test data: 0.970 (full dataset AUC training: 0.980, AUC test: 0.974; full dataset with bias file AUC training: 0.976, AUC test: 0.970; systematic sampling 1000 m AUC training: 0.982, AUC test: 0.963). The Jackknife test for assessing variable contribution revealed that the parameters that most affected the dispersal of the bug are the digital terrain model, urban land use and distance from houses and from streets (Figure S4). Therefore, the high suitability of the Adige valley across Trentino can probably be explained by its appropriate elevation, as well as by the prevalent agricultural-urban interfaces in Trento (Figure 7). For a polyphagous species that browses across landscapes tracking crop phenology, these diverse rural boundaries may facilitate BMSB population growth by offering diverse host plants that meet its nutritional requirements, in addition to natural and human-made overwintering structures [18,44,64,67]. Studies on BMSB haplotype diversity in North America and Europe revealed that Italy housed the second most diverse population of the bug, with 2–8 haplotypes represented in Emilia-Romagna, Piemonte and Veneto regions. This suggests that there is an ongoing invasion in Italy, with frequent re-introductions of the bug from several localities [68]. Through the Adige valley passes one of the main traffic corridors (Brennero) linking Italy to the rest of Europe, thus BMSB populations in this valley are thought to be periodically augmented by stratified diffusion via human transportation (i.e., movement of plants, goods and contaminated cargo) from the heavily infested Central Italian regions.

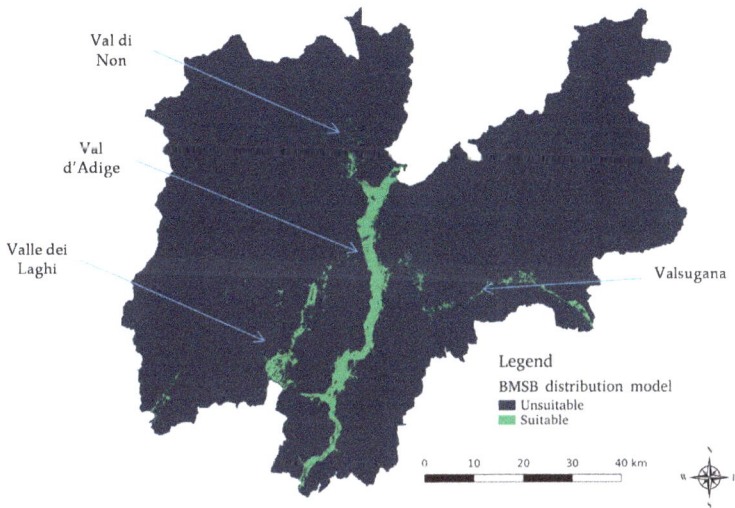

Figure 7. A suitability map of Trentino generated by MaxEnt, illustrating the areas suitable for the establishment of BMSB. ROC analysis was performed on the results of the 500 m sub-sampled model allowing the identification of the best cut-off value at 0.26 with a classification accuracy of 0.92. The logistic output of the map from MaxEnt was reclassified according to the best cut-off in two classes: unsuitable habitat below 0.26 of the logistic output and suitable habitat above that value.

To the West of the Adige, lies another suitable area which is Valle dei Laghi, nearby Garda Lake. This is a predominantly touristic zone characterized by a unique Mediterranean climate; aspects that constitute major driving forces for the passive flow and establishment of BMSB.

North of the map, a scattered suitability is projected for Val di Non, one of the most important apple growing regions of Trentino. Given that BMSB is a chill-intolerant species and mortality due to cold stress commences at temperatures as high as 4 °C [69], the ascending altitudinal gradient in Val di Non might be hampering the capacity of BMSB to spend the winter and overcome cumulative cold temperatures in this region. This might explain the unsuitability for BMSB establishment in this critical zone for the time being. On another note, the registration of several valid BugMap reports from this area may be due to a possible source to sink population dynamic, with Val di Non populations being replaced each spring by migrants from the southern, highly suitable Adige valley. Such a behavior has been previously observed in Alberta for the diamondback moth *Plutella xylostella* Linnaeus [70]. Therefore, we propose intensification of the monitoring activities in this region and appeal for citizens and farmers to stay on guard for early-season inoculations, as spring-adults are easier to manage than late-season populations [71]. The impacts of climate change on the distribution of BMSB have been assessed and a northward expansion of its suitable range is projected for Europe, as well as an increase of the number of annual generations [72]. These predictions indicate that in the absence of adequate control measures and lack of co-evolved natural enemies, BMSB will increasingly jeopardize agricultural areas around the world.

The suitable Eastern strip in Trentino corresponds to Valsugana. This area might be appropriate for BMSB establishment due to its richness in small fruit production and various horticultural crops. In addition, its confinement to the East by the BMSB infested Friuli and Veneto regions poses a ceaseless re-introduction risk of new individuals.

The global model produced by Zhu et al. [37] is helpful in understanding the suitable areas for BMSB establishment. They indicated that the whole Italian peninsula is suitable for setting up breeding populations, whereas our regional fine-scale model accounts for the extreme altitudinal variation and land morphology in Trentino, while offering monitoring and management support for affected areas.

3.3.2. Nationwide Involvement

A total of 431 reports were obtained from the whole of Italy, of which 306 were validated as accurate sightings of BMSB. Traditional monitoring methods were only employed in Trentino, and the contribution of citizens was also mostly focused in this same freshly invaded North Italian region. Therefore, 244 reports originated from Trentino, while the remaining 62 sightings were registered via BugMap in the other Italian regions (Figure 8) namely: Veneto, Lombardia, Friuli, Piemonte, Liguria, Emilia Romagna, Toscana, Lazio and Basilicata.

The serendipitous reports from different parts of Italy indicate that although the application was not advertised there, protagonists from the general public exhibited awareness, excitement and motivation for addressing national ecological issues. This behavior suggests a rising environmental democracy, the notion of making science more accessible to the public and scientists more aware of local knowledge and public enthusiasm [73]. Moreover, the registration of BMSB in regions where it has not been previously detected represents an early warning that can be communicated to phytosanitary services and collaborators. The latter entities are then advised to establish monitoring strategies and start the employment of preventive methods to limit the spread and potential damage that can be inflicted by this pest.

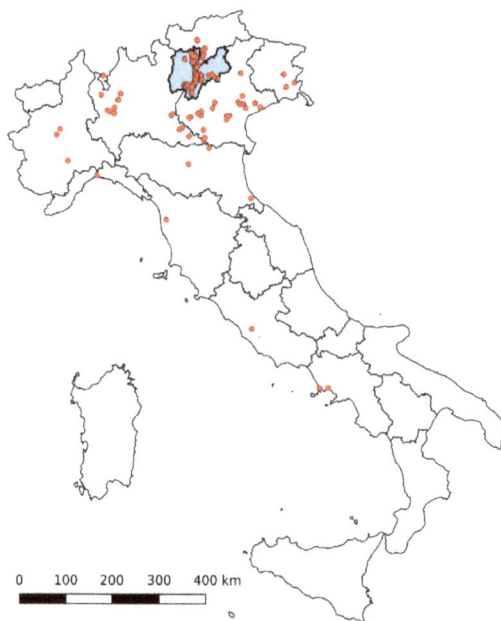

Figure 8. A map of Italy showcasing BMSB reports from the targeted Trentino (light blue) along with scattered registrations (red dots) from various Italian regions, offering an early warning for unsuspecting communities.

4. Conclusions

This study demonstrated that the coupling of volunteer-collected data with traditional ecological surveys is indeed instrumental for the improvement of existing and future monitoring programs and is worthy of the term 'monitoring 2.0' (Figure 9). Harnessing the capabilities of citizens helped uncover the invasion pattern and potential dissemination of the brown marmorated stink bug in Trentino. Although young and in its early stages, BugMap has proven efficient in stimulating scientific literacy and aided in raising public awareness regarding local ecological and economic endeavors; a cornerstone towards a more active scientific citizenship. Future activities will aim at (1) elucidating the nutritional requirements of BMSB by projecting BugMap reports onto GIS layers of plant species in the urban area of Trento; (2) refining the MaxEnt models by including abiotic parameters such as temperature and humidity; (3) expanding the BugMap domain to farther geographical boundaries and other invasive species; (4) utilizing BugMap data to assess the accuracy of the different invasive species modeling approaches currently employed and (5) developing an image recognition algorithm for identifying and validating the increasing flux of BugMap reports.

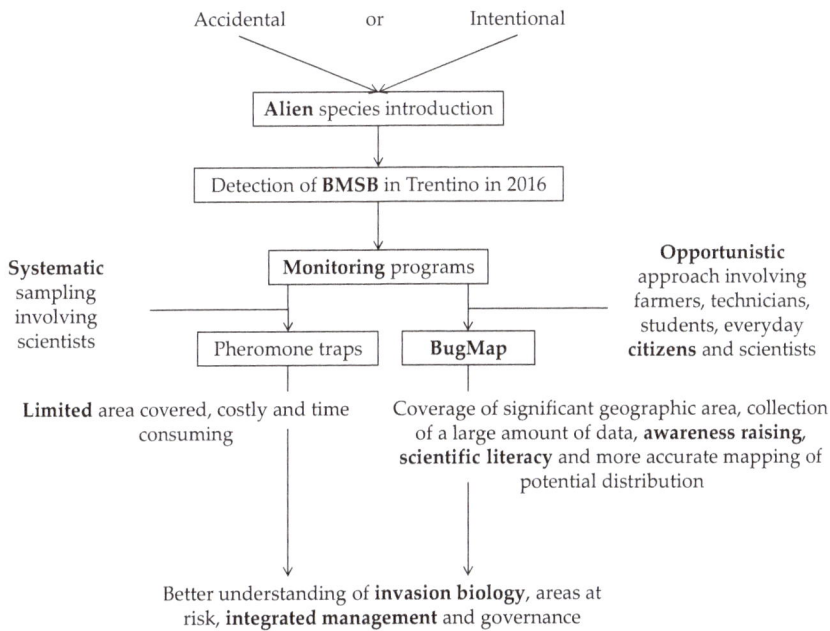

Figure 9. A summary of the workflow adopted in Trentino for tracking the spread of the invasive BMSB, termed "Monitoring 2.0", that potentially constitutes a model for future programs.

Supplementary Materials: The following are available online at http://www.mdpi.com/2220-9964/7/5/171/s1, Figure S1: BugMap flyer, Figure S2: BugMap real time tracking, Figure S3: Citizen monitoring and traditional monitoring, Figure S4: Jackknife test for variable contribution, Table S1: land use classes, Table S2: Travel-time bias.

Author Contributions: R.M., C.T., V.M., G.A. and M.C. conceived and designed the experiments; R.M. and C.T. performed the experiments, analyzed the data and generated the MaxEnt models; S.C. and D.A. designed the mobile application. R.M., C.T. and A.I. wrote the manuscript; A.E. managed the science communication campaign.

Acknowledgments: The authors wish to extend their gratitude to all the citizens who participated in this inquiry; their collaboration was essential for shaping and completing this project. We also thank all the technicians who took part in the monitoring activities and the BugMap validating experts. This work has been partly funded by E-STaR, Bando "I comunicatori STAR della scienza"-legge provinciale 2 agosto 2005, n. 14, art 22, the Autonomous Province of Trento and by APOT, Associazione Produttori Orto-Frutticoli Provincia di Trento.

Conflicts of Interest: The authors declare no conflict of interest. The funding sponsors had no role in the design of the study; in the collection, analyses, or interpretation of data; in the writing of the manuscript, and in the decision to publish the results.

References

1. Oxford English Dictionary. "Art, n.1." OED Online. Oxford University Press: Oxford, UK, January 2018. Available online: www.oed.com/viewdictionaryentry/Entry/11125 (accessed on 21 January 2018).
2. Dickinson, J.L.; Zuckerberg, B.; Bonter, D.N. Citizen Science as an ecological research tool: Challenges and benefits. *Annu. Rev. Ecol. Evol. Syst.* **2010**, *41*, 149–172. [CrossRef]
3. Danielsen, F.; Burgess, N.D.; Balmford, A. Monitoring matters: Examining the potential of locally-based approaches. *Biodivers. Conserv.* **2005**, *14*, 2507–2542. [CrossRef]
4. Measham, T.G. Building Capacity for Environmental Management: Local knowledge and rehabilitation on the Gippsland Red Gum Plains. *Aust. Geogr.* **2007**, *38*, 145–159. [CrossRef]
5. Brossard, D.; Lewenstein, B.; Bonney, R. Scientific knowledge and attitude change: The impact of a citizen science project. *Int. J. Sci. Educ.* **2005**, *27*, 1099–1121. [CrossRef]
6. Lepczyk, C.A. Integrating published data and citizen science to describe bird diversity across a landscape. *J. Appl. Ecol.* **2005**, *42*, 672–677. [CrossRef]
7. Greenwood, J.J.D. Citizens, science and bird conservation. *J. Ornithol.* **2007**, *148*, 77–124. [CrossRef]
8. Cooley, J. The distribution of periodical cicada (*Magicicada*) Brood I in 2012 with previously unreported disjunct populations (Hemiptera: Cicadadae, *Magicicada*). *Bull. Entomol. Soc. Am.* **2015**, *61*, 51–56. [CrossRef]
9. Palmer, J.R.B.; Oltra, A.; Collantes, F.; Delgado, J.A.; Lucientes, J.; Delacour, S.; Bengoa, M.; Eritja, R.; Bartumeus, F. Citizen science provides a reliable and scalable tool to track disease-carrying mosquitoes. *Nat. Commun.* **2017**, *8*. [CrossRef] [PubMed]
10. Hulme, P.E.; Bacher, S.; Kenis, M.; Klotz, S.; Kuhn, I.; Minchin, D.; Nentwig, W.; Olenin, S.; Panov, V.; Pergl, J.; et al. Grasping at the routes of biological invasions: A framework for integrating pathways into policy. *J. Appl. Ecol.* **2008**, *45*, 403–414. [CrossRef]
11. Cini, A.; Anfora, G.; Escudero-Colomar, L.A.; Grassi, A.; Santosuosso, U.; Seljak, G.; Papini, A. Tracking the invasion of the alien fruit pest *Drosophila suzukii* in Europe. *J. Pest Sci.* **2014**, *87*, 559–566. [CrossRef]
12. Geslin, B.; Gauzens, B.; Baude, M.; Dajoz, I.; Fontaine, C.; Henry, M.; Ropars, L.; Rollin, O.; Thébault, E.; Vereecken, N.J. Massively introduced managed species and their consequences for plant-pollinator interactions. In *Networks of Invasion: Empirical Evidence and Case Studies*, 1st ed.; Bohan, D.A., Dumbrell, A.J., Massol, F., Eds.; Academic Press: Cambridge, MA, USA, 2017; Volume 57, pp. 147–199.
13. Leskey, T.C.; Nielsen, A.L. Impact of the invasive brown marmorated stink bug in North America and Europe: History, biology, ecology, and management. *Annu. Rev. Entomol.* **2018**, *63*, 599–618. [CrossRef] [PubMed]
14. Fogain, R.; Graff, S. First records of the invasive pest, *Halyomorpha halys* (Hemiptera: Pentatomidae), in Ontario and Quebec. *J. Entomol. Soc. Ont.* **2011**, *142*, 45–48.
15. Wermelinger, B.; Wyniger, D.; Forster, B. First records of an invasive bug in Europe: *Halyomorpha halys* Stål (Heteroptera: Pentatomidae), a new pest on woody ornamentals and fruit trees? *Bull. Soc. Entomol. Suiss.* **2008**, *81*, 1–8.
16. EPPO Global Database. Available online: https://gd.eppo.int (accessed on 13 January 2018).
17. Bariselli, M.; Bugiani, R.; Maistrello, L. Distribution and damage caused by *Halyomorpha halys* in Italy. *Bull. OEPP* **2016**, *46*, 332–334. [CrossRef]
18. Nielsen, A.L.; Hamilton, G.C. Life history of the invasive species *Halyomorpha halys* (Hemiptera: Pentatomidae) in northeastern United States. *Ann. Entomol. Soc. Am.* **2009**, *102*, 608–616. [CrossRef]
19. Oregon Department of Agriculture: Pest Alert: Brown Marmorated Stink Bug *Halyomorpha halys*. Available online: http://www.oregon.gov/ODA/Pages/default.aspx (accessed on 4 December 2017).
20. Inkley, D. Characteristics of home invasion by the brown marmorated stink bug (Hemiptera: Pentatomidae). *J. Entomol. Sci.* **2012**, *47*, 125–130. [CrossRef]
21. Cesari, M.; Maistrello, L.; Piemontese, L.; Bonini, R.; Dioli, P.; Lee, W.; Park, C.G.; Partsinevelos, G.K.; Rebecchi, L.; Guidetti, R. Genetic diversity of the brown marmorated stink bug *Halyomorpha halys* in the invaded territories of Europe and its patterns of diffusion in Italy. *Biol. Invasions* **2017**, 1–20. [CrossRef]
22. Trentino Agricoltura. Available online: http://www.trentinoagricoltura.it/Trentino-Agricoltura/Settori2/Ortofrutticoltura/Mela (accessed on 9 January 2018).

23. Roura-Pascual, N.; Krug, R.M.; Richardson, D.M.; Hui, C. Spatially-explicit sensitivity analysis for conservation management exploring the influence of decisions in invasive alien plant management. *Divers. Distrib.* **2010**, *16*, 426–438. [CrossRef]

24. Jackson, S.T.; Overpeck, J.T. Responses of plant populations and communities to environmental changes of the late Quaternary. *Paleobiology* **2000**, *26*, 194–220. [CrossRef]

25. Anderson, R.P.; Lew, D.; Peterson, A.T. Evaluating predictive models of species' distributions: Criteria for selecting optimal models. *Ecol. Model.* **2003**, *162*, 211–232. [CrossRef]

26. Merow, C.; Smith, M.J.; Silander, J.A., Jr. A practical guide to MaxEnt for modeling species' distributions: What it does, and why inputs and settings matter. *Ecocraphy* **2013**, *36*, 1058–1069. [CrossRef]

27. Hahn, N.G.; Kaufman, A.J.; Rodriguez-Saona, C.; Nielsen, A.L.; Laforest, J.; Hamilton, G.C. Exploring the spread of the brown marmorated stink bug in New Jersey through the use of crowdsourced reports. *Am. Entomol.* **2016**, *62*, 36–45. [CrossRef]

28. Silverton, J. A new dawn for citizen science. *Trends Ecol. Evolut.* **2009**, *24*, 467–471. [CrossRef] [PubMed]

29. Hamilton, R.M. Remote Sensing and GIS Studies on the Spatial Distribution and Management of Japanese Beetle Adults and Grubs. Ph.D. Dissertation, Purdue University, West Lafayette, IN, USA, 2003.

30. Dminić, I.; Kozina, A.; Bažok, R.; Barčić, J.I. Geographic information system (GIS) and entomological research: A Review. *J. Food Agric. Environ.* **2010**, *8*, 1193–1198.

31. Istat-Istituto Nazionale di Statistica. Available online: http://www.istat.it/en/ (accessed on 16 February 2018).

32. Provincia di Trento. Available online: http://www.statweb.provincia.tn.it/ (accessed on 20 February 2018).

33. Capinha, C.; Anastácio, P. Assessing the environmental requirements of invaders using ensembles of distribution models. *Divers. Distrib.* **2011**, *17*, 13–24. [CrossRef]

34. Autonomous Province of Trento. Available online: http://www.territorio.provincia.tn.it/portal/server.pt/community/cartografia_di_base/260/cartografia_di_base/19024 (accessed on 6 August 2017).

35. Philips, S.J.; Anderson, R.P.; Schapire, R.E. Maximum entropy modeling of species geographic distributions. *Ecol. Model.* **2006**, *190*, 231–359. [CrossRef]

36. Elith, J.; Philips, S.J.; Hastie, T.; Dudík, M.; Chee, Y.E.; Yates, C.J. A statistical explanation of MaxEnt for ecologists. *Divers. Distrib.* **2011**, *17*, 43–57. [CrossRef]

37. Zhu, G.; Bu, W.; Gao, Y.; Liu, G. Potential geographic distribution of brown marmorated stink bug invasion (*Halyomorpha halys*). *PLoS ONE* **2012**, *7*, e31246. [CrossRef] [PubMed]

38. Fourcade, Y.; Engler, J.O.; Rödder, D.; Secondi, J. Mapping Species Distributions with MAXENT Using a Geographically Biased Sample of Presence Data: A Performance Assessment of Methods for Correcting Sampling Bias. *PLoS ONE* **2014**, *9*, e97122. [CrossRef] [PubMed]

39. GRASS Development Team. *Geographic Resources Analysis Support System (GRASS) Software, Version 7.0;* Open Source Geospatial Foundation: Chicago, IL, USA, 2015.

40. Phillips, S.J.; Dudík, M.; Elith, J.; Graham, C.H.; Lehmann, A.; Leathwick, J.; Ferrier, S. Sample selection bias and presence-only distribution models: Implications for background and pseudo-absence data. *Ecol. Appl.* **2009**, *19*, 181–197. [CrossRef] [PubMed]

41. R Core Team. *R: A Language and Environment for Statistical Computing;* R Foundation for Statistical Computing: Vienna, Austria, 2015.

42. Cantor, S.B.; Sun, C.C.; Tortolero-Luna, G.; Richards-Kortum, R.; Follen, M. A comparison of C/B ratios from studies using receiver operating characteristic curve analysis. *J. Clin. Epidemiol.* **1999**, *52*, 885–892. [CrossRef]

43. Liu, C.; Berry, P.; Dawson, T.; Pearson, R. Selecting thresholds of occurrence in the prediction of species distributions. *Ecography* **2005**, *28*, 385–393. [CrossRef]

44. Wallner, A.M.; Hamilton, G.C.; Nielsen, A.L.; Hahn, N.; Green, E.J.; Rodriguez-Saona, C.R. Landscape factors facilitating the invasive dynamics and distribution of the brown marmorated stink bug, *Halyomorpha halys* (Hemiptera: Pentatomidae), after arrival in the United States. *PLoS ONE* **2014**, *9*, e95691. [CrossRef] [PubMed]

45. Maistrello, L.; Dioli, P.; Bariselli, M.; Mazzoli, G.L.; Giacalone-Forini, I. Citizen science and early detection of invasive species: Phenology of first occurrences of *Halyomorpha halys* in Southern Europe. *Biol. Invasions* **2016**, *18*, 3109–3116. [CrossRef]

46. Sargent, C.; Martinson, H.M.; Raupp, M.J. Traps and trap placement may affect location of brown marmorated stink bug (Hemiptera: Pentatomidae) and increase injury to tomato fruits in home gardens. *Environ. Entomol.* **2014**, *43*, 432–438. [CrossRef] [PubMed]

47. Rossi-Stacconi, M.V.; Rupinder, K.; Mazzoni, V.; Ometto, L.; Grassi, A.; Gottardello, A.; Rota-Stabelli, O.; Anfora, G. Multiple lines of evidence for reproductive winter diapause in the invasive pest *Drosophila suzukii*: Useful clues for control strategies. *J. Pest Sci.* **2016**, *89*, 689–700. [CrossRef]

48. Holtz, T.; Kamminga, K. Qualitative Analysis of the Pest Risk Potential of the Brown Marmorated Stink Bug (BMSB), *Halyomorpha halys* (Stål), in the United States. United States Department of Agriculture: APHIS 2010. Available online: https://www.michigan.gov/documents/mda/BMSB_Pest_Risk_Potential_-_USDA_APHIS_Nov_2011_344862_7.pdf (accessed on 6 January 2018).

49. Tindall, K.V.; Fothergill, K.; McCormack, B. *Halyomorpha halys* (Hemiptera: Pentatomidae): A first Kansas record. *J. Kansas Entomol. Soc.* **2012**, *85*, 169. [CrossRef]

50. Anfora, G.; Trento University, Trento, Trentino, Italy. Personal Communication, 2018.

51. Crall, A.W.; Newman, G.J.; Jarnevich, C.S.; Stohlgren, T.J.; Waller, D.M.; Graham, J. Improving and integrating data on invasive species collected by citizen scientists. *Biol. Invasions* **2010**, *12*, 3419–3428. [CrossRef]

52. Kolar, C.S.; Lodge, D.M. Progress in invasion biology: Predicting invaders. *Trends Ecol. Evolut.* **2001**, *16*, 199–204. [CrossRef]

53. Keane, R.M.; Crawley, M.J. Exotic plant invasions and the enemy release hypothesis. *Trends Ecol. Evol.* **2002**, *17*, 164–170. [CrossRef]

54. Nielsen, A.L.; Holmstrom, K.; Hamilton, G.C.; Cambridge, J.; Ingerson–Mahar, J. Use of black light traps to monitor the abundance, spread, and flight behavior of *Halyomorpha halys* (Hemiptera: Pentatomidae). *J. Econ. Entomol.* **2013**, *106*, 1495–1502. [CrossRef] [PubMed]

55. Blossey, B.; Notzold, R. Evolution of increased competitive ability in invasive nonindigenous plants: A hypothesis. *J. Ecol.* **1995**, *83*, 887–889. [CrossRef]

56. Lee, K.A.; Klasing, K.C. A role for immunology in invasion biology. *Trends Ecol. Evolut.* **2004**, *19*, 523–529. [CrossRef] [PubMed]

57. Bergmann, E.J.; Venugopal, P.D.; Martinson, H.M.; Raupp, M.J.; Shrewsbury, P.M. Host plant use by the invasive Halyomorpha halys (Stål) on woody ornamental trees and shrubs. *PLoS ONE* **2016**, *11*, e0149975. [CrossRef] [PubMed]

58. Philips, C.R.; Kuhar, T.P.; Dively, G.P.; Hamilton, G.; Whalen, J.; Kamminga, K. Seasonal abundance and phenology of *Halyomorpha halys* (Hemiptera: Pentatomidae) on different pepper cultivars in the Mid-Atlantic (United States). *J. Econ. Entomol.* **2016**, *110*, 192–200. [CrossRef]

59. Nielsen, A.L.; Fleischer, S.; Hamilton, G.C.; Hancock, T.; Krawczyk, G.; Lee, J.C.; Ogburn, E.; Pote, J.M.; Raudenbush, A.; Rucker, A.; et al. Phenology of brown marmorated stink bug described using female reproductive development. *Ecol. Evol.* **2017**, *7*, 6680–6690. [CrossRef] [PubMed]

60. Haye, T.; Abdallah, S.; Gariepy, T.; Wyniger, D. Phenology, life table analysis and temperature requirements of the invasive brown marmorated stink bug, *Halyomorpha halys*, in Europe. *J. Pest Sci.* **2014**, *87*, 407–418. [CrossRef]

61. Morrison, W.R.; Lee, D.H.; Short, B.D.; Khrimian, A.; Leskey, T.C. Establishing the behavioral basis for an attract-and-kill strategy to manage the invasive *Halyomorpha halys* in apple orchards. *J. Pest Sci.* **2016**, *89*, 81–96. [CrossRef]

62. Bergh, J.C.; Morrison, W.R.; Joseph, S.V.; Leskey, T.C. Characterizing spring emergence of adult *Halyomorpha halys* using experimental overwintering shelters and commercial pheromone traps. *Entomol. Exp. Appl.* **2017**, *162*, 336–345. [CrossRef]

63. Wiman, N.G.; Walton, V.M.; Shearer, P.W.; Rondon, S.I. Electronically monitored labial dabbing and stylet 'probing' behaviors of brown marmorated stink bug, *Halyomorpha halys*, in simulated environments. *PLoS ONE* **2014**, *9*, e113514. [CrossRef] [PubMed]

64. Acebes-Doria, A.L.; Leskey, T.C.; Bergh, J.C. Host plant effects on *Halyomorpha halys* (Hemiptera: Pentatomidae) nymphal development and survivorship. *Environ. Entomol.* **2016**, *45*, 663–670. [CrossRef] [PubMed]

65. Joseph, S.V.; Nita, M.; Leskey, T.C.; Bergh, J.C. Temporal effects on the incidence and severity of brown marmorated stink bug (Hemiptera: Pentatomidae) feeding injury to peaches and apples during the fruiting period in Virginia. *J. Econ. Entomol.* **2015**, *108*, 592–599. [CrossRef] [PubMed]

66. McCarthy, K.P.; Fletcher, R.J.; Rota, C.T.; Hutto, R.L. Predicting species distributions from samples collected along roadsides. *Conserv. Biol.* **2012**, *26*, 68–77. [CrossRef] [PubMed]

67. Venugopal, P.D.; Coffey, P.L.; Dively, G.P.; Lamp, W.O. Adjacent habitat influence on stink bug (Hemiptera: Pentatomidae) densities and the associated damage at field corn and soybean edges. *PLoS ONE* **2014**, *9*, e109917. [CrossRef] [PubMed]

68. Morrison, W.R.; Milonas, P.; Kapantaidaki, D.E.; Cesari, M.; Della, E.; Guidetti, R.; Haye, T.; Maistrello, L.; Moraglio, S.T.; Piemontese, L.; et al. Attraction of *Halyomorpha halys* (Hemiptera: Pentatomidae) haplotypes in North America and Europe to baited traps. *Sci. Rep.* **2017**, *7*, 16941. [CrossRef] [PubMed]

69. Cira, T.M.; Venette, R.C.; Aigner, J.; Kuhar, T.; Mullins, D.E.; Gabbert, S.E.; Hutchison, W.D. Cold tolerance of *Halyomorpha halys* (Hemiptera: Pentatomidae) across geographic and temporal scales. *Environ. Entomol.* **2016**, *45*, 484–491. [CrossRef] [PubMed]

70. Dosdall, L.M. Evidence for successful overwintering of diamondback moth, *Plutella xylostella* (L.) (Lepidoptera: Plutellidae), in Alberta. *Can. Entomol.* **1994**, *126*, 183–185. [CrossRef]

71. Leskey, T.C.; Short, B.D.; Lee, D.H. Efficacy of insecticide residues on adult *Halyomorpha halys* (Stål) (Hemiptera: Pentatomidae) mortality and injury in apple and peach orchards. *Pest Manag. Sci.* **2014**, *70*, 1097–1104. [CrossRef] [PubMed]

72. Kistner, E.J. Climate Change Impacts on the Potential Distribution and Abundance of the Brown Marmorated Stink Bug (Hemiptera: Pentatomidae) with Special Reference to North America and Europe. *Environ. Entomol.* **2017**, *46*, 1212–1224. [CrossRef] [PubMed]

73. Conrad, C.C.; Hilchey, K.G. A review of citizen science and community-based environmental monitoring: Issues and opportunities. *Environ. Monit. Assess.* **2011**, *176*, 273–291. [CrossRef] [PubMed]

International Journal of
Geo-Information

MDPI

Article

Obstacles and Opportunities of Using a Mobile App for Marine Mammal Research

Courtney H. Hann [1,*], Lei Lani Stelle [2], Andrew Szabo [3] and Leigh G. Torres [4]

[1] College of Earth Ocean and Atmospheric Science, Oregon State University, 104 CEOAS Administration Building, Corvallis, OR 97333, USA
[2] Department of Biology, University of Redlands, 1200 E. Colton Ave., Redlands, CA 92373, USA; leilani_stelle@redlands.edu
[3] Alaska Whale Foundation, PO Box 1927, Petersburg, AK 99833, USA; andyszabo@gmail.com
[4] Department of Fisheries and Wildlife, Marine Mammal Institute, Oregon State University, 2030 SE Marine Science Drive, Newport, OR 97365, USA; Leigh.Torres@oregonstate.edu
* Correspondence: courtshann@gmail.com; Tel.: +1-760-712-7315

Received: 30 March 2018; Accepted: 28 April 2018; Published: 3 May 2018

Abstract: This study investigates the use of a mobile application, Whale mAPP, as a citizen science tool for collecting marine mammal sighting data. In just over three months, 1261 marine mammal sightings were observed and recorded by 39 citizen scientists in Southeast Alaska. The resulting data, along with a preliminary and post-Whale mAPP questionnaires, were used to evaluate the tool's scientific, educational, and engagement feasibility. A comparison of Whale mAPP Steller sea lion distribution data to a scientific dataset were comparable (91% overlap) given a high enough sample size ($n = 73$) and dense spatial coverage. In addition, after using Whale mAPP for two weeks, citizen scientists improved their marine mammal identification skills and self-initiated further learning, representing preliminary steps in developing an engaging citizen science project. While the app experienced high initial enthusiasm, maintaining prolonged commitment represents one of the fundamental challenges for this project. Increasing participation with targeted recruitment and sustained communication will help combat the limitations of sample size and spatial coverage. Overall, this study emphasizes the importance of early evaluation of the educational and scientific outcomes of a citizen science project, so that limitations are recognized and reduced.

Keywords: citizen science; marine mammal; opportunistic data; Alaska; spatial bias; sample size; volunteer; education; recruitment

1. Introduction

An estimated 37% of marine mammals are at risk of extinction [1] due to long-term, broad-scale, and adverse geopolitical issues including pollution, habitat loss, shipping, and global climate change [2,3]. Monitoring the impacts of these pressures on the distribution patterns of wide-ranging marine mammals requires data across large temporal and spatial scales [4,5]. Current monitoring efforts lack the spatial extent and frequency required to detect abrupt and fast marine mammal declines [6], and therefore limit efforts to manage and protect marine mammals from various pressures. Furthermore, many applied and basic ecological questions are left unanswered as they occur at large geographic and temporal scales beyond the reach of traditional research methods [7].

One approach to combat the scarcity of such data is to embrace citizen science, a non-standardized alternative method of data collection [8]. Citizen science projects are often focused on collecting multi-year and regional-scale data [9,10] primarily to estimate species distributions [9,11–13]. Citizen science volunteered geographic information projects gather a diverse set of data on both website and mobile application platforms [14–16] to accomplish various tasks, ranging from mapping wheel chair accessibility locations [17] to monitoring environmental health in a city [18]. Further, citizen science research represents a low cost approach for gathering large spatial scale data, enabling more data collection than what could feasibly be accomplished by a scientist with equivalent monetary and time budgets [9,11,13]. In addition, these projects often provide educational benefits including increased content knowledge and perspective on science [12,19,20]. One such example is the avian citizen science project eBird that contributed to over 150 peer-reviewed articles to date, increased scientific understanding of avian migration ecology, informed conservation policy, and provided an invaluable educational tool for users [21]. Similarly, these methods can be applied to marine mammal research as well. Examples of marine mammal citizen science research include: the American Cetacean Society surveys [22,23] annual volunteer cetacean counts in Hawaii [24], specific species projects [25,26], and analyses of historical whaling records [5,27–30].

While citizen science projects can provide a vast array of scientific and educational benefits, the methods and data should be properly evaluated and interpreted with caution. Limitations including data fragmentation, uncertainty regarding data accuracy, and limited applicability for research, must be overcome to have a successful project [19,31]. In addition, unlike traditional scientific studies that use standardized transects and survey methods, citizen science projects often do not regulate the "survey area" or range covered by an individual data collector. This discrepancy can be seen in studies that show data over-reported in high use areas [32] and for uncommon species [33]. Sampling error can also occur when observers differ in their ability to detect, identify and quantify species or events [32], leading to variation in data accuracy [10,34,35]. Sample size, coupled with species' ecological and detection differences can alter the performance of distribution models [36]. Consequently, inadequate evaluation of these biases associated with citizen science data will lead to false results that fail to accurately describe the species distribution.

This study examined some of the opportunities and obstacles faced when using the citizen science Android-based mobile application (app) called Whale mAPP (www.whalemapp.org) to record opportunistic marine mammal sighting data in Southeast Alaska. In sum, the Whale mAPP project aims to apply tailored citizen science methods to monitoring marine mammals within the study area of Southeast Alaska. Overall, Southeast Alaska offers a potentially ideal location for collecting these data due to a high summer ship-based tourism [37] and marine mammal abundance [3,38,39]. Concurrently, long-term studies describing marine mammal distributions are resource intensive and therefore limited [38–41], making robust and reliable marine mammal data in this region valuable.

To test how well Whale mAPP could generate such a dataset, citizen scientists were recruited to use the app in the summer of 2014. Objectives included evaluating the potential educational and scientific benefits and limitations of Whale mAPP for the purpose of improving this and future citizen science projects. To achieve the educational objectives, citizen scientists completed a questionnaire before and after using the mobile app to assess participants' motivations, general experience, and educational outcomes of using the app. Technological glitches and participant retention added additional insight.

To evaluate data quality, spatial, user, and sample size biases were measured. A Steller sea lion case study was used to measure the robustness of the resulting species data by comparing the Whale mAPP data to data collected using traditional survey methods [41]. By examining the Whale mAPP citizen science project through both a scientific and educational lens, this study emphasizes the importance of evaluating all components of a citizen science project.

Overall, Whale mAPP is used as a case study to demonstrate the value of project assessment for improving citizen science recruitment, retention, and data quality. This study also describes common citizen science data limitations including user retention, spatial bias, sample size, and technology errors. Thus, evaluating these restrictions and designing a project to reduce biases is key for supporting a successful citizen science project.

2. Materials and Methods

Citizen scientists used Whale mAPP, a mobile app designed for collecting opportunistic marine mammal data [42], in Southeast Alaska from 20 June to 30 September 2014. The app was tailored to this region, limiting possible species selection to those commonly found in the area: mysticetes (humpback whale *Megaptera novaeangliae*, minke whale *Balaenoptera acutorostrata*, fin whale *Balaenoptera physalus*, gray whale *Eschrichtius robustus*), odontocetes (killer whale *Orcinus orca*, Dall's porpoise *Phocoenoides dalli*, harbor porpoise *Phocoena phocoena*, Pacific white-sided dolphin *Lagenorhynchus obliquidens*), pinnipeds (Steller sea lion *Eumetopias jubatus*, harbor seal *Phoca vitulina*, California sea lion *Zalophus californianus*), and other (sea otter *Enhydra lutris*) [3,39].

Once activated, Whale mAPP automatically recorded the user's location every 30 s, resulting in a record of the survey track line (Figure 1). Upon sighting a marine mammal, users selected the binocular icon (Figure 1) and were asked to identify the species using a drop-down menu. Additionally, users were asked to record the number of individuals (categorical options: 1, 2, 3, . . . to 10, then 11–15, 16–20, 21–30, etc.), distance and direction to the animal(s) (categorical options: N, NE, E, SE, S, SW, W, NW, and 0–500 m, 500 m–1 km, 1–2 km, 2–5 km, 5+ km), the animal's behavior (categorical options: feeding, logging, milling, socializing, thermal regulation, travelling, other, unknown behavior), weather conditions (cloud cover and wind), and a single five star confidence rating scoring the user's confidence in the accuracy of the data entered for each sighting (1 star being the lowest confidence, 5 stars being the highest confidence) (Figure 1). Because volunteers could, but were not required to take a photo of the marine mammal, their species identification accuracy was assessed based on their confidence ranking. The assumption was made that a higher confidence rating coincided with greater accuracy. Upon completion, a recorded sighting was noted on the map with an icon on top of the black track line (Figure 1). Since many areas of Southeast Alaska do not support cellular reception, Whale mAPP stored data locally on a base map of Southeast Alaska that was accessible offline. Once cellular reception was available, the data were automatically transmitted to a geodatabase.

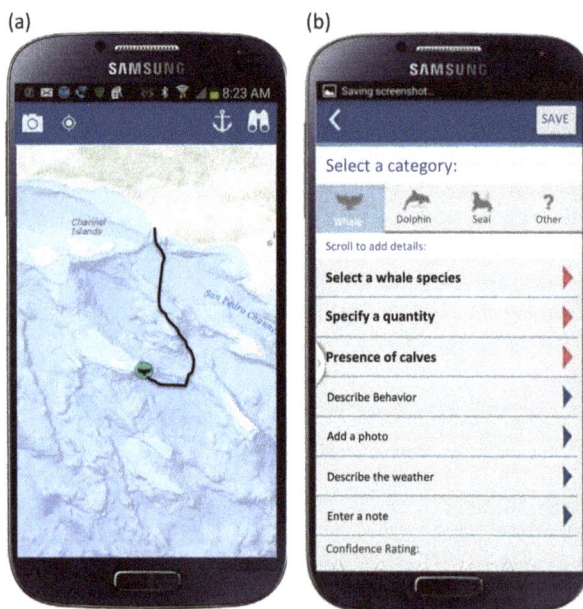

Figure 1. (**a**) Whale mAPP map display showing the map, track line, and sighting record icon; (**b**) Whale mAPP display for recording a marine mammal sighting.

A summary of the processes and topics evaluated in this study are provided as a flowchart (Figure 2). Citizen scientists were recruited at the Alaska Whale Foundation's Center for Coastal Conservation in Warm Springs Bay, Southeast Alaska (57.09° N, −134.84° W). Every person who entered the center was told about the project, marine mammals in the area, and asked if they were interested in using Whale mAPP. Potential participants included guests and crew aboard nature tourism cruises, private vessel operators, and commercial fishermen. All participants were provided with a user manual and marine mammal identification guide with illustrations completed by Pieter Folken (full body drawings) and Courtney Hann (all other illustrations) (Figure 3). While all participants were offered additional training, the length and duration varied based on their available time and interest. Additional training walked the citizen scientist through a mock scenario of how to use Whale mAPP, discussed scientific technique and rationale, and clarified data collection methods. Outside of this study, Whale mAPP citizen scientists rarely receive in-person training and instead rely on online instructions for using the app. Therefore, the in-person training was not required for participants in this study. To maximize user participation, citizen scientists were allowed to collect data following their preferred routes of travel [9,11]. A simple data entry platform with clear data protocols [43] and a standardized method [44] were implemented to minimize error and ensure the public could collect and submit accurate data.

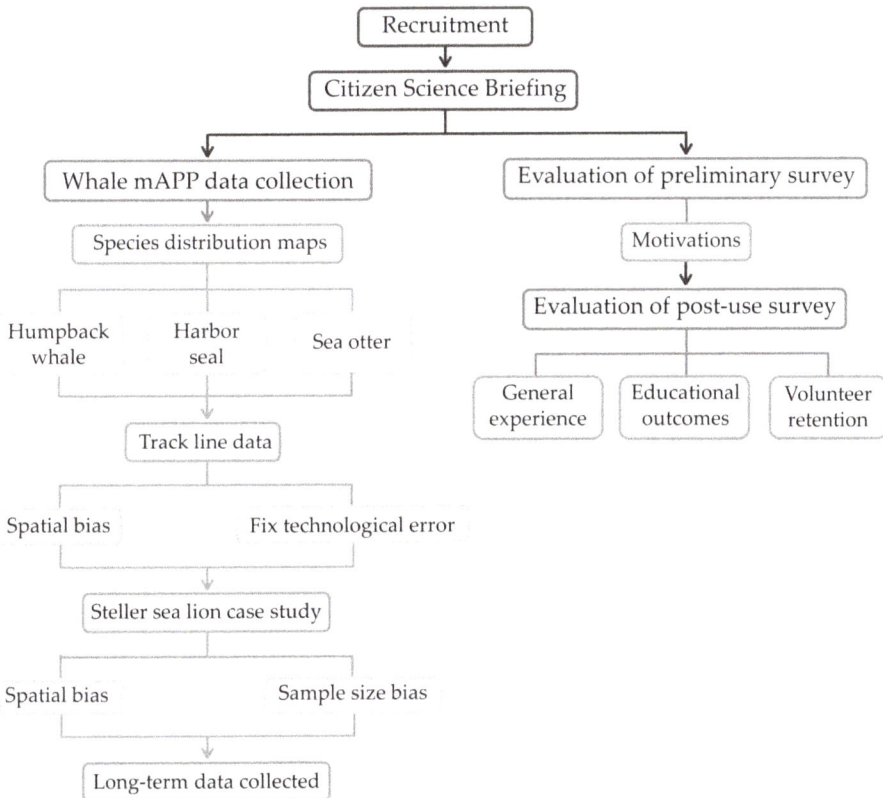

Figure 2. Flowchart of methods with the left hand column showing how the Whale mAPP data were used and the right hand column describing how the questionnaire survey data were used.

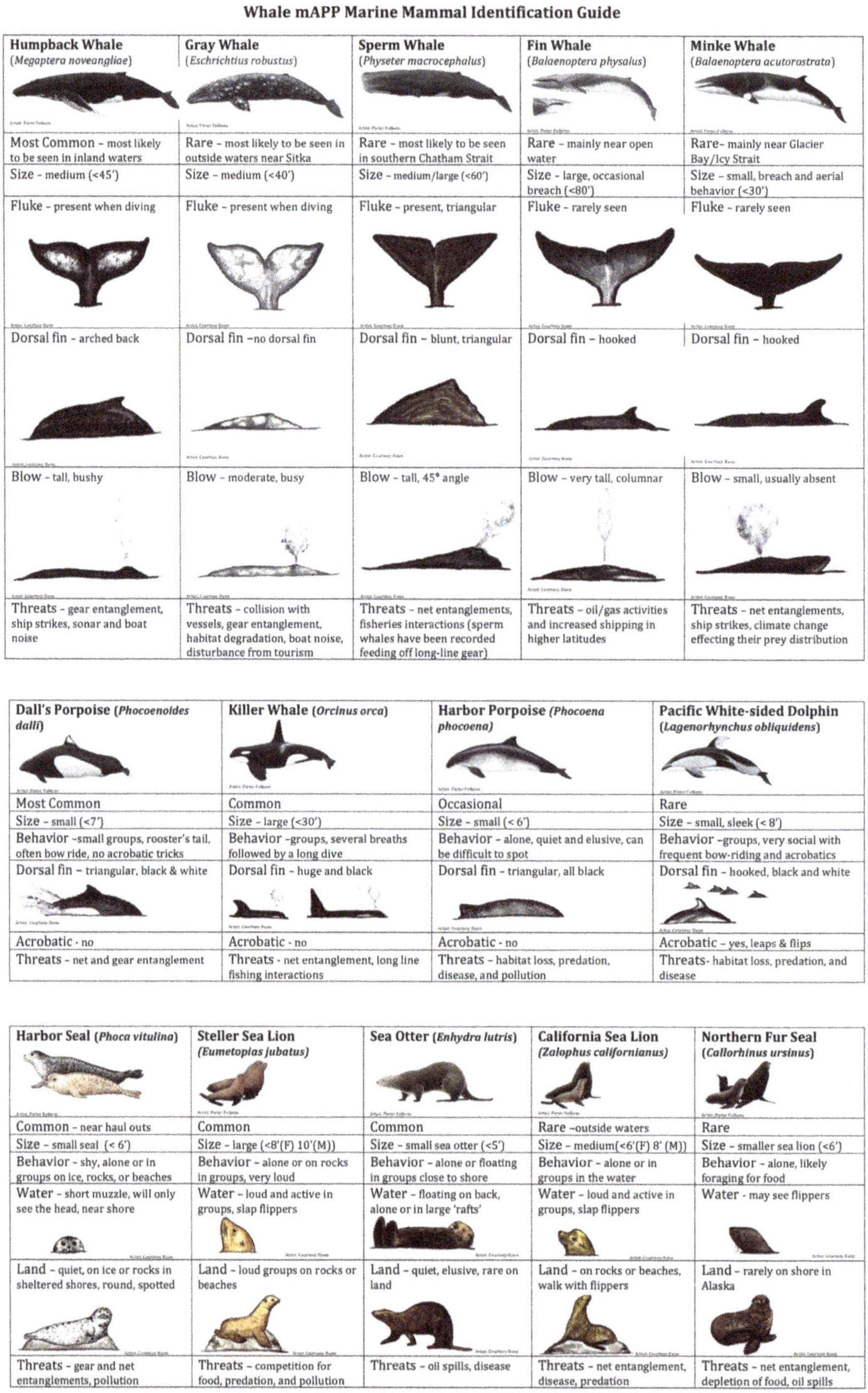

Whale mAPP Marine Mammal Identification Guide

Humpback Whale (*Megaptera noveangliae*)	Gray Whale (*Eschrichtius robustus*)	Sperm Whale (*Physeter macrocephalus*)	Fin Whale (*Balaenoptera physalus*)	Minke Whale (*Balaenoptera acutorastrata*)
Most Common – most likely to be seen in inland waters	Rare – most likely to be seen in outside waters near Sitka	Rare – most likely to be seen in southern Chatham Strait	Rare – mainly near open water	Rare- mainly near Glacier Bay/Icy Strait
Size – medium (<45')	Size – medium (<40')	Size – medium/large (<60')	Size – large, occasional breach (<80')	Size – small, breach and aerial behavior (<30')
Fluke – present when diving	Fluke – present when diving	Fluke – present, triangular	Fluke – rarely seen	Fluke – rarely seen
Dorsal fin – arched back	Dorsal fin –no dorsal fin	Dorsal fin – blunt, triangular	Dorsal fin – hooked	Dorsal fin – hooked
Blow – tall, bushy	Blow – moderate, busy	Blow – tall, 45° angle	Blow – very tall, columnar	Blow – small, usually absent
Threats – gear entanglement, ship strikes, sonar and boat noise	Threats – collision with vessels, gear entanglement, habitat degradation, boat noise, disturbance from tourism	Threats – net entanglements, fisheries interactions (sperm whales have been recorded feeding off long-line gear	Threats – oil/gas activities and increased shipping in higher latitudes	Threats – net entanglements, ship strikes, climate change effecting their prey distribution

Dall's Porpoise (*Phocoenoides dalli*)	Killer Whale (*Orcinus orca*)	Harbor Porpoise (*Phocoena phocoena*)	Pacific White-sided Dolphin (*Lagenorhynchus obliquidens*)
Most Common	Common	Occasional	Rare
Size – small (<7')	Size – large (<30')	Size – small (< 6')	Size – small, sleek (< 8')
Behavior –small groups, rooster's tail, often bow ride, no acrobatic tricks	Behavior –groups, several breaths followed by a long dive	Behavior – alone, quiet and elusive, can be difficult to spot	Behavior –groups, very social with frequent bow-riding and acrobatics
Dorsal fin – triangular, black & white	Dorsal fin – huge and black	Dorsal fin – triangular, all black	Dorsal fin – hooked, black and white
Acrobatic - no	Acrobatic - no	Acrobatic - no	Acrobatic – yes, leaps & flips
Threats – net and gear entanglement	Threats - net entanglement, long line fishing interactions	Threats – habitat loss, predation, disease, and pollution	Threats- habitat loss, predation, and disease

Harbor Seal (*Phoca vitulina*)	Steller Sea Lion (*Eumetopias jubatus*)	Sea Otter (*Enhydra lutris*)	California Sea Lion (*Zalophus californianus*)	Northern Fur Seal (*Callorhinus ursinus*)
Common – near haul outs	Common	Common	Rare –outside waters	Rare
Size – small seal (< 6')	Size – large (<8'(F) 10'(M))	Size – small sea otter (<5')	Size – medium(<6'(F) 8' (M))	Size – smaller sea lion (<6')
Behavior – shy, alone or in groups on ice, rocks, or beaches	Behavior – alone or on rocks in groups, very loud	Behavior – alone or floating in groups close to shore	Behavior – alone or in groups in the water	Behavior – alone, likely foraging for food
Water – short muzzle, will only see the head, near shore	Water – loud and active in groups, slap flippers	Water – floating on back, alone or in large 'rafts'	Water – loud and active in groups, slap flippers	Water – may see flippers
Land – quiet, on ice or rocks in sheltered shores, round, spotted	Land – loud groups on rocks or beaches	Land – quiet, elusive, rare on land	Land – on rocks or beaches, walk with flippers	Land – rarely on shore in Alaska
Threats – gear and net entanglements, pollution	Threats – competition for food, predation, and pollution	Threats – oil spills, disease	Threats – net entanglement, disease, predation	Threats – net entanglement, depletion of food, oil spills

Figure 3. Marine mammal identification guide for common marine mammals in Southeast Alaska.

2.1. Participant Surveys

Participant motivations, general experience, and educational outcomes were evaluated using two questionnaires: one completed prior to using Whale mAPP and the second completed after using the app for more than two weeks. Questions included marine mammal identification test, ranking knowledge of various topics, selecting why they chose to use Whale mAPP and why/if they stopped, noting the percent of time spent doing various activities while using Whale mAPP, noting actions taken outside of using Whale mAPP to learn about marine mammals, and ranking various statements regarding their experience using Whale mAPP. Questionnaires were designed using Qualitrics software and distributed in person and by email. All statistical analyses were performed with Microsoft Excel 14.2.0 and R [45]. Permission to collect data on volunteers over the age of 18 was granted by the Oregon State University 5234 Institutional Review Board, study number 6273.

Two informal learning goals were also considered in conjunction with this study: (1) participants' developing interest in science; and (2) participants' understanding of science knowledge [46]. The first goal, interpreting participants' developing interest in science, was assessed by identifying user interests and any actions they took to learn more about marine mammals. Consequently, to evaluate user motivation, participants responded to questions on why they chose to participate in the project, if they would continue to use Whale mAPP, and their enjoyment level from participating. From this, user attention and variability was determined by recording the time spent scanning the water and looking for marine mammals compared to focusing on other activities.

The second goal, evaluating participants' understanding of science knowledge, focused on interpreting the user's growth in marine mammal content knowledge. To assess content knowledge, participants responded to questions regarding marine mammal identification and knowledge. Then, responses from the post-use survey were compared to those of the preliminary survey to identify improvement in identification skills and content knowledge. Since the data were not normally distributed, paired Wilcoxon rank sum tests using the R package 'coin' were used to determine significance change between the preliminary and post questionnaire responses [47].

2.2. Mapping and Evaluating Whale mAPP Data

Marine mammal distribution maps were produced using a lattice-based density estimator [48]. Given the irregular boundaries and abundance of islands in Southeast Alaska, a lattice-based density estimator was used to generate estimates of core- (25% density), intermediate- (50% density), and broad-use (95% density) areas of species distribution. Similar methods were used to estimate marine species distributions in other areas with complex shorelines and islands [49,50]. The probability that the random walk stayed in the same location, M, was set in accordance with previous studies to be 0.5 [48,49]. The estimation of the optimal smoothing parameter, k, was determined using cross-validation in the package 'latticeDensity' for each species [51]. Node spacing of 50 m was used because it was sufficient for delineating the coastlines while still allowing computer computation of the complex study area. All analyses were conducted in R [45].

To demonstrate a potential use of Whale mAPP data for marine mammal research, the collected data on Steller sea lions was compared to results from Womble et al.'s (2005) standardized survey of haul out locations in Southeast Alaska. Haul out sites are areas where pinnipeds, in this case Steller sea lions, temporarily leave the water and 'haul out' on land. These are generally established locations and commonly used to describe Steller sea lion distribution [41]. To minimize inter-annual variation, we aggregated the scientifically collected data from March through May 2001 and 2002 [41] for comparison to Whale mAPP data between June and September 2014. This dataset and species were chosen as the case study because the dataset represents the most comprehensive published dataset that surveyed all of Southeast Alaska for Steller sea lions [41]. In addition, haul out sites are easily identifiable and comparable between datasets.

Whale mAPP Steller sea lion data accuracy were measured by counting the number of scientifically collected haul out locations [41] recorded by Whale mAPP users. To determine how sample size influenced the accuracy of Whale mAPP data, a discovery curve was generated that plots the percent of scientifically collected haul out sites [41] recorded by Whale mAPP users with every 5-unit sighting increase in Whale mAPP sample size. For each five-unit sample size increase, ten random Whale mAPP sub-samples were generated.

For the Steller sea lion case study, 'high survey effort' was differentiated to quantify how spatial bias affected the quality of citizen science data. To accomplish this, track line data of Whale mAPP user paths were mapped to examine spatial biases. Because 90% of total track line data were represented in the 75% lattice-based density contour of survey effort, the 75% lattice-based density contour was used to define 'high track line effort'. The 'high Steller sea lion sighting effort' was defined by the 50% lattice-based contour because these data included more than 90% of all Steller sea lion data. The final 'high survey effort' was the compilation of the 'high track line effort' and 'high Steller sea lion sighting effort' spatial layers. While, minimum survey effort was determined by combining all track line data with all Whale mAPP sighting data. Ultimately, this process created one layer representing all locations that Whale mAPP users traveled to at least once.

3. Results and Evaluations

Of the 216 people encountered at the recruitment center during the summer, 73.5% were interested in using Whale mAPP; however, of those only 44.7% possessed an Android device to download the app. Of the resulting 39 participants who followed through in using the app, 18 were private vessel owners, 18 were involved with nature tourism cruises, and three were commercial fishermen. From 20 June to 30 September 2014, these participants logged over 800 h to record 1261 marine mammal sightings and 10,892 km of track line data from the northern portion of Southeast Alaska to Seattle, Washington. In sum, 52.9% of sightings were of humpback whale ($n = 665$), 11.5% were sea otters ($n = 146$), 11.3% were harbor seals ($n = 143$), 9.2% were Steller sea lions ($n = 117$), 6.0% were killer whales ($n = 76$), 3.9% were Dall's porpoises ($n = 50$), 3.3% were harbor porpoises ($n = 43$), 1.0% were Pacific white-sided dolphins ($n = 12$), and <1% were California sea lions ($n = 5$), minke whales ($n = 2$), elephant seals ($n = 2$), fin whale ($n = 1$), and gray whales ($n = 1$).

3.1. Questionnaire Results

In sum, 29 (74%) Whale mAPP users completed the pre-use questionnaire, and of those, 24 (83%) followed up with the post-use questionnaire. A majority of volunteers, 78.3% reported they enjoyed using Whale mAPP, 73.9% found it easy to use and 82.6% would recommend it. When asked about their interests prior to using Whale mAPP, citizen scientists reported pre-existing interests in marine mammals, the ocean, using Whale mAPP, collecting marine mammal data, and to a lesser degree technology (Figure 4). When asked whether they took any self-initiated actions to learn about marine mammals as a result of using the app, 56.5% reported talking to a peer, 47.8% read a book, 34.8% talked to a scientist, and 8.7% reported going on a wildlife tour, talking to family, or going to a museum.

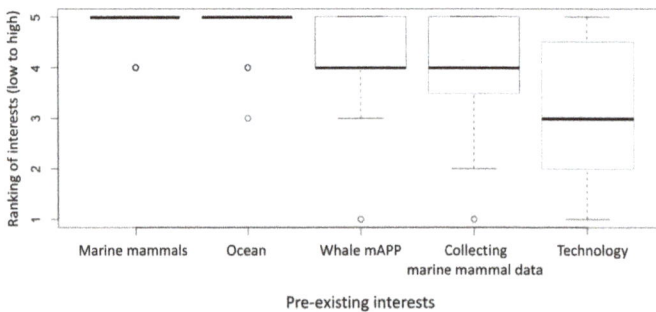

Figure 4. Boxplot of volunteer rankings of pre-existing interest on a scale of 1 (low) to 5 (high).

Volunteers were primarily motivated to use Whale mAPP due to their interest in marine mammals (n = 17), collecting data (n = 13), science (n = 13), citizen science (n = 11), their tourist company's association with Whale mAPP (n = 6), technology (n = 3), and/or for another reason (n = 2).

Marine mammal identification skills significantly improved after using Whale mAPP (Wilcoxon rank sum test, Z = 1.83, p-value = 0.035), with an increase in average test score rising from 76.3% ± 20.4% to 86.1% ± 19.3%. Yet, citizen scientists' content knowledge of marine mammal conservation topics did not change after using Whale mAPP for at least two weeks.

When the study ended in September 2014, 27.3% noted they were still using Whale mAPP. Of these, 83.3% were involved with tourism cruises as either a captain or naturalist staff. The remaining 16.7% were private vessel owners. Furthermore, all of these volunteers possessed very strong pre-existing interests for the ocean and marine mammals, and very strong to strong pre-existing interest in monitoring marine mammals and Whale mAPP.

The remaining 72.7% of citizen scientists reported that they stopped using the app because they either left Southeast Alaska (68.4%), felt it was too time consuming (15.8%), encountered technology problems (10.5%), and/or for another reason (5.3%). Of those that stopped using the app because they left Southeast Alaska, 64.3% were private vessel owners and 35.7% were involved with nature tourism cruises. Furthermore, around 26.1% of volunteers thought the app required too much data entry. Long-term commitment to data collection was limited as the number of Whale mAPP sightings from Southeast Alaska to Seattle dropped from 1256 in 2014 to nine in 2015, 38 in 2016, and none in 2017. One naturalist originally recruited in 2014 also collected data in 2016. All other data collected in 2015 and 2016 were by citizen scientists not recruited during the 2014 field season.

User effort also differed between individuals as the top five recorders collected 16.6%, 14.7%, 11%, 9.2%, and 6.6% of sighting data. The other 34 participants recorded the remaining 41.9% approximately evenly. In addition, nine users collected ~73% of track line data, a value that may be due to user choice or app malfunctions as only 30% of track line data were correctly saved (participants did not know if their track line data was correctly uploaded or not). Around half of participants (52.2%) did not purposefully go out to use Whale mAPP; rather they used it when already travelling to a destination.

All participants spent the majority of their time scanning the water for marine mammals while using the app; however, users also reported simultaneously driving a vessel and talking for more than a third of the time (Figure 5).

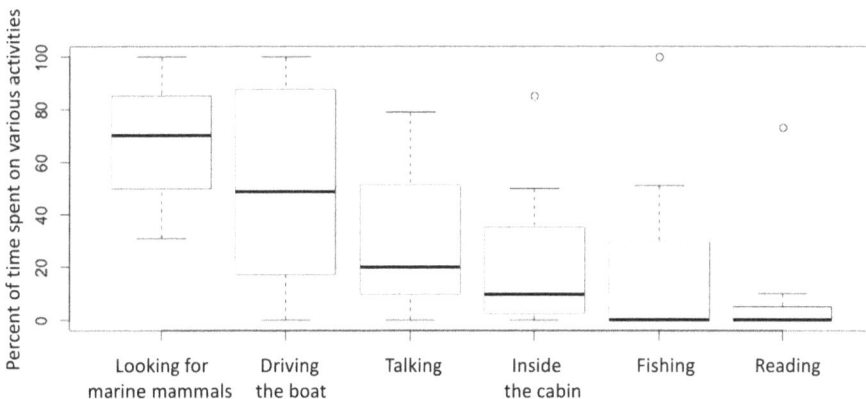

Figure 5. Percent of time citizen scientists spent doing various activities while using Whale mAPP.

3.2. Applicability to Marine Mammal Research

For the lattice-based density method, the total water area used to estimate home ranges was 15,630 km^2, with each node located 0.05 km apart. Only sightings north of latitude 54°, where a majority (91.6%) of the sightings were located, were analyzed. Due to the structure of Whale mAPP, only presence data were recorded. It cannot be assumed that all marine mammals encountered on each track line were recorded. This is true for all marine mammal surveys, as even trained observers miss animals due to weather conditions, animal diving, or change. Data removed from the analysis included duplicate track line data ($n = 10$), data with unidentifiable users, data noted as a "mistake" by the recorder, and sightings that were revisions to previous recordings ($n = 7$). Additionally, due to low confidence rating (less than 3 rating) and poor visibility (<3 mile visibility), 47 sightings were removed, primarily for humpback whales ($n = 22$), but proportionally more for cryptic harbor ($n = 4$) and Dall's ($n = 4$) porpoises. An average of $77.4 \pm 11.5\%$ of removed sighting were due to poor visibility. Data with confidence ratings lower than a three were removed because this ranking indicated that citizen scientists had lower than 75% confidence in the accuracy of those data. Data with visibility of less than three miles were removed because Southeast Alaska can have sudden, thick fog that significantly reduces visibility to a few hundred meters. Therefore, visibility can greatly impact the number of marine mammals a citizen scientist can spot from a boat.

To illustrate the potential of Whale mAPP data to inform marine mammal distribution patterns, three lattice-based density maps were created for humpback whales ($n = 665$, $k = 1$), sea otters ($n = 146$, $k = 3$) and harbor seals ($n = 143$, $k = 3$) (Figure 6). Humpback whale distributions were comparable to other scientific datasets along common travel routes, but data gaps were present in the southern part of Southeast Alaska (south of $56.5°$ N) and the more remote northern sections of Glacier Bay and Icy Strait [38,39]. Comparison of Whale mAPP results regarding sea otter distribution to previous work [52] suggests their range has expanded from the outer western edges to throughout Southeast Alaska since early 1990. Harbor seal distribution was difficult to compare to other studies, because no detailed published studies were found outside of Glacier Bay or specific Southeast Alaskan inlets [53–55]. Thus, this lack of pre-existing data highlights the added value of citizen science data.

3.2.1. Steller Sea Lion Case Study

Altogether, Whale mAPP users recorded 54.2% of Steller sea lion haul out sites identified by the scientifically collected data [41], which improved to 72.2% accuracy when the comparison was limited to the areas surveyed at least once by Whale mAPP users (Figure 7). Accuracy increased further to 90.9% when the comparison was focused on high Whale mAPP survey effort areas, a common measurement of spatial bias (orange areas in Figure 7).

Furthermore, the discovery curve shows a linear increase in the percent of scientifically collected haul out sites [41] identified by Whale mAPP users with an increase in sample size of Whale mAPP Steller sea lion sightings (Figure 8). This data demonstrates how sample size can impact citizen science data accuracy and needs to be considered when interpreting results.

3.2.2. Spatial Bias

Approximately 70% of the track line data collected by 11 citizen scientists were not recorded due to a technological glitch in which the track line data were not uploaded to the cloud geodatabase. The remaining 30% ($n = 120$ track lines) were used to estimate areas of high and low survey effort. Twenty-eight, or ~71.8% of participants, contributed to the vessel track line data. On average, each user travelled 80 km and recorded 3.5 ± 4.0 marine mammal sightings per track line. Technological errors that led to the loss of track line data have subsequently been fixed in Whale mAPP, as these data are essential for interpreting spatial bias.

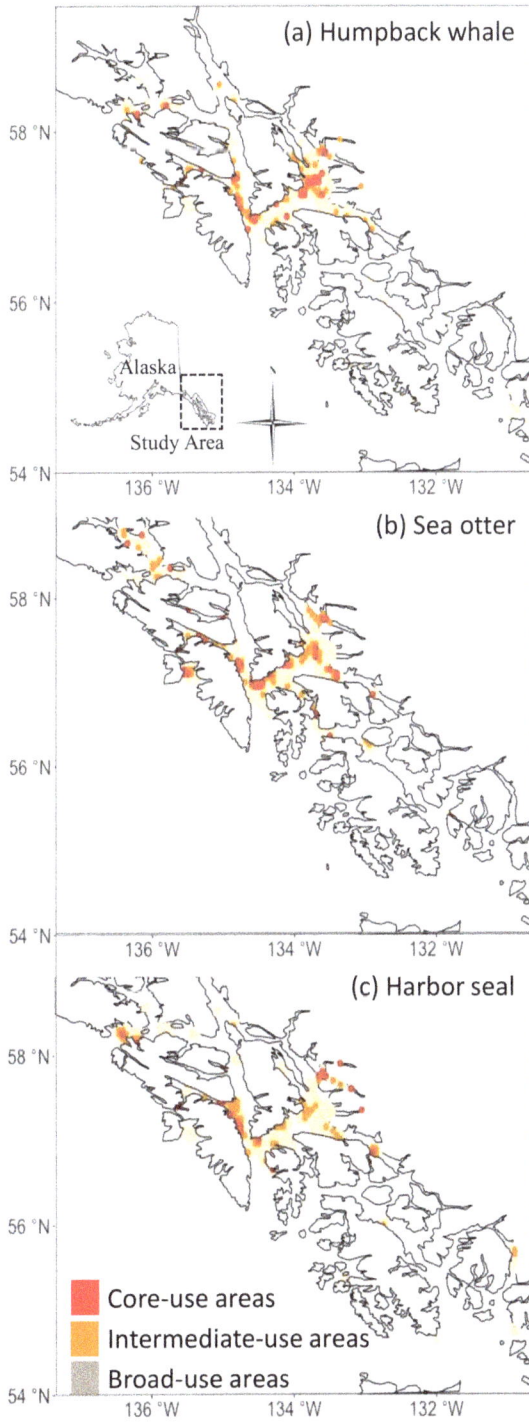

Figure 6. (**a**) Humpback whale; (**b**) sea otter; and (**c**) harbor seal distribution maps.

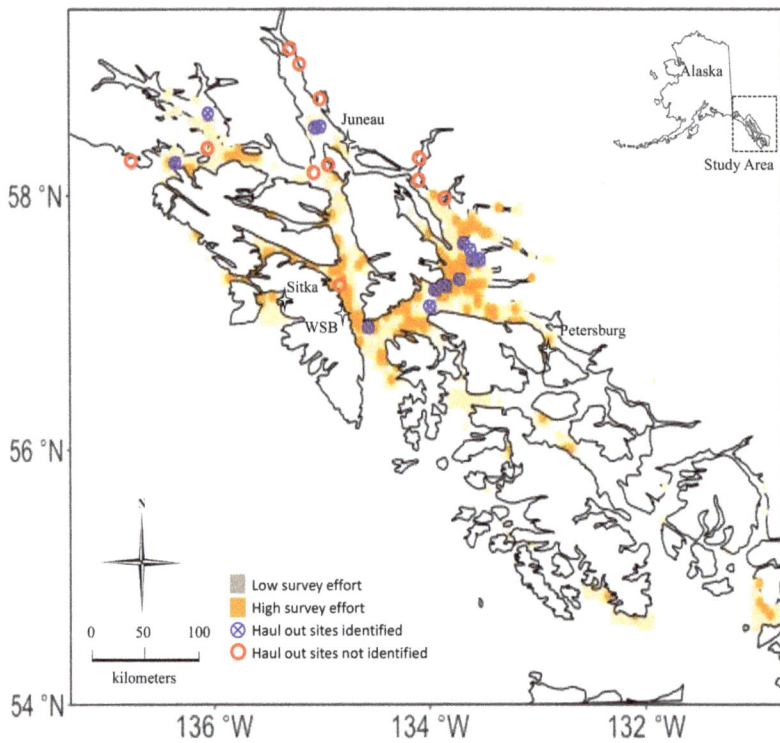

Figure 7. Map showing scientifically collected haul out sites [41] identified with Whale mAPP data in low and high survey effort areas.

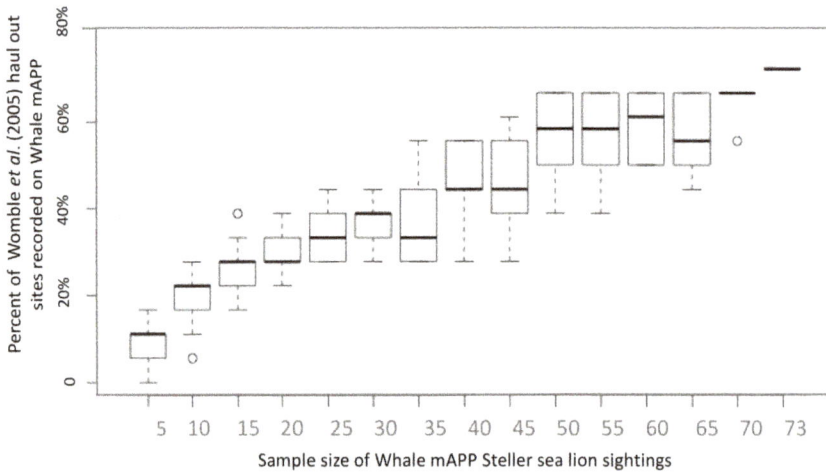

Figure 8. Steller sea lion discovery curve showing an increase in the percent of scientifically collected haul out sites [41] recorded through Whale mAPP with every five-unit increase in Whale mAPP Steller sea lion sighting sample size.

Higher survey effort (within the 75% density contour) occurred near the recruitment center (Warm Springs Bay), other towns, and along common travel routes such as Chatham Strait, Peril Strait, Fredrick Sound, and Stephens Passage (Figure 9). Reduced coverage occurred in remote areas, waterways far from the recruitment center, and/or restricted access areas, such as along offshore facing coastlines, Lynn Canal, Glacier Bay, and south by Wrangell. Variation in survey effort makes the data difficult to extrapolate for all of Southeast Alaska. Therefore, this caveat needs to be considered when determining Whale mAPP data quality and interpreting the results.

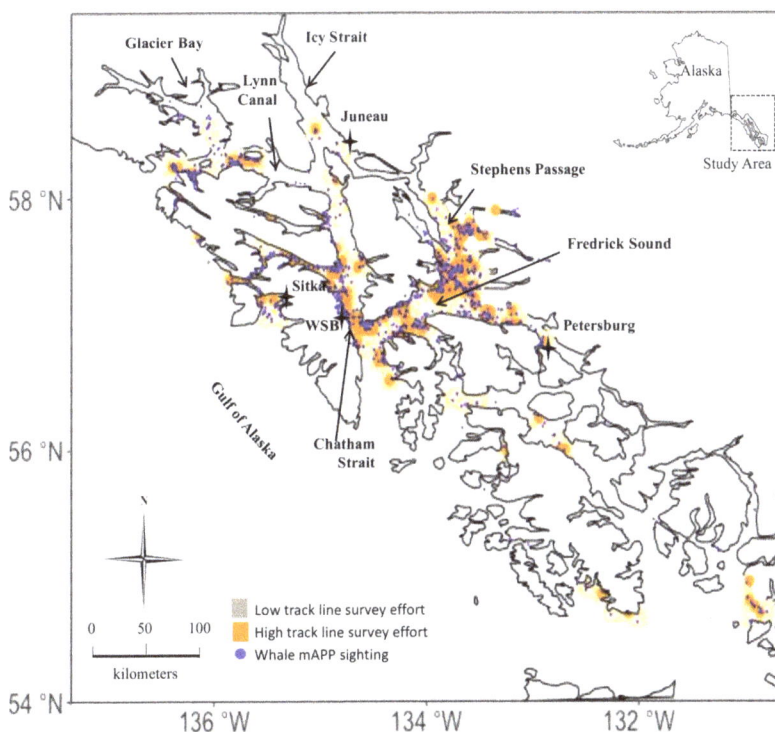

Figure 9. Map showing low and high track line survey effort and Whale mAPP-sighting data.

4. Discussion

With an increasing need for monitoring marine mammal populations in the face of broad scale environmental changes, citizen science research presents a low-cost solution to collect the vast spatial and temporal data required for robust regional scale population monitoring. This citizen science study cost less than $10,000 for the mobile application revisions, field work equipment, and graduate student time; and it covered a spatial area that had only been surveyed several times in Southeast Alaska, yet had never been surveyed for all cetaceans, pinnipeds, and sea otter species [38–41]. While the Whale mAPP citizen science project received immense positive feedback and can provide inexpensive and potentially useful scientific data, it still has biases and constraints. Thus, this study emphasizes the importance of acknowledging and measuring the limitations of each citizen science project, so that data collection methods and structure can adapt for improved education and scientific benefits.

This initial review demonstrates the potential of a marine mammal citizen science app to support both scientific and education objectives. For instance, Whale mAPP data identified 90.9% of Steller sea lion haul out sites identified by previous scientific surveys [41] in areas of high survey coverage.

In addition, short-term educational benefits, such as improved marine mammal identification skills and self-initiated learning, represent an added benefits of the application.

Citizen scientists' engagement level and retention play a vital role in the success of any citizen science project. Mobile application citizen science projects often receive funding for the initial app development and recruitment phase [15], but may find it challenging to acquire the long term funds to support continued app management, volunteer recruitment, and volunteer communication. This study stands testament to the effects of supplying resources for one field season, only to have the data collection cease without persistent recruitment and retention efforts. While, the initial excitement and data collection of Whale mAPP was promising, long-term participation beyond two weeks was limited to 27.3% of the citizen scientists. The remaining 72.7% of volunteers stopped using Whale mAPP because they left the area, believed the app required too much time, or encountered technological problems. This indicates that perhaps volunteers did not realize Whale mAPP works globally, were only interested in recording data for the Southeast Alaska project, or no longer had an opportunity to use the mobile app because they were no longer on the water. Furthermore, an even sharper decline in the number of marine mammal records in the years after recruitment indicate that perhaps this transient area is not an ideal location to recruit returning volunteers or that more effort is required to maintain communication and create a stimulating Whale mAPP community. Since the goal of many citizen science projects is to be self-sustaining, this study clearly demonstrated that this goal could not be accomplished with only one season of recruitment effort. For Whale mAPP, only one naturalist working on a nature tourism boat continued to use the mobile app two years after the initial recruitment summer. After three years, no originally recruited volunteers continued to use the app. Thus, to support and develop a long-term successful citizen science project, continual effort needs to be directed into recruiting and retaining volunteers, communicating with those volunteers, updating the mobile application, fixing technological problems, and funding the project so that resources are retained after one field season.

Targeted recruitment may also represent a viable option. Many crowdsourcing volunteers are actually not amateurs, but rather self-selected professionals and experts who elect to participate [56]. The five citizen scientists who contributed over 58% of the data likely represent these self-selecting experts. This trend is not uncommon, as there are often just a few citizen scientists who contribute a majority of the data [33]. Thus, finding these knowledgeable candidates will be key to collecting high quality data and retaining dedicated citizen science volunteers. Results from this study suggest Whale mAPP recruitment should focus on captain or naturalist staff on nature tourism vessels, especially those who express interest in marine mammals and the project. Since 83.3% of citizen scientists who continued using Whale mAPP were involved with nature tourism cruises and expressed very strong pre-existing interests in the ocean and marine mammals, and to a slightly lesser degree monitoring marine mammals and Whale mAPP, targeting this same audience would likely yield better results. This more specialized approach will likely connect to even more interested and dedicated citizen scientists with a stronger pre-existing interests [12,20,30,31] than recruiting volunteers from a nature outreach center, which likely resulted in a broadly targeted audience. Furthermore, if staff working on nature tourism vessels are targeted, the resulting participants would likely follow set travel routes, facilitating easier temporal data comparison. While a more standardized travel route might aid in data analyses, it may also lead to data gaps in narrow channels, marine protected areas, or locations not travelled by these vessels (Figure 9).

Previous citizen science studies also emphasize the importance of creating a community of volunteers who enjoy working together and identify with the projects' goals. This community can also play a role in driving the future direction and modifications of the citizen science project [30,57]. Putting more energy into the citizen scientists' experience through the Whale mAPP website, forum, or email notification may also help improve sustained volunteer commitment. Additionally, increasing data collection masks inevitable errors and improves data quality [7], a common concern for citizen science projects [10]. Like many things in life, having a supportive community can motivate people to continue contributing to a mutual project.

Another strategy to increase volunteer recruitment would be to provide an iOS version of Whale mAPP. Around half of the people encountered at the recruitment center did not own an Android, and therefore could not participate. By either creating an iOS version of Whale mAPP or providing Android tablets for dedicated volunteers, the project could reach this still untapped audience.

While still cheaper than running an entire field season to collect comparable data, this effort is not cost-free. Promoting Whale mAPP to the general public and target audiences, developing an iOS version, and providing tablets to dedicated users requires staff time and resources. Furthermore, sustaining new volunteers demands long-term communication, collaboration, and commitment. Securing long-term funding for such a project, although challenging, will likely by the only means to continue the quality of the citizen science project that collects this inclusive marine mammal data.

Results suggest this project is worth investing in because Whale mAPP provides an opportunity to learn more about many marine mammals by contributing to sustained marine mammal research and outreach. This outcome was one of the motivating reasons for why people chose to use Whale mAPP. Moreover, as a result of using Whale mAPP, participants, on average, improved their marine mammal identification skills, a positive learning outcome for many citizen science projects [12]. A more challenging, and often long-term, educational benefit is inspiring volunteers to engage in further learning outside of the citizen science project [19]. Few citizen science projects have reportedly accomplished this task [12]. The first steps to this goal are seen in the Whale mAPP volunteers who took action to learn more about marine mammals by talking to peers, reading about marine mammals, and talking to scientists. There are many methods for bolstering the educational components of a citizen science project. However, methods need to address various user types, from teaching marine mammal identification, to novice users to providing a more stimulating learning environment for the self-initiated participants with previous knowledge. Perhaps one way to address this gap could be for Whale mAPP to enhance citizen scientists' engagement in the scientific process and education on marine mammals by developing interactive maps that focus on connecting spatiotemporal processes (i.e., current, depth, time of year, etc.) with marine mammal distributions. Overall, providing more enhanced services via the application or website, or actively through ann onsite Whale mAPP steward, would likely enhance the reach of the project and prolong interests of Whale mAPP citizen scientists.

In addition to increasing science literacy and knowledge, citizen science projects also need to consider how researchers can use the data and what caveats are associated with the dataset. One common data limitation is spatial bias. Variability in citizen science survey effort is not uncommon, as a majority of the effort is often focused around common, human populated areas [11,58–60]. This spatial bias toward populated areas is not necessarily a drawback, as urban areas frequently require regular monitoring because they can be more impacted by anthropogenic stressors [61,62]. Based on available track line data, these same trends of high survey effort near towns and common travel routes were present in the Whale mAPP data as well. Because citizen scientists often inadvertently collect more data in these regions, Whale mAPP could provide a unique and useful method to examine the impacts of anthropogenic stressors on marine mammals. An alternative method to combat spatial bias may be to highlight areas of low effort in the app's map display and encourage volunteers to travel to those low survey areas. Either way, survey effort and bias need to be considered when interpreting the citizen science data.

Results from comparing Steller sea lion citizen science data to traditionally collected data illustrate the importance of spatial coverage and sample size. For Steller sea lions, Whale mAPP data accuracy in co-locating the haul out sites identified by a previous scientific study [41] improved from ~72% to 91% when the comparison was limited to high survey effort areas (Figure 7). Data quality also improved with an increased sample size (Figure 8). A Steller sea lion citizen science sample size of 73 recordings was adequate to predicting 72.2% of haul out sites identified with traditional research methods [41], an accuracy value higher than results comparing eBird and iNaturalist bird lists to National Park

Service records [14]. These two results demonstrate that increased sample size and spatial coverage is crucial for scientific usefulness of data. Sample size is important for all research methods, including assessments of species distributions where low sample size effects both citizen science [14,35] and traditional scientific studies [63]. In sum, a large enough Whale mAPP data sample size and spatial coverage enable this citizen science method to become scientifically valuable for gathering broad scale and marine mammal data.

User training can also contribute to data quality. Sufficient training and accurate protocols are required to reinforce data consistency and accuracy [15]. Future work should specifically test the effectiveness of the Whale mAPP protocol and marine mammal identification guide by shadowing Whale mAPP citizen scientists while they use the app and associated materials, noting where error and/or frustration occur while simultaneously using Whale mAPP to compare the data recorded by the citizen scientists to that of a scientist. This study would provide a more thorough evaluation of the current Whale mAPP protocols, supporting future improvements to the user guidelines and the resulting data quality.

5. Conclusions

Overall, Whale mAPP received positive user feedback and produced valuable scientific data for the Steller sea lion case study. Whale mAPP data quality, like many citizen science data, improved with a larger sample size [21,33] and spatial coverage [64]. This will most likely be achieved by recruiting self-selecting experts, such as naturalists and captains working on nature tourism cruises, and developing the educational and engagement components of Whale mAPP beyond the level of improved marine mammal identification, and into something more beneficial for expert volunteers. Committing future funding to Whale mAPP user recruitment, especially of nature tourism staff, and building a Whale mAPP citizen science learning and networking community would contribute substantially to the success of the project. Subsequently, more consistent data could be collected, improving spatial coverage and sample size, thereby reducing limitations with the resulting Whale mAPP data. With these efforts, citizen scientists equipped with Whale mAPP could provide the broad scale, long term and continuous marine mammal monitoring data needed to identify at risk populations and fill many data gaps currently present in marine mammal conservation and management research.

Author Contributions: C.H.H., L.L.S., and A.S. conceived the experiments; C.H.H. designed the experiments; C.H.H. performed the experiments; C.H.H. and L.G.T. analyzed the data; A.S. contributed field material and the recruitment center; C.H.H. led manuscript preparation, and all authors contributed to writing.

Acknowledgments: This study was made possible due to the generous contribution of 39 citizen scientists who collected the marine mammal data. The Alaska Whale Foundation provided financial assistance and use of the Center for Coastal Conservation as a recruitment center during the 2014 field season. The Mamie Markham Research Award provided financial support for data analysis and writing of the manuscript. Funding for the development of Whale mAPP was generously provided by a grant from the California Coastal Commission's Whale Tail License Plate Fund. The project was initiated with the support of Lei Lani Stelle through a LENS fellowship funded by the Keck Foundation, Melodi King of Smallmelo Geographic Information Services, and Crown Chimp Design & Development. Special thank you to Josef Tecumseh Stitts for his immense support and assistance throughout the process.

Conflicts of Interest: Andrew Szabo works for the Alaska Whale Foundation, one of the funding sponsors. Neither the Alaska Whale Foundation nor the Mamie Markham Research Award had a role in the design of the study; in the collection, analyses, or interpretation of data; in the writing of the manuscript, and in the decision to publish the results.

References

1. Davidson, A.D.; Boyer, A.G.; Kim, H.; Pompa-Mansilla, S.; Hamilton, M.J.; Costa, D.P.; Ceballos, G.; Brown, J.H. Drivers and hotspots of extinction risk in marine mammals. *Proc. Natl. Acad. Sci. USA* **2012**, *109*, 3395–3400. [CrossRef] [PubMed]
2. Hazen, E.L.; Jorgensen, S.; Rykaczewski, R.R.; Bograd, S.J.; Foley, D.G.; Jonsen, I.D.; Shaffer, S.A.; Dunne, J.P.; Costa, D.P.; Crowder, L.B.; et al. Predicted habitat shifts of Pacific top predators in a changing climate. *Nat. Clim. Chang.* **2013**, *3*, 234–238. [CrossRef]

3. Allen, B.M.; Angliss, R.P. *Alaska Marine Mammal Stock Assessments, 2014*; Agencies and Staff of the US Department of Commerce: Seattle, WA, USA, 2015; p. 11.

4. Magurran, A.E.; Baillie, S.R.; Buckland, S.T.; Dick, J.M.; Elston, D.A.; Scott, E.M.; Smith, R.I.; Somerfield, P.J.; Watt, A.D. Long-term datasets in biodiversity research and monitoring: Assessing change in ecological communities through time. *Trends Ecol. Evol.* **2010**, *25*, 574–582. [CrossRef] [PubMed]

5. Torres, L.G.; Smith, T.D.; Sutton, P.; MacDiarmid, A.; Bannister, J.; Miyashita, T. From exploitation to conservation: Habitat models using whaling data predict distribution patterns and threat exposure of an endangered whale. *Divers. Distrib.* **2013**, *19*, 1138–1152. [CrossRef]

6. Taylor, B.L.; Martinez, M.; Gerrodette, J.; Barlow, J. Lessons from monitoring trends in abundance of marine mammals. *Mar. Mammal Sci.* **2007**, *23*, 157–175. [CrossRef]

7. Dickinson, J.L.; Zuckerberg, B.; Bonter, D.N. Citizen science as an ecological research tool: Challenges and benefits. *Annu. Rev. Ecol. Evol. Syst.* **2010**, *41*, 149–172. [CrossRef]

8. Parsons, E.; Baulch, S.; Bechshoft, T.; Bellazzi, G.; Bouchet, P.; Cosentino, A.; Godard-Codding, C.; Gulland, F.; Hoffmann-Kuhnt, M.; Hoyt, E.; et al. Key research questions of global importance for cetacean conservation. *Endanger. Species Res.* **2015**, *27*, 113–118. [CrossRef]

9. Kelling, S.; Gerbracht, J.; Fink, D.; Lagoze, C.; Wong, W.K.; Yu, J.; Damoulas, T.; Gomes, C. A human/computer learning network to improve biodiversity conservation and research. *AI Mag.* **2012**, *34*, 10. [CrossRef]

10. Thiel, M.; Penna-Díaz, M.A.; Luna-Jorquera, G.; Salas, S.; Sellanes, J.; Stotz, W. Citizen scientists and marine research: Volunteer participants, their contributions, and projection for the future. *Oceanogr. Mar. Biol. Annu. Rev.* **2014**, *52*, 257–314.

11. Goffredo, S.; Pensa, F.; Neri, P.; Orlandi, A.; Gagliardi, M.S.; Velardi, A.; Piccinetti, C.; Zaccanti, F. Unite research with what citizens do for fun: "recreational monitoring" of marine biodiversity. *Ecol. Appl.* **2010**, *20*, 2170–2187. [CrossRef] [PubMed]

12. Crall, A.W.; Jordan, R.; Holfelder, K.; Newman, G.J.; Graham, J.; Waller, D.M. The impacts of an invasive species citizen science training program on participant attitudes, behavior, and science literacy. *Public Underst. Sci.* **2013**, *22*, 745–764. [CrossRef] [PubMed]

13. Hochachka, W.M.; Fink, D.; Hutchinson, R.A.; Sheldon, D.; Wong, W.K.; Kelling, S. Data-intensive science applied to broad-scale citizen science. *Trends Ecol. Evol.* **2012**, *27*, 130–137. [CrossRef] [PubMed]

14. Clemens, J.; Jacobs, Z.A.; Zipf, A. Completeness of citizen science biodiversity data from a volunteered geographic information perspective. *Geo-Spat. Inf. Sci.* **2017**, *20*, 3–13. [CrossRef]

15. Ferster, C.J.; Coops, N.C. Assessing the quality of forest fuel loading data collected using public participation methods and smartphones. *Int. J. Wildland Fire* **2014**, *23*, 585–590. [CrossRef]

16. Fritz, S.; Fonte, C.C.; See, L. The role of citizen science in earth observation. *Multidiscip. Digit. Publ. Inst.* **2017**, *9*, 357. [CrossRef]

17. Mobasheri, A.; Deister, J.; Dieterich, H. Wheelmap: The wheelchair accessibility crowdsourcing platform. *Open Geospat. Data Softw. Stand.* **2017**, *2*, 27. [CrossRef]

18. Castell, N.; Kobernus, M.; Liu, H.Y.; Schneider, P.; Lahoz, W.; Berre, A.J.; Noll, J. Mobile technologies and services for environmental monitoring: The Citi-Sense-MOB approach. *Urban Clim.* **2014**, *14*, 370–382. [CrossRef]

19. Conrad, C.C.; Hilchey, K.G. A review of citizen science and community-based environmental monitoring: Issues and opportunities. *Environ. Monit. Assess.* **2011**, *176*, 273–291. [CrossRef] [PubMed]

20. Raddick, M.J.; Bracey, G.; Gay, P.L.; Lintott, C.J.; Murray, P.; Schawinski, K.; Szalay, A.S.; Vandenberg, J. Galaxy zoo: Exploring the motivations of citizen science volunteers. *Astron. Educ. Rev.* **2010**, *9*, 010103. [CrossRef]

21. Sullivan, B.L.; Aycrigg, J.L.; Barry, J.H.; Bonney, R.E.; Bruns, N.; Cooper, C.B.; Damoulas, T.; Dhondt, A.A.; Dietterich, T.; Farnsworth, A.; et al. The eBird enterprise: An integrated approach to development and application of citizen science. *Biol. Conserv.* **2014**, *169*, 31–40. [CrossRef]

22. Rugh, D.J.; Shelden, K.E.; Schulman-Janiger, A. Timing of the gray whale southbound migration. *J. Cetacean Res. Manag.* **2001**, *3*, 31–40.

23. Shelden, K.E.; Rugh, D.J.; Schulman-Janiger, A. Gray whales born north of Mexico: Indicator of recovery or consequence of regime shift? *Ecol. Appl.* **2004**, *14*, 1789–1805. [CrossRef]

24. Tonachella, N.; Nastasi, A.; Kaufman, G.; Maldini, D.; Rankin, R.W. Predicting trends in humpback whale (*Megaptera novaeangliae*) abundance using citizen science. *Pac. Conserv. Biol.* **2012**, *18*, 297–309. [CrossRef]

25. Bruce, E.; Albright, L.; Sheehan, L.; Blewitt, M. Distribution patterns of migrating humpback whales (*Megaptera novaeangliae*) in Jervis Bay, Australia: A spatial analysis using geographical citizen science data. *Appl. Geogr.* **2014**, *54*, 83–95. [CrossRef]

26. Carlson, B.S.; Sims, C.; Brunner, S. Cook Inlet Beluga Whale, *Delphinapterus leucas*, observations near Anchorage, Alaska between 2008 and 2011: Results from a citizen scientist project. *Mar. Fish. Rev.* **2015**, *77*, 115–130. [CrossRef]

27. Gregr, E.J. Insights into North Pacific right whale *Eubalaena japonica* habitat from historic whaling records. *Endanger. Species Res.* **2011**, *15*, 223–239. [CrossRef]

28. Smith, T.D.; Reeves, R.R.; Josephson, E.A.; Lund, J.N. Spatial and seasonal distribution of American whaling and whales in the age of sail. *PLoS ONE* **2012**, *7*, e34905. [CrossRef] [PubMed]

29. Hann, C.H.; Smith, T.D.; Torres, L.G. A sperm whale's perspective: The importance of seasonality and seamount depth. *Mar. Mammal Sci.* **2016**, *32*, 1470–1481. [CrossRef]

30. Nov, O.; Arazy, O.; Anderson, D. Dusting for science: Motivation and participation of digital citizen science volunteers. In Proceedings of the 2011 iConference, Seattle, WA, USA, 8–11 February 2011; ACM: New York, NY, USA, 2011; pp. 68–74.

31. Koss, R.S.; Miller, K.; Wescott, G.; Bellgrove, A.; Boxshall, A.; McBurnie, J.; Bunce, A.; Gilmour, P.; Ierodiaconou, D. An evaluation of Sea Search as a citizen science programme in Marine Protected Areas. *Pac. Conserv. Biol.* **2009**, *15*, 116–127. [CrossRef]

32. Bird, T.J.; Bates, A.E.; Lefcheck, J.S.; Hill, N.A.; Thomson, R.J.; Edgar, G.J.; Stuart-Smith, R.D.; Wotherspoon, S.; Krkosek, M.; Stuart-Smith, J.F.; et al. Statistical solutions for error and bias in global citizen science datasets. *Biol. Conserv.* **2014**, *173*, 144–154. [CrossRef]

33. Paul, K.; Quinn, M.S.; Huijser, M.P.; Graham, J.; Broberg, L. An evaluation of a citizen science data collection program for recording wildlife observations along a highway. *J. Environ. Manag.* **2014**, *139*, 180–187. [CrossRef] [PubMed]

34. Bray, G.S.; Schramm, H.L. Evaluation of a statewide volunteer angler diary program for use as a fishery assessment tool. *N. Am. J. Fish. Manag.* **2011**, *21*, 606–615. [CrossRef]

35. Galloway, A.W.; Tudor, M.T.; HAEGEN, W.M.V. The reliability of citizen science: A case study of Oregon white oak stand surveys. *Wildl. Soc. Bull.* **2006**, *34*, 1425–1429. [CrossRef]

36. Reese, G.C.; Wilson, W.R.; Hoeting, J.H.; Flather, C.H. Factors affecting species distribution predictions: A simulation modeling experiment. *Ecol. Appl.* **2005**, *15*, 554–564. [CrossRef]

37. Walter, B.; Hladick, C.; Cioni-Haywood, B. *Commercial Passenger Vessel Excise Tax: Community Needs, Priorities, Shared Revenue, and Expenditures (Fiscal Years 2007 to 2016)*; Alaska Department of Commerce, Community, and Economic Development: Seattle, WA, USA, 2017. Available online: https://www.commerce.alaska.gov/web/Portals/6/pub/TourismResearch/00%20FULL%20CPV%20RPT%2016%202017.pdf?ver=2017-03-23-160339-903 (accessed on 12 April 2018).

38. Calambokidis, J.; Falcone, E.A.; Quinn, T.J.; Burdin, A.M.; Clapham, P.J.; Ford, J.K.B.; Gabriele, C.M.; LeDuc, R.; Mattila, D.; Rojas-Bracho, L.; et al. *SPLASH: Structure of Populations, Levels of Abundance and Status of Humpback Whales in the North Pacific*; Cascadia Research Collective to USDOC: Seattle, WA, USA, 2009.

39. Dahlheim, M.E.; White, P.A.; Waite, J.M. Cetaceans of Southeast Alaska: Distribution and seasonal occurrence. *J. Biogeogr.* **2009**, *36*, 410–426. [CrossRef]

40. Dahlheim, M.E.; Zerbini, A.N.; Waite, J.M.; Kennedy, A.S. Temporal changes in abundance of harbor porpoise (*Phocoena phocoena*) inhabiting the inland waters of Southeast Alaska. *Fish. Bull.* **2015**, *113*, 242–256. [CrossRef]

41. Womble, J.N.; Willson, M.F.; Sigler, M.F.; Kelly, B.P.; VanBlaricom, G.R. Distribution of Steller sea lions Eumetopias jubatus in relation to spring-spawning fish in SE Alaska. *Mar. Ecol. Prog. Ser.* **2005**, *294*, 271–282. [CrossRef]

42. Stelle, L.L.; King, M.; Hann, C.H. Chapter 8 Whale mAPP: Engaging Citizen Scientists to Contribute and Map Marine Mammal Sightings. In *Ocean Solutions, Earth Solutions*; White, D.J., Ed.; Esri Press: Redland, MD, USA, 2016; pp. 151–170.

43. Couvet, D.; Jiguet, F.; Julliard, R.; Levrel, H.; Teyssedre, A. Enhancing citizen contributions to biodiversity science and public policy. *Interdiscip. Sci. Rev.* **2008**, *33*, 95–103. [CrossRef]

44. Silvertown, J. A new dawn for citizen science. *Trends Ecol. Evol.* **2009**, *24*, 467–471. [CrossRef] [PubMed]

45. R Core Team. *R: A Language and Environment for Statistical Computing*; R Foundation for 343 Statistical Computing; R Core Team: Vienna, Austria, 2015. Available online: http://www.R-project.org/ (accessed on 2 May 2018).

46. Bell, P.; Lewenstein, B.; Shouse, A.W.; Feder, M.A. *Learning Science in Informal Environments: People, Places, and Pursuits in Report of the National Research Council of the National Academies*; The National Academic Press: Washington, DC, USA, 2009.

47. Hothorn, T.; Hornik, K.; Mark, A.; van de Wiel, A.Z. Implementing a Class of Permutation Tests: The coin Package. *J. Stat. Softw.* **2008**, *28*, 1–23. [CrossRef]

48. Barry, R.P.; McIntyre, J. Estimating animal densities and home range in regions with irregular boundaries and holes: A lattice-based alternative to the kernel density estimator. *Ecol. Model.* **2011**, *222*, 1666–1672. [CrossRef]

49. Citta, J.J.; Quakenbush, L.T.; Okkonen, S.R.; Druckenmiller, M.L.; Maslowski, W.; Clement-Kinney, J.; George, J.C.; Brower, H.; Small, R.J.; Ashjian, C.J.; et al. Ecological characteristics of core-use areas used by Bering–Chukchi–Beaufort (BCB) bowhead whales, 2006–2012. *Prog. Oceanogr.* **2015**, *136*, 201–222. [CrossRef]

50. Legare, B.; Kneebone, J.; DeAngelis, B.; Skomal, G. The spatiotemporal dynamics of habitat use by blacktip (*Carcharhinus limbatus*) and lemon (*Negaprion brevirostris*) sharks in nurseries of St. John, United States Virgin Islands. *Mar. Boil.* **2015**, *162*, 699–716. [CrossRef]

51. Barry, R. latticeDensity: Density Estimation and Nonparametric Regression on Irregular Regions. R Package Version 1.0.7 2012. Available online: http://CRAN.R-project.org/package=latticeDensity (accessed on 2 May 2018).

52. Kvitek, R.G.; Bowlby, C.E.; Staedler, M. Diet and foraging behavior of sea otters in southeast Alaska. *Mar. Mammal Sci.* **1993**, *9*, 168–181. [CrossRef]

53. Calambokidis, J.; Taylor, B.L.; Carter, S.D.; Steiger, G.H.; Dawson, P.K.; Antrim, L.D. Distribution and haul-out behavior of harbor seals in Glacier Bay, Alaska. *Can. J. Zool.* **1987**, *65*, 1391–1396. [CrossRef]

54. Mathews, E.A.; Pendleton, G.W. Declines in harbor seal (*Phoca vitulina*) numbers in Glacier Bay national park, Alaska, 1992–2002. *Mar. Mammal Sci.* **2006**, *22*, 167–189. [CrossRef]

55. Karpovich, S.A.; Skinner, J.P.; Mondragon, J.E.; Blundell, G.M. Combined physiological and behavioral observations to assess the influence of vessel encounters on harbor seals in glacial fjords of southeast Alaska. *J. Exp. Mar. Boil. Ecol.* **2015**, *473*, 110–120. [CrossRef]

56. Brabham, D.C. The myth of amateur crowds: A critical discourse analysis of crowdsourcing coverage. *Inf. Commun. Soc.* **2012**, *15*, 394–410. [CrossRef]

57. Cooper, S.; Khatib, F.; Treuille, A.; Barbero, J.; Lee, J.; Beenen, M.; Leaver-Fay, A.; Baker, D.; Popović, Z. Predicting protein structures with a multiplayer online game. *Nature* **2010**, *466*, 756. [CrossRef] [PubMed]

58. Bart, J.; Hofschen, M.; Peterjohn, B.G. Reliability of the breeding bird survey: Effects of restricting surveys to roads. *Auk* **1995**, *112*, 758–761.

59. Lawler, J.J.; O'Connor, R.J. How well do consistently monitored breeding bird survey routes represent the environments of the conterminous United States? *Condor* **2004**, *106*, 801–814. [CrossRef]

60. Niemuth, N.D.; Dahl, A.L.; Estey, M.E.; Loesch, C.R. Representation of landcover along breeding bird survey routes in the Northern Plains. *J. Wildl. Manag.* **2007**, *71*, 2258–2265. [CrossRef]

61. Halpern, B.S.; Walbridge, S.; Selkoe, K.A.; Kappel, C.V.; Micheli, F.; D'agrosa, C.; Bruno, J.F.; Casey, K.S.; Ebert, C.; Fox, H.E.; et al. A global map of human impact on marine ecosystems. *Science* **2008**, *319*, 948–952. [CrossRef] [PubMed]

62. Gilarranz, L.J.; Mora, C.; Bascompte, J. Anthropogenic effects are associated with a lower persistence of marine food webs. *Nat. Commun.* **2016**, *7*, 10737. [CrossRef] [PubMed]

63. Fitzpatrick, M.C.; Preisser, E.L.; Ellison, A.M.; Elkinton, J.S. Observer bias and the detection of low-density populations. *Ecol. Appl.* **2009**, *19*, 1673–1679. [CrossRef] [PubMed]

64. Jackson, M.M.; Gergel, S.E.; Martin, K. Citizen science and field survey observations provide comparable results for mapping Vancouver Island White-tailed Ptarmigan (*Lagopus leucura sazatilis*) distributions. *Biol. Conserv.* **2015**, *181*, 162–172. [CrossRef]

MDPI
St. Alban-Anlage 66
4052 Basel
Switzerland
Tel. +41 61 683 77 34
Fax +41 61 302 89 18
www.mdpi.com

ISPRS International Journal of Geo-Information Editorial Office
E-mail: ijgi@mdpi.com
www.mdpi.com/journal/ijgi

www.ingramcontent.com/pod-product-compliance
Lightning Source LLC
Chambersburg PA
CBHW051849210326
41597CB00033B/5828